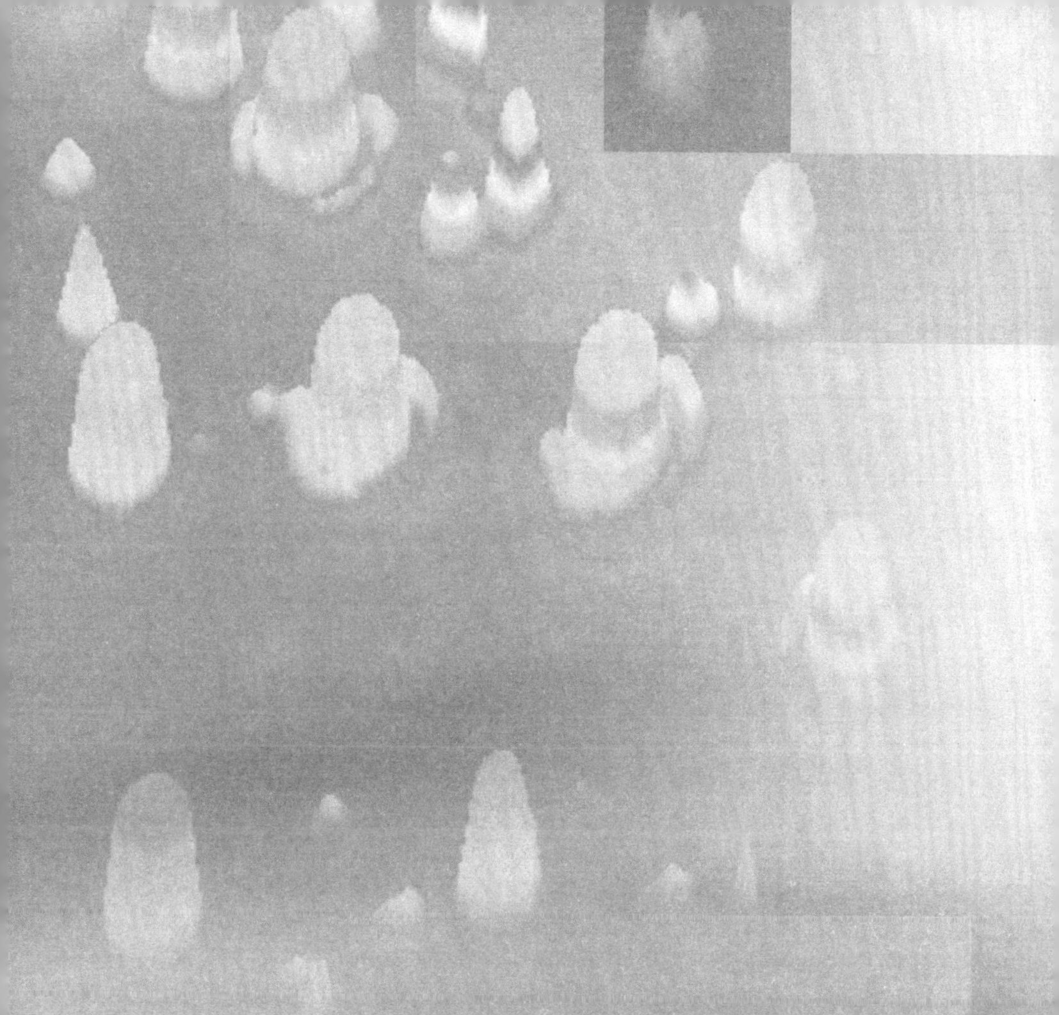

Tuning Semiconducting and Metallic Quantum Dots

Tuning Semiconducting and Metallic Quantum Dots

Spectroscopy and Dynamics

edited by

Christian von Borczyskowski

Eduard Zenkevich

Pan Stanford Publishing

Published by

Pan Stanford Publishing Pte. Ltd.
Penthouse Level, Suntec Tower 3
8 Temasek Boulevard
Singapore 038988

Email: editorial@panstanford.com
Web: www.panstanford.com

British Library Cataloguing-in-Publication Data
A catalogue record for this book is available from the British Library.

Tuning Semiconducting and Metallic Quantum Dots: Spectroscopy and Dynamics

ISBN 978-981-4745-24-6 (Hardcover)
ISBN 978-981-4745-25-3 (eBook)

MIX
Paper from responsible sources
FSC
www.fsc.org
FSC® C013056

Printed and bound in Great Britain by
TJ International Ltd, Padstow, Cornwall

For Sabine and Tamara

Contents

Preface

During the last few decades, nanomaterials have become an intensively growing field of interest, with respect to both applications and fundamental science. These materials, especially from semiconductors and metals, are equally important for applications such as nanoelectronics, sensoric applications, and life sciences. Nowadays, this field can be characterized by several material classes, including the related preparation or manufacturing processes. According to the widely varying features of these materials, scientific approaches in analyzing, understanding, and modeling are inherently interdisciplinary and cover a still growing range of physical, chemical, and biological disciplines. Common to many of these materials is the fact that their properties depend on the dimensions in the nanometer range: size matters! According to the extremely small size, a large surface-to-volume ratio is an inherent property of all these materials. This opens access to both surface and core features. However, this perspective is not merely a linear superposition of the respective properties but constitutes completely new ones. Naturally, interfaces play a central role. Though this is a central aspect of nanomaterials, a detailed understanding and control of inherent properties of surfaces and interfaces, such as structure–function relationships, has to be investigated with respect to the specific class or even subclass of nanomaterials.

Moreover, in an analogy to quantum well structures, dimensions can be further downsized to quantum dots from semiconductors or metals. At a critical dimension, bulk properties (such as optical ones) deviate from those of the bulk and become size dependent. This results in considerably changed features intermediate between both the original solid and its constituting atoms (or molecules). In this respect (optical) properties can be tuned by the dimension of the (nano)material.

In this book, we will concentrate on the optical properties of colloidal semiconductor quantum dots and to some extent also those of metal particles. Since there is already a considerable number of publications reviewing the size dependence of optical properties of such nanomaterials, we prefer to concentrate on interfaces or surfaces and the perspectives of tuning of the related electronic energies and dynamics.

The experimental approach is based on static and dynamic optical spectroscopy, both on ensembles of quantum dots and on single ones. The latter approach is extremely sensitive and opens a wide field for detailed understanding and sensoric applications in material and life sciences. This has been recently acknowledged by the 2014 Nobel Prize in Chemistry awarded to E. Betzig, S. Hell, and W. E. Moerner for the invention and implementation of super-resolution optical microscopy.

Experimental reports in this book, some of them even unpublished, are based on results obtained in our groups during the last 10 years and especially include luminescence intermittency (blinking dynamics) in single quantum dots. The latter is a fascinating phenomen, often observed but far from being understood at a microscopic level.

The book is organized in the following way: We have virtually divided it into two parts: "Electronic States of Quantum Dots" (Chapters 1–3) and "Single Quantum Dots" (Chapters 4–6). Each part starts with an introductory chapter followed by several case studies in separate chapters. The introductory parts summarize relevant basic aspects of the related subjects and are to a certain extent a kind of review without being exhaustive in the respect. Correspondingly, the bibliography at the end of each chapter contains some leading papers, recent reviews, and books in which the readers may find some additional specific references relevant to their subjects of interest.

In "Electronic States of Quantum Dots," we report on several not yet reported spectroscopic features of CdSe excitons or trap states and related temperature-dependent exciton–phonon coupling. Both subjects pay special attention to the influence of the heterogeneous interfaces, including organic ligands on colloidal quantum dots.

The chapters in "Single Quantum Dots," besides a basic introduction to the related optical properties and detection schemes, concentrate on a review of experiments on single metal (Ag) nanoclusters. The central part is the tuning of photoluminescence intermittency (blinking) and spectral diffusion of CdSe quantum dots with special emphasis on interface-related effects. We present a new model for blinking processes.

Altogether, we aim with this book for a thorough understanding of subtle effects of interfaces and surfaces, which noticeably tune optical properties with respect to both optical energies and (interfacial) dynamics. We feel that we contribute to a deeper understanding of these to a large extent heterogeneous nanostructures. We make as often as possible use of a combination of experiments on ensemble and single nanostructures, an approach which has so far not been intensively used in literature. In the end our contributions might in the case of blinking processes be useful for either suppressing such effects for highly sensitive sensoric applications or even controling blinking for applications in stochastically based super-resolution microscopy.

Christian von Borczyskowski
Chemnitz, Germany
Eduard Zenkevich
Minsk, Belarus
September 2016

Acknowledgments

All results from our groups in Chemnitz and Minsk have been obtained by graduate students, PhD students, and postdocs. We appreciate all the work which they have contributed to the development of this field during the last 10 years.

The book contains many results which have not yet been published: Ines Trenkmann (parts of Chapter 6), Cornelius Krasselt and Clemens Göhler (parts of Chapters 4 and 6), and Stefan Krause (parts of Chapter 5).

Writing a book like this would not have been possible without many fruitful discussions and help in writing and layout. Especially, we would like to mention Danny Kowerko (experimental part of Chapter 4), Frank Cichos and Michel Orrit (modified version of blinking models in Chapter 4), and Clemens Göhler, Thomas Blaudeck, and Cornelius Krasselt for figure preparations. Stefan Krause prepared the graph of Ag nanoparticles on the cover page, which is gratefully acknowledged.

We would also like to acknowledge and appreciate financial support throughout the years from varies institutions: (i) Belarussian State Program for Scientific Research "Convergence 3.2.08: Photophysics of Bioconjugates, Semiconductor and Metallic Nanostructures and Supramolecular Complexes and Their Biomedical Applications" (Chapters 2 and 3); (ii) INTAS Project 03-50-4540, "Optical Active Assemblies of Colloid Quantum Dots and Tetrapyrrole Compounds: Laser Induced Relaxation Processes and Opto-Electronic Functionalities" (Chapters 2 and 3); *(iii)* DAAD Grant No A/08/08573, "Single Particle Spectroscopy of Individual Nanoobjects Fixed on Heterogeneous Surfaces" (Chapters 1, 2, and 3); (iv) DFG Project GR 2695/8-1, "Optical Probe for Semiconductor Nanoparticles" (Chapters 4 and 6); (v) DFG Research Unit FOR 388, "Laboratory Astrophysics" (Chapter 4); (vi) DFG Graduate College GRK 82, "Accumulation of Single Molecules to Nanostructures (Accumol)" (Chapters 4 and 6); and (vii) DFG Research Unit FOR

877, "From Local Constraints to Macroscopic Transport" (Chapters 1, 2, and 3).

Quantum dot spectroscopy is based on for over more than 25 years on a stimulating and fruitful cooperation between research groups from Minsk (Zenkevich) and Chemnitz (von Borczyskowski). The scientific and personal exchange was strongly supported over these years by the Volkswagen-Foundation, INTAS, DAAD, and especially within the framework of the DFG Research Unit FOR 877.

Chapter 1

Size Matters: Optical Properties of Nanoparticles

Christian von Borczyskowski[a] **and Eduard Zenkevich**[b]
[a]*Institute of Physics, Center for Nanostructured Materials and Analytics, Technische Universität Chemnitz, Reichenhainer Str. 70, D-09107 Chemnitz, Germany*
[b]*Department of Information Technologies and Robototechnique, National Technical University of Belarus, Prospect Nezavisimosti 65, 220013 Minsk, Belarus*
borczyskowski@physik.tu-chemnitz.de, zenkev@tut.by

1.1 Introduction

One of the fascinations about nanoparticles or nanocrystals (NCs) built from a few to several thousand atoms is the fact that they are from a quantum mechanical point of view a class of materials between atoms and solids showing properties of both of them. Though a deep understanding of their properties emerged only during the last few decades, the related versatility has been used already since centuries. Probably the most obvious feature is the dependence of properties, like the color, on the size of NCs, both for semiconducting

Tuning Semiconducting and Metallic Quantum Dots: Spectroscopy and Dynamics
Edited by Christian von Borczyskowski and Eduard Zenkevich

ISBN 978-981-4745-24-6 (Hardcover), 978-981-4745-25-3 (eBook)
www.panstanford.com

and for metal particles. A lot of features can nowadays be understood in terms of quantum confinement resembling in its most simple approximation, the "particle in a box" model taught in undergraduate courses on quantum mechanics. NCs falling into this category are thought of as 1D quantum objects and have been also named "quantum dots" (QDs). We will use this name, or QDs, throughout the book. The fundamental optical excitations in a semiconducting QD depend on electrons, holes, and excitons in a 3D (confining) potential. Systematic research began in the early 1980s with the identification of quantum confinement in small semiconductor NCs. Different from scaling down nanostructures lithographically as is the present standard in micro- and nanotechnology, QDs are in most cases formed via bottom-up approaches in glasses or polymers or from solutions. In general, semiconductor nanostructures show an amazing variety of interesting properties different from conventional solid state materials. A recent comprehensive review outlines the prospects of nanoscience with NCs [Kovalenko et al., 2015].

Though quantum confinement and its consequences already open an enormously wide field with respect to both fundamental research and applications, QDs are not only in between atoms and solids but depend to a large extent on surface properties. This can be considered not as a disadvantage but just the opposite since elementary processes such as charge accumulation or catalysis depend essentially on interface properties. In this book, we will pay special attention from an empirical point of view to surface or interface features. Especially we will discuss how they can be tuned by various means.

Figure 1.1 shows schematically what has to be taken into account when analyzing a QD, including influences outside the intrinsically confining potential. We have indicated that a crystal structure may not be perfect under realistic conditions. In any case a crystal structure includes facets and surface domains, depending on crystal symmetry. Besides the QD core (e.g., CdSe as a II–VI semiconductor), QDs may contain shells from other semiconductor materials (e.g., ZnS) and/or organic surfactants (ligands). The latter are essential when QDs are synthesized from solution to stop further growth or to protect from aggregation (Oswald ripening). Such a structure already constitutes two to three interfaces. In general, QDs are embedded in a matrix (liquid, polymer, glass). Moreover, in the case of thin films, a supporting substrate might also have to be taken into account. All these subsystems have to be followed since though the electronic wavefunctions are confined in a potential, they might tunnel through confining barriers, "exploring" the corresponding environment on a nanoscale. These considerations are essential, especially with respect to applications.

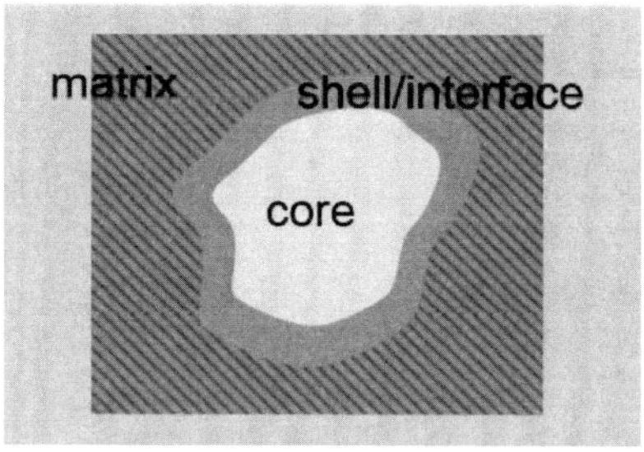

Figure 1.1 Schematic illustration of a (deformed) semiconductor core capped by semiconductor and/or (organic) surfactant shells. This combined quantum system is embedded in a matrix, which is treated as a continuum and supported by a substrate.

1.2 Quantum Confinement

1.2.1 Basic Concepts

A whole bunch of textbooks describes the treatment of excited states of (semiconductor) QDs, taking into account quantum confinement [Woggon, 1997; Gaponenko, 1998; Gaponenko, 2010; Rogach, 2008; Gavrilenko, 2011]. Therefore we will restrict ourselves here to the basic principles without going into detail. The band energy of a particle in a quasi-infinite bulk semiconductor is described by the dispersion relation

$$E(\vec{k}) = E_G + \frac{\hbar^2 \vec{k}^2}{2m^*} \tag{1.1}$$

with band-gap energy E_G and the reciprocal effective mass $1/m^*$ at $k = 0$ according to

$$\frac{1}{m^*} = \frac{1}{\hbar^2}\frac{d^2E}{dk^2}\,. \tag{1.2}$$

The band structure is characterized in Fig. 1.2 for II–VI semiconductors (e.g., CdSe or ZnS) with a direct band gap.

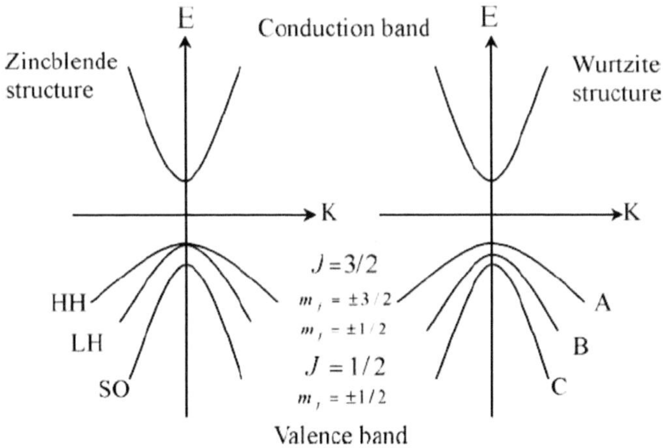

Figure 1.2 Simplified band-gap scheme of direct-gap II–VI semiconductors (zinc-blende and wurtzite type). HH and LH correspond to heavy and light hole valence subbands, respectively, while SO is the spin–orbit split-off band. The conduction band is related to the metal atoms, while the valence band is related to the chalcogenides. From Issac, A. (2006). *Photoluminescence Intermittency of Semiconductor Quantum Dots in Dielectric Environments*, PhD Thesis, TU Chemnitz, Germany, with kind permission from Abey Issac.

For semiconductors like CdS, CdSe, ZnS, or ZnSe, the conduction band is formed from s orbitals of the metal ions Cd or Zn, whereas the valence band corresponds to p orbitals of the chalcogenides S or Se. While the conduction band can in most cases be approximated by parabolic potentials, the valence bands are partly degenerate and more complex, depending on the crystal structure (zinc-blende or wurtzite). We will describe the details after introducing quantum confinement.

Without quantum size effects the (optically generated) electron–hole pair is a hydrogen-like bound state, the so-called exciton. The interacting hole and electron (electron–hole pair) are described by the Hamiltonian in Eq. 1.3, which applies when the size of an NC is comparable to the critical length parameters, that is, the de Broglie wavelength λ and the exciton Bohr radius a_B of the quasi-particles: electron, hole, and exciton. It is described as

$$H = -\frac{\hbar^2}{2m_e^*}\nabla_e^2 - \frac{\hbar^2}{2m_h^*}\nabla_h^2 - \frac{e^2}{\varepsilon_2 |r_e - r_h|} \tag{1.3}$$

where $m_{e(h)}^*$ is the effective mass of the electron (hole), respectively, and the static dielectric constant of the bulk semiconductor is ε_2 (~9.5 for CdSe). The exciton is similarly, as a hydrogen atom, characterized by the exciton Bohr radius a_B according to

$$a_B = \frac{\hbar^2 \varepsilon_2}{e^2}\left[\frac{1}{m_e^*} + \frac{1}{m_h^*}\right]. \tag{1.4}$$

a_B is in the range of 1–10 nm for typical II–VI semiconductors. In this range the QD is with respect to lattice constants of a macroscopic solid, but quantum size effects become important for the quasi-particles. Figure 1.3 shows schematically how the various scales (lattice constant, Bohr radius, crystal size) are related.

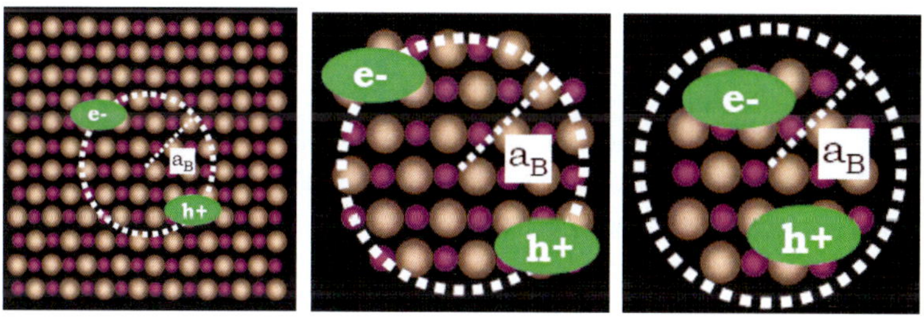

Figure 1.3 Schematic presentations of a Wannier exciton with a constant Bohr radius a_B of an electron (e−)–hole (h+) pair (exciton) upon variation of the nanocrystal size D with a constant lattice constant. Left: $D >> a_B$; middle: $D \approx a_B$; right: $D < a_B$.

To proceed, confining potentials (barriers) both for the electron and the hole have to be added to the Hamiltonian, which results in

$$H = -\frac{\hbar^2}{2m_e^*}\nabla_e^2 - \frac{\hbar^2}{2m_h^*}\nabla_h^2 - \frac{e^2}{\varepsilon_2 |r_e - r_h|} + V_e(r_e) + V_h(r_h). \tag{1.5}$$

Now ε_2 has to be replaced by the one of the confined QDs (~8 for CdSe QDs). Additionally, a surface polarization potential $\delta V(r_e, r_h, \varepsilon_2, \varepsilon_m)$ depending on the embedding matrix with a dielectric constant ε_m has to be added. A qualitative description of what is expected for excitonic energies E and density of states $N(E)$ upon solution of the

Eigenwert problem is shown in Fig. 1.4 when proceeding from bulk material to a QD .

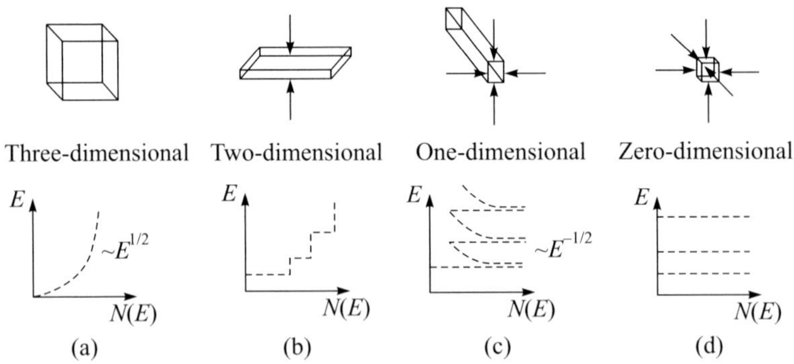

Figure 1.4 Schematic representation of the density of states for semiconductor structures of (a) 3D ideal bulk, (b) 2D confined thin film, (c) 1D quantum wire, and (d) 0D quantum dot (Yoffe, 2002).

Several solutions either via an effective mass approximation (EMA) or treatments within the linear combination of atomic orbitals (LCAO) approach have been published to calculate excitonic energies and optical transition probabilities [Gaponenko, 1998; Brus; 1984; Franceschetti and Zunger, 1997; Woggon, 1997; Wang and Zunger, 1998; Franceschetti et al., 1999; Burdett, 1984; Wang et al., 1987; Einevoll, 1992; Gómez et al., 2006]. It is beyond the scope of this short introduction to review the theoretical progress during the last two decades from one of the first perturbation approaches [Brus, 1984] up to first-principle theories [Franceschetti et al., 1999]. A qualitative insight of the impact of the quantum size effect on the energies of the lowest exciton shows Eq. 1.6 as a function of the radius R of the QD, following the perturbation approach [Brus, 1984] according to

$$E_{10} = \frac{\hbar^2 \pi^2}{2R^2}\left[\frac{1}{m_e^*} + \frac{1}{m_h^*}\right] - \frac{1.8e^2}{\varepsilon_2 R} \tag{1.6}$$

to which the bulk band-gap energy E_G has to be added. A further example of the excitonic energies as a function of the diameter of a ZnS-capped CdSe QD is shown in Fig. 1.5 and adapted to absorption spectra (Table 1.1).

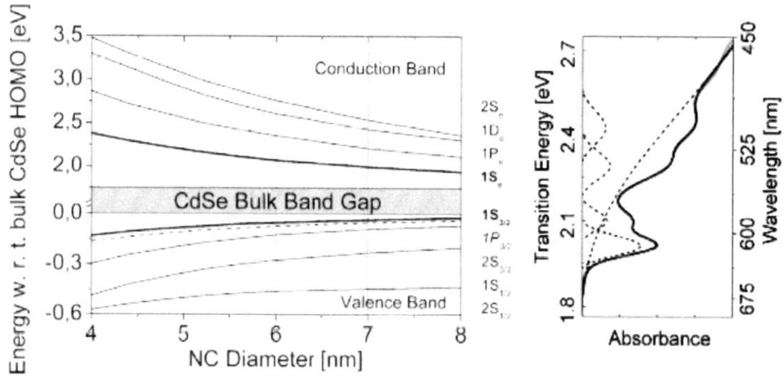

Figure 1.5 Left: Calculated excitonic energies of electron–hole pair in ZnS-capped CdSe nanocrystals as a function of their diameter [Brus, 1984] Right: Fitted absorption spectra of CdSe/ZnS ($d \approx 6$ nm) [Blaudeck, 2007] (from Blaudeck, T. (2007). *Self-Assembly of Functionalized Porphyrin Molecules on Semiconductor Nanocrystal Surfaces*, PhD Thesis, TU Chemnitz, Germany, with kind permission from Thomas Blaudeck) and assignment (Table 1.1) [Brus, 1984].

1.2.2 Wavefunction Tunneling

Though quantum confinement is very effective, potential barriers are in real systems not infinite. From this it follows that the confined wavefunctions of an electron and hole have finite amplitude outside the potential box. Due to the larger effective mass of the hole, as compared to the one of the electron, the hole is more strongly confined. However, in principle, there is always a finite amplitude of the respective wavefunctions beyond the surface of the QD. Consequently, the interaction with surface states and the embedding environment is nonnegligible and often gives rise to a reduction of the quantum yield of the photoluminescence (PL). Dabbousi et al. (1997) have calculated the amplitude of the radial part of the wavefunction for both the electron and the hole outside the QD under various conditions. Figure 1.6 shows that the amplitudes of the hole vanish outside the QD, while the one of the electrons has still considerable amplitude at the outer ZnS interface but decays exponentially beyond that interface. This exponential decay function depends on the kind of the matrix, especially on the related dielectric properties [Rabani et al., 1999; Leatherdale and Bawendi, 2001]. Blaudeck et al. (2008) have probed the wavefunction beyond the outer interface for a variety of QDs, making use of PL quenching induced by surface-attached organic dye molecules [Borczyskowski and Zenkevich, 2014].

It also becomes evident from Fig. 1.6 that capping the CdSe core by a thin layer of a different II–VI semiconductor material (in this case two monolayers of ZnS) imposes, if properly chosen, additional barriers for an electron and hole to reduce tunneling beyond the core material.

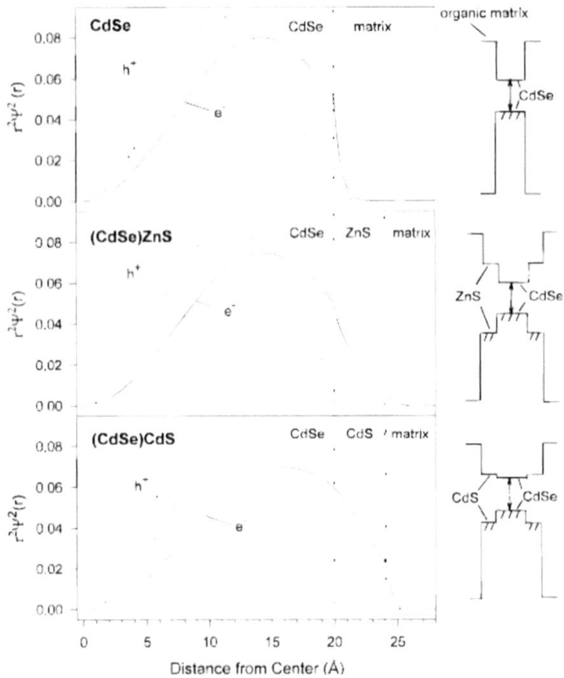

Figure 1.6 Radial probability functions for the lowest-energy ($1S^{3/2}$-Se) electron and hole wavefunctions for (top) 2 nm diameter CdSe QDs, (middle) CdSe/ZnS QDs capped by a 0.4 nm thick ZnS shell, and (bottom) CdSe/CdS with a 0.4 nm thick CdS shell. The sketches on the right show the approximate band offsets between various components. Reprinted with permission from Dabbousi, B. O., Rodriguez-Viejo, J., Mikulec, F. V, Heine, J. R., Mattoussi, H., Ober, R., Jensen, K. F., and Bawendi, M. G. (1997). (CdSe) ZnS core-shell quantum dots: synthesis and characterization of a size series of highly luminescent nanocrystallites, *Phys. Chem. B*, **101**, 9463. Copyright (1997) American Chemical Society.

The figure shows with respect to the electron wavefunction that the ZnS shell imposes a higher barrier as compared to the CdS shell.

Since tunneling of the wavefunction occurs, not only the barrier height is of importance but also the barrier thickness, as shown in Fig. 1.7. With recent improvements in the barrier design of thick CdS shells (giant QDs), it was even possible to suppress almost completely blinking processes [Vela et al., 2010, Chen et al., 2013; Qin et al., 2014].

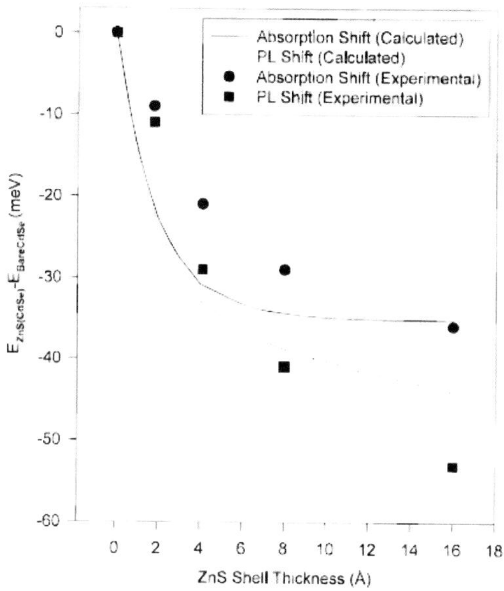

Figure 1.7 Energy difference between CdSe (2 nm radius) and CdSe/ZnS QDs as a function of ZnS shell thickness. Shifts in photoluminescence (PL) are given in solid squares and those in absorption as solid circles. Calculated energy shifts in absorption (PL) are plotted versus shell thickness as solid (dashed) lines. Reprinted with permission from Dabbousi, B. O., Rodriguez-Viejo, J., Mikulec, F. V, Heine, J. R., Mattoussi, H., Ober, R., Jensen, K. F., and Bawendi, M. G. (1997). (CdSe)ZnS core-shell quantum dots: synthesis and characterization of a size series of highly luminescent nanocrystallites, *Phys. Chem. B*, **101**, 9463. Copyright (1997) American Chemical Society.

When designing capping shells one has to take into account that lattice constants vary from compound to compound, which causes strain at the interface. Strain will give rise to defects, reducing the quantum efficiency. This can be partly overcome by designing gradual shell growing [Xie et al., 2005]. The lattice mismatch between CdSe and ZnS (CdS) is 12% (4%) [Xie et al., 2005; Wei et al., 2000]. On the other hand, strain can be used to tune optical and electronic properties of QDs [Smith et al., 2009]. The electronic structure of ZnSe/CdS QDs has also been engineered by interface alloying [Boldt et al., 2014].

1.2.3 Density Functional Theory

So far, we have discussed theoretical approaches to calculate quantum confinement based on EMA and related models. It is difficult in these methods to take surface properties other than by considering the impact of dielectric properties of the embedding environment into account. As we have pointed out in the beginning the surface-to-volume ratio is very high in QDs. Atoms at the surface carry unsaturated (dangling) bonds. Special efforts are necessary to take surface properties into account at a microscopic level. Recent progress in density functional theory (DFT) enables calculations for QDs as large as 1000 atoms.

Figure 1.8 shows such a calculation of valence and band-edge states for CdS as a function of the number of Cd atoms [Frenzel et al., 2007]. Calculations have been also performed by passivating dangling bonds of Cd by hydrogen. While without saturation many midgap states are created, they become much less upon passivation. Even then the band edge is not a clear cutoff energy, as suggested by EMA approaches. Though calculated crystals are considerably smaller than most of the experimentally investigated ones, these and similar calculations pave the road for a microscopic understanding of the surface- and interface-related optical properties. The work has been extended to CdSe/CdS core-shell systems for zinc-blende and wurtzite structures [Frenzel et al., 2011].

Other groups have calculated via ab initio methods (time domain DFT) the influence of organic surfactants (ligands) and the influence of solvents on the density of states, optical oscillator strengths, and absorption spectra. These calculations also provide access to hot-electron and exciton–phonon relaxation [Prezhdo, 2009]. Interesting is the calculation of the influence of the removal of a single ligand molecule, depending on the specific surface site [Fischer et al., 2012; Kilina et al., 2009]. Figure 1.9 shows the structure optimization for two different ligands and the specific sites which show different sensitivities on the density of states to ligand removal. Certainly, crystals are small, but calculations show the potential to investigate subtle processes at QD surfaces and interfaces. Several of these aspects have been discussed in more detail recently in relation with investigations on the influence of one dye molecule on CdSe/ZnS QD PL spectra and dynamics [Zenkevich et al., 2015].

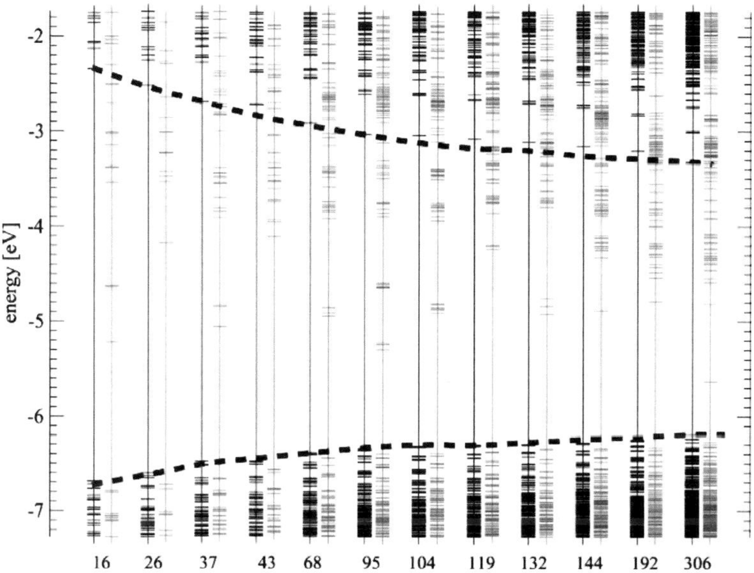

Figure 1.8 Eigenvalue spectra of the molecular orbitals closest to the Fermi level. The energy levels of the hydrogen-saturated nanoparticles are depicted in black, and those of the corresponding unsaturated particles are depicted in gray. The number of Cd atoms is given below for each set of eigenvalues representing the particle size. Reprinted with permission from Frenzel, J., Joswig, J. O., and Seifert, G. (2007). Optical excitations in cadmium sulfide nanoparticles, *J. Phys. Chem. C*, **111**, 10761–10770. Copyright (2007) American Chemical Society.

Surface traps play a crucial role in controlling the quantum yield and blinking processes. Voznyy (2011) had performed ab initio calculations of electronic properties of Cd-rich CdSe QDs with covalently bound carboxylic acid. Surface traps originating from ligands can form near the top of the valence band. Some of the ligands are found to be mobile on the surface, and this mobility is accompanied by a spectral diffusion of the associated trap energy levels. This provides a microscopic approach to explain the spectral diffusion, lifetime fluctuations, and blinking of QDs, as will be discussed in more detail in Chapters 4 and 6. Figure 1.10 shows a schematic view of the relation between spatial and spectral diffusion.

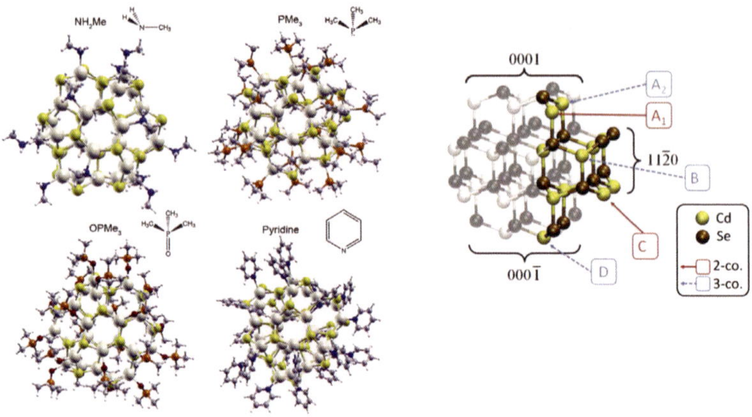

Figure 1.9 Optimized geometries of four ligated QD clusters. The ligands used are methylamine (NH_2Me), trimethylphosphine (PMe_3), trimethylphosphine oxide ($OPMe_3$), and pyridine, as shown on the insets. The right-hand panel shows the initial $Cd_{33}Se_{33}$ configuration obtained from the bulk CdSe crystal. One-third of the crystal is in color, highlighting its threefold symmetry. The two- and three-coordinated atoms on the surface of the cluster are listed in red and blue, respectively. Miller–Bravais indices for surfaces of the cluster are shown in black. Geometry optimization of the cluster leads to a reconstruction of the surface so that all surface atoms become at least three-coordinated. Reproduced from Fischer, S. A., Crotty, A. M., Kilina, S. V., Ivanov, S. A., and Tretiak, S. (2012). Passivating ligand and solvent contributions to the electronic properties of semiconductor nanocrystals nanoscale, *J. Am. Chem. Soc.*, **4**, 904–914 with permission of The Royal Society of Chemistry.

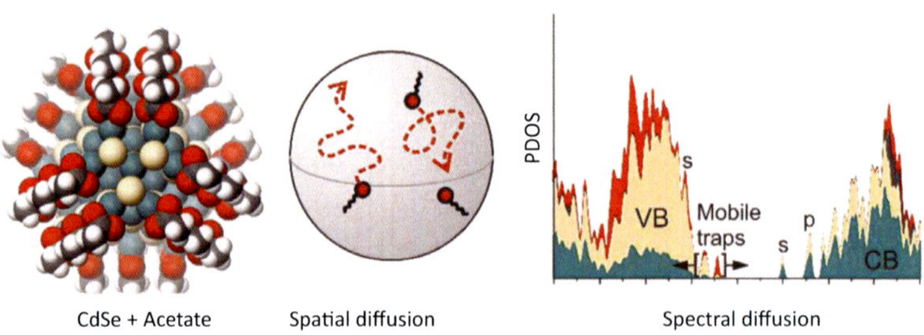

Figure 1.10 Schematic view of the relation between spatial diffusion of surface ligands and calculated spectral diffusion. From Voznyy, O. (2011). Mobile surface traps in CdSe nanocrystals with carboxylic acid ligands, *J. Phys. Chem. C*, **115**, 15927–15932, with kind permission from Oleksandr Voznyy.

Finally, Gómez-Campos and Califano [2012] calculated the trapping of holes to surface trap sites located on the surface of the core or the shell or at the core/shell interface in CdSe QDs with both organic and inorganic passivation, which is important for carrier trapping with respect to blinking processes but also for charge-related devices.

1.3 Optical Transitions

1.3.1 Assignment of Optical Transitions

Single-band EMA has been successfully applied to conduction bands of II–VI semiconductor QDs (of metal s-type orbitals, for example, Zn^{2+}) but mainly fails for valence bands of p-type character (e.g., Se^{2-}). Figure 1.2 has already shown that valence band (hole states) are much more complex than conduction band states. The electronic configuration of Se^{2-} is $[Ar]4s^23d^{10}4p^6$. Spin–orbit coupling becomes important, and the description is based on the total angular momentum. The spin-related p states can be written as |3/2, ±3/2>, |3/2, ±1/2>, and |1/2, ±1/2>. The conduction band of CdSe is to first approximation composed of Cd^{2+} orbitals with an electronic configuration [Kr] $4d^{10}5s^0$. As s orbitals are single-degenerate, electron states are of the $1s_e$, $1p_e$, $1d_e$, or $2s_e$ character. Optical transitions between hole and electron states will now follow selection rules of electric dipole–allowed transitions, as shown for the energetically lowest transitions in Fig. 1.11. Naturally, such assignments depend on the symmetry group of the QD that is its crystallographic structure. More detailed information is presented in Chapter 2.

An experimental absorption spectrum of a CdSe/ZnS QD is shown in Fig. 1.5, including the corresponding assignment of transitions given in Table 1.1. The size-dependent ground state of an exciton is assigned to the separated electron–hole pair. Its energy and size dependence depend on the degree of (weak or strong) confinement [Thoai et al., 1990]. The fine structure of the pair state ($1S_{3/2}$, $1S_e$) assigned to transition 1 in the given example of Fig. 1.11 is split in nonspherical QDs [Nirmal et al., 1995; Chamarro et al., 1996]. As is shown in more detail in Chapter 2 the related optical transitions can be classified as allowed or forbidden, depending on the crystal symmetry (crystal field). The lowest exciton state of the fine structure of CdSe QDs is dark (long lived). Since the splitting is very small (~10 meV), thermal equilibration will allow for PL on temperature increase.

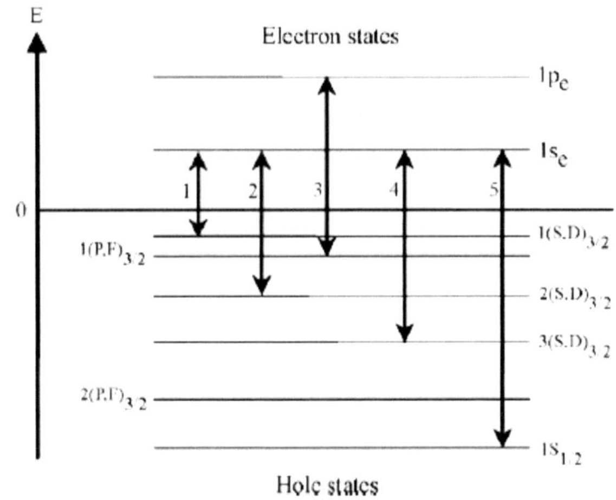

Figure 1.11 Dipole-allowed transitions among the lowest valence and conduction band states, taking into account valence state mixing (zinc-blende type), according to Woggon (1996), Adapted from Issac, A. (2006). *Photoluminescence Intermittency of Semiconductor Quantum Dots in Dielectric Environments*, PhD Thesis, TU Chemnitz, Germany, with kind permission from Abey Issac.

Table 1.10 Fit of the optical absorption spectrum of CdSe/ZnS (d_{core} = 5.1 nm; 2 ML ZnS) by four optical transition Gaussian bands according to amplitudes A_i, widths w_i, and center wavelengths λ_i. Assignment of transitions according to Ekimov et al. (1993)

i	Transition	$\lambda_{c,i}$ [nm]	w_i [nm]	A_i [1]
1	1S(e)-1S$_{3/2}$(h)	613.6 ± 0.2	20.1 ± 0.3	21.1 ± 0.3
2	1S(e)-2S$_{3/2}$(h)	588.0 ± 0.4	19.0 ± 0.6	8.9 ± 0.4
3	1P(e)-1P$_{3/2}$(h)	538.6 ± 0.4	23.1 ± 0.6	12.0 ± 0.5
4	1S(e)-3S$_{1/2}$(h)	509.0 ± 0.5	23.8 ± 0.7	10.5 ± 0.5
5		361.7 ± 1.8	214.0 ± 2.6	796 ± 15

1.3.2 Relaxation of Excited Quantum Dots

In the previous section we have shown which excitonic energies are coupled with each other by dipole-allowed optical transitions. This applies for absorption but not in the same way for PL emission following an absorption process since fast nonradiative transitions occur from higher to lower states. In general, the population of an excited exciton state will be followed by fast relaxation (subpicosecond) processes (hot-electron relaxation) to the lowest (fine structure splitted) exciton state from which photons are emitted (depending on the temperature according to thermal equilibration of the fine structure level population) with a radiative rate k^{r} (k_{fl}). Figure 1.12 summarizes these and other nonradiative relaxation processes among exciton states. Intraband relaxation takes place both for electrons and holes (dotted lines). For electrons, this relaxation process is on a timescale of a few hundred femtoseconds and that for holes is a few picoseconds. In competition to the radiative rate (k_{fl}) nonradiative processes (k_{ic}) such as exciton–phonon coupling (see Chapter 3) and relaxation to trap or surface states take place on various timescales. These states return to the ground state via nonradiative (or weakly radiative) rates. Such surface and/or trap states are heavily involved in luminescence intermittency [Nirmal et al., 1996; Cichos et al., 2007] (or reversible photobleaching [Chung and Bawendi, 2004; Cichos et al., 2004]). Luminescence intermittency (or blinking) is one of the major topics of this book and will be discussed in great detail in Chapters 4 and 6.

1.3.3 Surface and Trap States

Calculations discussed in Section 1.2.3 show that the band edges are split into discrete levels. An assignment as trap, surface, or exciton states might become obsolete since calculations show that these states also might be electric dipole allowed. Experiments show that both ensembles and single QDs show in many cases intensity, lifetime, and energy fluctuations which cannot in all cases be merely assigned to charging, charge separation, or Auger processes. For a more detailed discussion, see also Chapters 4 and 6. Issac (2006) has discussed trap states under different aspects. A trap can also be considered as a localized electronic state that captures charge carriers. Generally one can characterize trap states by deep and shallow traps (see Chapter 2). Whether the latter can be clearly discriminated against real QD states might be a matter of semantics. Following Issac (2006) several types of traps might be considered:

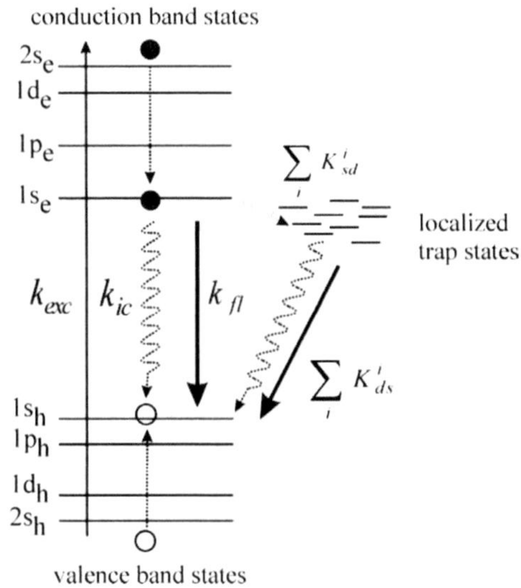

Figure 1.12 Scheme of the relaxation processes among exciton states. Excitation with a rate k_{exc} generates excitons which relax to the band edges via fast nonradiative relaxation (dotted arrows). Radiative recombination of the electro-hole pair takes place with a rate k_{fl}. k_{ic} represents nonradiative relaxation processes from the band-edge states. Trapping of charges occurs to spatially and energetically distributed trap states. Radiative decay from trap states in parallel to strong, dark back reactions results in defect luminescence. From Issac, A. (2006). *Photoluminescence Intermittency of Semiconductor Quantum Dots in Dielectric Environments*, PhD Thesis, TU Chemnitz, Germany, with kind permission from Abey Issac.

- Crystal imperfections such as point defects due to missing or misplaced atoms, chemical impurities, and large structural imperfections such as dislocations or grain boundaries [Bube, 1987].
- Surface-related traps [Hache et al., 1991; Lueth, 2001] are associated with surface reconstruction, stoichiometric defects (Cd^{2+}, Zn^{2+}, Se^{2-}, and S^{2-} vacancies in the case of CdSe/ZnS QDs), dangling bonds, external adatoms, or molecules (like oxygen or dyes) and interfacial defects due to lattice mismatch [Němec et al., 2000].

- Self-trapped states: Carriers might be trapped in states localized at the QD–matrix interface [Tian et al., 1994; Wang et al., 1991; Zhang et al., 1994] or outside the QD in the surrounding matrix [MacDonald et al., 1993; Woggon et al., 1992; Kaschke et al., 1990; Issac et al., 2005]. Some aspects of self-trapped charges will be discussed in Chapter 4.

The crucial parameters describing the process of localization and trapping are the relation between the spatial overlap of the exciton wavefunction with that of trap potential as well as the depth of local potential fluctuations. The trapping process decreases the overlap of electron–hole wavefunctions and therefore increases the recombination time. The presence of traps not only changes the recombination process but can also cause lattice distortion, alternate the potential barrier, or activate chemical reaction, as has been shown by calculations [Voznyy, 2011] experimentally [de Mello Donega et al., 2006; Zenkevich et al. 2015] and is discussed in more detail by Borczyskowski and Zenkevich (2014).

The presence and creation of deep traps have been shown in an investigation of CdSe QDs upon exchange of dodecylamine on tri-(n-octyl)phosphinoxid (TOPO)-capped QDs by 3-mercaptopropionic acid [Baker and Kamat, 2010], which amplifies deep trap emission, always abundant to a certain extent in uncapped QDs, without changing the absorption considerably (see Fig. 1.13). This way the color can be tuned across a wide range.

1.3.4 Multiexcitons and Charged Quantum Dots

Multiexciton effects on optical properties of QDs are extensively studied both experimentally [Klimov, 2007] and theoretically [Prezhdo, 2008] to name only a few and are beyond the scope of this book. However, we mention multiexcitons for two reasons. First, one of the models for luminescence intermittency in QDs is (as also discussed in Chapter 4) that a charge (electron) is photoejected, leaving the QD positively charged, and that the newly excited exciton decays radiationless, following an Auger relaxation process (particle–particle interaction) [Nirmal et al., 1996]. Second, the PL of QDs, both the one of bulk material as well as the one of single dots, is in most cases multiexponential [Schlegel et al., 2002; Zenkevich et al., 2015]. One reason for this finding might be due to the creation of charged QDs (as discussed in Chapter 6).

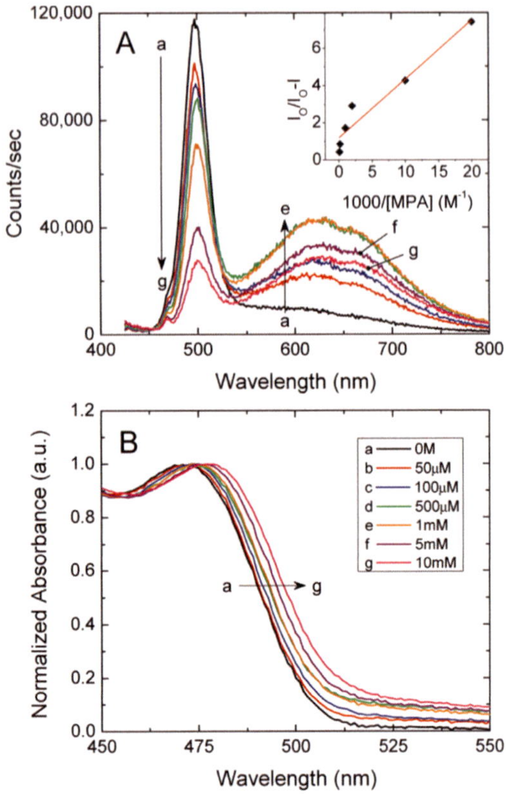

Figure 1.13 (A) Photoluminescence spectra and (B) absorbance spectra (normalized to the excitonic peak) of CdSe quantum dots (2.6 nm diameter) with different concentrations of MPA in solution: (a) 0 M, (b) 50 μM, (c) 100 μM, (d) 500 μM, (e) 1 mM, (f) 5 mM, and (g) 10 mM. The inset of (A) is a double reciprocal plot representing the quenching of the band-edge emission at 497 nm. Reprinted with permission from Baker, D. R., and Kamat, P. V. (2010). Tuning the emission of CdSe quantum dots by controlled trap enhancement, *Langmuir*, **26**, 11272–11276. Copyright (2010) American Chemical Society.

Figure 1.14 shows a schematic excitonic energy diagram with possible optical transitions [Akimov et al., 2003]. Due to interactions among the particles and quasi-particles under confinement conditions, fast relaxation is observed, which can be investigated by optical pump-probe experiments. Since several excitation processes are involved, generation of multiexcitonic states requires high excitation intensities, though arguments have been presented that multiexcitations can be accumulated [Peterson and Nesbitt, 2009].

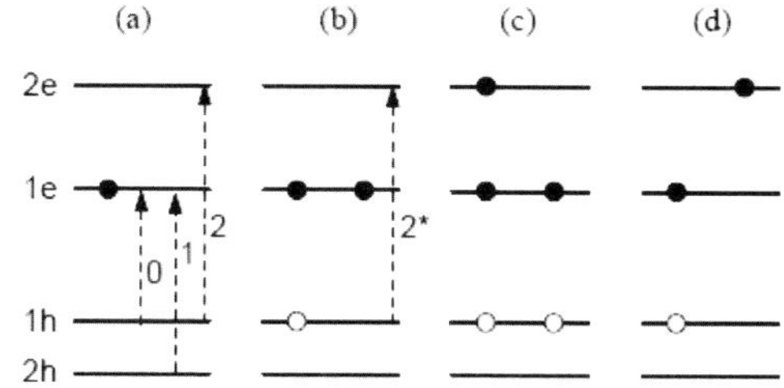

Figure 1.14 Excitonic energy diagrams showing with allowed optical transitions showing (a) a negatively charged CdSe QD, (b) a trion, (c) a charged biexciton, and (d) an excited trion. A biexciton consist of two holes and two electrons. Reprinted from Akimov, I. A., Hundt, A., Flissikowski, T., Kratzert, P., and Henneberger, F. (2003). Energy spectrum of negatively charged single quantum dot: trion and charged biexciton states, *Physica E*, **17**, 31–34, Copyright (2003), with permission from Elsevier.

Theoretical investigations [Califano et al., 2007] report on calculations for energies and radiative transition rates, as shown exemplarily in Figs. 1.15 and 1.16.

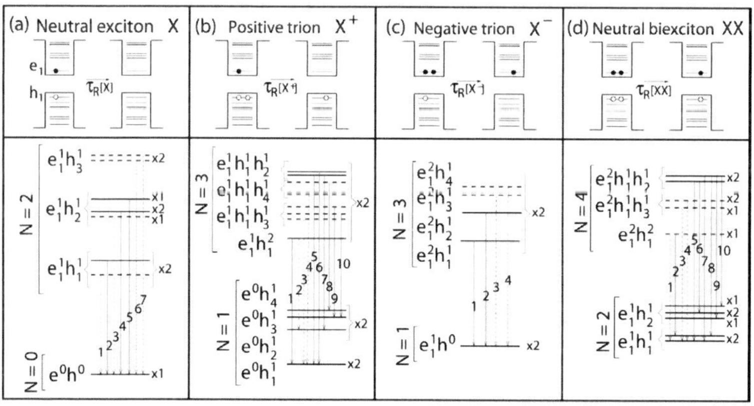

Figure 1.15 Schematics of calculated single-particle (upper panels) and excitonic (lower panels) levels for (a) neutral exciton X, (b) positive trion X^+, (c) negative trion X^-, and (d) neutral biexciton XX. While the order of the excitonic levels in the lower panel is accurate, the energy scale is schematic. The total number of particles is denoted by $N = n + m$. The solid (dashed) horizontal lines indicate bright (dark) excitons, whereas the vertical arrows indicate transitions between levels. The numbers on these arrows label the transitions shown in Fig. 1.16. Reprinted (figure) with permission from [Califano, M., Franceschetti, A., and Zunger, A. (2007). Lifetime and polarization of the radiative decay of excitons, biexcitons, and trions in CdSe nanocrystal quantum dots, *Phys. Rev. B*, **75**, 115401.] Copyright (2007) by the American Physical Society.

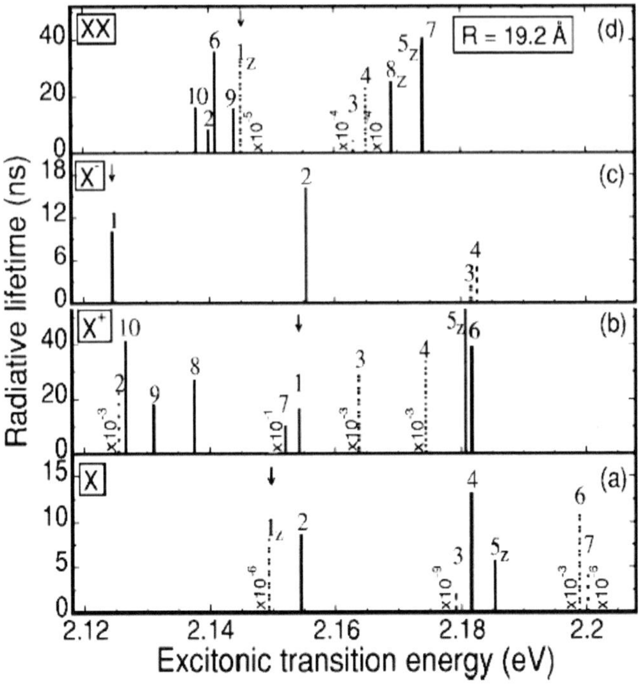

Figure 1.16 Radiative lifetimes at zero temperature as a function of emission energy for the various excitonic states considered here. Bright (solid lines) and dark (dashed lines) transitions are labeled according to Fig. 1.15. The subscript *z* denotes transitions polarized along the *z* axis. The arrows mark the position of the fundamental transition (labeled 1 in Fig. 1.15) from the *N*-particle ground state to the (*N* – 2)-particle ground state. Reprinted (figure) with permission from [Califano, M., Franceschetti, A., and Zunger, A. (2007). Lifetime and polarization of the radiative decay of excitons, biexcitons, and trions in CdSe nanocrystal quantum dots, *Phys. Rev. B*, **75**, 115401.] Copyright (2007) by the American Physical Society.

Multiple excitons can be created at a sufficient excitation rate in a volume smaller than the excitonic volume $V \approx a_B^3$. A QD may carry two electron–hole pairs within the lifetime of the excited state. The energy of one of the pairs is transferred to the remaining electron (hole), which is termed Auger relaxation, and may cause charging of a QD known as autoionization. Due to the high density of states far from the band edge, creation of a second electron–hole pair becomes possible even in an ionized QD. The electron–hole pair couples to the remaining charge carrier, resulting in very fast nonradiative decays

(~100 ps) [Roussignol et al., 1987], which is much faster than the radiative lifetime (~20 ns). Due to these processes a charged QD appears as a dark QD, though still absorbing light. Naturally, the strength of the interparticle coupling depends on the degree of confinement and might be less pronounced, for example, in CdSe/CdS QDs as compared to CdSe/ZnS ones, since the wavefunction of the electron is much more extended due to a reduced barrier for the conduction band [Zhao et al., 2010].

1.4 Synthesis, Crystal Structure, and Ligands

Since this book concentrates on the tuning of optical properties of *as-prepared* QDs, focusing on spectroscopy and self-assembly [Zenkevich and Borczyskowski, 2016], we will give only a short introduction into the synthesis and crystal structure. Ligands will be treated only in the sense that we will just stress their importance during crystal growth and their role in preventing aggregation of size-selected QDs. Recently a comprehensive review has been published, which outlines with respect to semiconducting QDs the present status of material manufacturing and tailoring [Kovalenko et al., 2015]. In those chapters of this book, which deal with semiconducting QDs, we concentrated on colloidal QDs and, among them, especially on CdSe. We will follow this concept in this section also.

Synthesis of colloidal semiconductor QDs has been reported by Brus and coworkers [Cheatnoy et al., 1986] for solution-phase synthesis of inorganic II–VI clusters. The route of organometallic crystallization in solution is nowadays the standard routine for a reliable synthesis of (photo)chemically stable QDs characterized by a high quantum yield and narrow size distribution, referred to as the benchmark synthesis [Eychmüller, 2000]. Among the organometallic approaches, the tri-(n-octyl)phosphine (TOP)/TOPO synthesis evolved as the standard routine for II–VI semiconductor QDs. It is based on pyrolysis of organometallic precursors injected into a hot coordinating organic liquid. For CdSe QDs, dimethylcadmium and bis(trimethylsilyl)selenium are often used as precursors and TOP and TOPO as solvents [Murray et al., 1993]. According to the processing CdSe QDs crystallize in one of two possible crystal structures:

cubic zinc-blende (face-centered cubic [fcc]) or hexagonal wurtzite (hexagonal closed packing [hcp]). After preparation QDs are usually stored in (water-free) organic liquids, for example, toluene with an excess of ligands.

Organic ligands play an essential role not only during synthesis but also during the further use since they prevent from ongoing aggregation if the QDs are stored in solution, and protect the QD surface from oxidation and attachment against possible PL quenchers. Depending on the solvent, ligands are (after crystal preparation) in dynamic equilibrium with excess ligands, concurrent ligands, and appropriately functionalized molecules. We have discussed the possible consequences recently [Zenkevich et al., 2015]. The relative binding of ligands varies for different facets of the CdSe surface and is in the case of TOPO a coordinative interaction between the lone pairs of the oxygen atom of phosphine and the Cd^{2+} ions of the conduction band, leaving hole-related Se^{2-} ions of the valence band unsaturated. Ab initio calculations further showed that the binding energies of the ligands control the relative growth rates of different facets and therefore significantly affect the resulting geometry of the QDs [Peng et al., 2000; Mann et al., 2002; Puzner et al., 2004]. DFT calculations also provide binding energies [Kilina et al., 2009; Schapotschnikow et al., 2002] and have suggested that ligand binding is more close to a (dielectrically controlled) electrostatic interaction.

In any case passivation of semiconductor surfaces (i) prevents from chemical reactions at the surface; (ii) influences effectively the energy landscape, quantum yield, and relaxation processes (e.g., monitored by intensity, energy, and PL decay time blinking) at the surface [Kilina et al., 2009]; and (iii) prevents from further QD growth or Ostwald ripening. However, eventually steric hindrance among spacious ligands (or attached spacious organic molecules) prevents from the passivation of all chemically accessible dangling bonds. In the case of TOPO, at most 50% bonds of the smaller QDs and 30% of the larger ones can be effectively accessed [Kowerko et al., 2010]. Figure 1.17 gives an impression of steric relations for two different types of ligands.

On the other hand, ligands control dynamic processes such as charge transfer outside the particle-in-a-box QD [Knowles et al., 2012]. PL emission energies can be tuned through ligand exchange [Brown et al., 2014]. Surface-related emission has been identified for CdSe QDs [Wang et al., 2003] and is also discussed in Chapter 2.

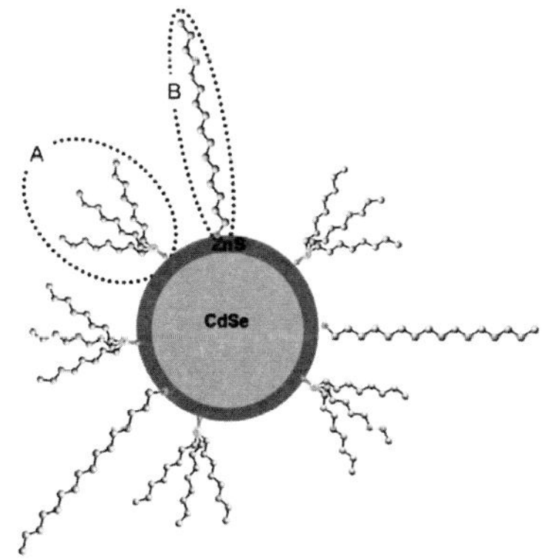

Figure 1.17 Schematic presentations (not on scale) of a CdSe/ZnS QD with TOPO (A) and hexadecylamine (HDA) ligands (B).

The relative weak binding of ligands allows for ligand substitution after crystal preparation [Ji et al., 2008]. This allows for exchange from conduction band to valence band passivation schemes and vice versa. Additionally, this way solubility (e.g., in water) and functionality might be changed, which is important for sensoric applications in biology. Finally, a basic characterization of a QD with respect to its optical properties is absorption and emission (PL) spectra, as shown in Fig. 1.18 for CdSe/ZnS in toluene.

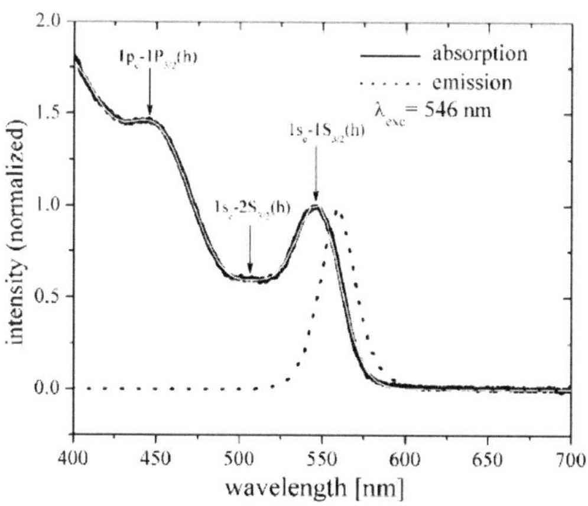

Figure 1.18 Absorption and PL emission of CdSe/ZnS in toluene at room temperature. Absorption bands to three different exciton states are indicated. From Issac, A. (2006). *Photoluminescence Intermittency of Semiconductor Quantum Dots in Dielectric Environments*, PhD Thesis, TU Chemnitz, Germany, with kind permission from Abey Issac.

At the very end we show a classical example of the effect of quantum confinement on optical spectra for CdSe QDs of different sizes, clearly showing the shifting excitonic energies in Fig. 1.19.

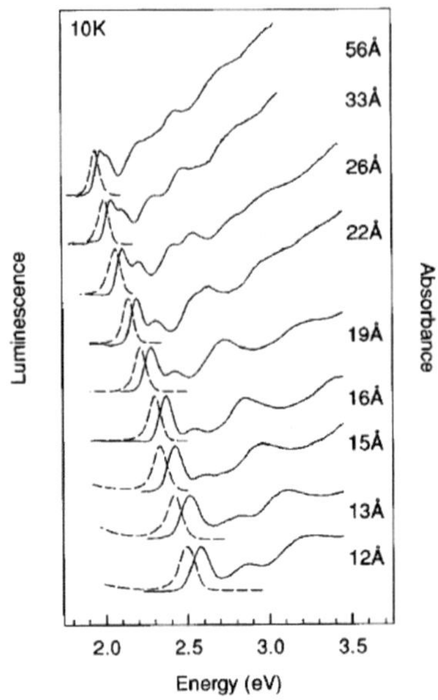

Figure 1.19 Normalized absorption and PL spectra for CdSe QDs between 12 and 56 Å in radius. The absorption spectra are indicated by solid lines and the corresponding PL spectra by dotted lines. Reprinted (figure) with permission from [Efros, A. L., Rosen, M., Kuno, M., Nirmal, M., Norris, D. J., and Bawendi, M. (1996). Band-edge exciton in quantum dots of semiconductors with a degenerate valence band: dark and bright exciton states, *Phys. Rev. B*, **54**, 4843.] Copyright (1996) by the American Physical Society.

1.5 Metal Nanoparticles

Metal nanoparticles are another class of quantum objects which show size-dependent optical properties. Silver (Ag), copper (Cu), and gold (Au) particles are the most commonly investigated systems. Nanoscale metals can be classified into three size domains—large nanoparticles, small nanoparticles, and clusters—corresponding to three different characteristic length scales with respect to the particle diameter R [Kreibig and Vollmer, 1995]. These regimes are (i) $R > \lambda$ (nearly bulk-like dielectric function), (ii) $R \approx$ free electron path length (strong size dependence), and (iii) $R \approx$ Fermi wavelength of an electron (breakdown of Mie theory and crossover to molecule-like behavior). For Au and Ag, these critical values are at about 50 nm and 0.5 nm, respectively.

Basically, there are two different approaches to describe the size-dependent physical features, namely a bottom-down and a bottom-up approach. The bottom-down approach is based on the size-dependent interaction of an electromagnetic (optical) field with a metal containing a free-electron gas. This approach makes use of the dielectric function of a gas of noninteracting charged particles (electrons) and can be considered as a plasma-based approach on the background of Mie theory constituting the contemporary, active research field of nanoplasmonics. The bottom-up approach starts with single metal atoms followed by cluster formation, making use of an atomic description of wavefunctions and describing the formation of aggregates in terms of a molecule-type binding. These very small clusters show both absorption and emission features in the optical range. The latter approach is limited to the lower size range and is especially important for fluorescence labeling applications in biology and eventually super-resolution microscopy.

1.5.1 Plasmonic Nanoparticles

Metal nanoparticles have been used through the centuries as dispersed metal colloids to create brilliant colors. There are several comprehensive descriptions of their optical properties [Kreibig and Vollmer, 1995; Maier, 2007; Gaponenko, 2010]. The optical properties are determined by the electrons in the conduction band upon excitation to energetically higher bands. The impact of the crystal lattice is taken into account in changing the mass of the free electron to an effective mass m_{eff}. Irradiating a small, spherical metal nanoparticle by light, the oscillating electric field causes the conduction electrons to oscillate coherently, as shown in Fig. 1.20. The oscillation frequency (plasma resonance of the so-called plasmon) is determined by Coulomb interaction between the electrons and the positively charged core atoms. This frequency depends on the electron density in the specific metal, the effective mass m_{eff} of the electron, the size and the shape of the NC, and the dielectric properties of the embedding medium. A (simplified)

Mie scattering theory describes for particles much smaller than the wavelength of light the plasmon resonance frequency [Kelly et al., 2003].

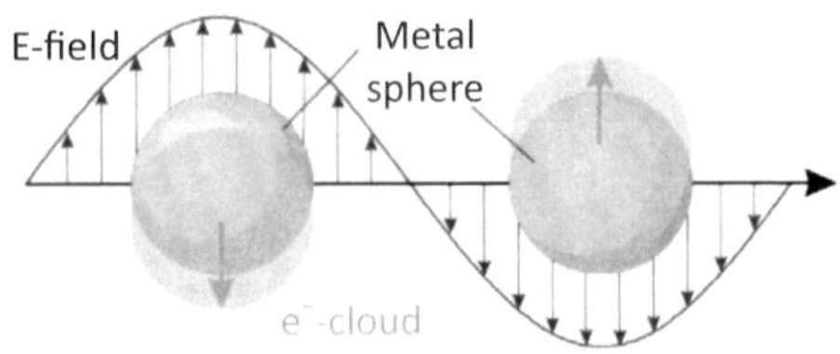

Figure 1.20 Schematic of plasmon oscillation for a sphere, showing the displacement of the conduction electron charge cloud relative to the nuclei. Reprinted with permission from Kelly, L., Coronado, E., Zhao, L. L., and Schatz, G. C. (2003). The optical properties of metal nanoparticles: the influence of size, shape, and dielectric environment, *J. Phys. Chem. B*, **107**, 668–677. Copyright (2003) American Chemical Society.

To calculate the influence of an electromagnetic field on the electron dynamics, the real and imaginary parts of the dielectric function of the nanoparticle have to be calculated. Interaction of an electromagnetic field with a metal particle results in absorption (extinction) and (Rayleigh) scattering. The frequency-dependent dielectric function of a free-electron gas can be written as

$$\varepsilon(\omega) = 1 - \frac{\omega_{\mathrm{p}}^2}{\omega^2 + i\gamma \cdot \omega}. \tag{1.7}$$

The plasmon resonance frequency ω_{p} corresponds to

$$\omega_{\mathrm{p}}{}^2 = (n\, e^2/\varepsilon_0 m_{\mathrm{eff}}) \tag{1.8}$$

where n is the electron density and γ a damping constant. In a nanoparticle the oscillations will be damped due to surface effects, depending on the radius R according to

$$\gamma(R) = \gamma_0 + \frac{A \cdot v_{\mathrm{F}}}{R}, \tag{1.9}$$

where γ_0 is the damping of the corresponding metal and A a phenomenological factor taking the specific interaction at the surface into account. V_{F} is the velocity of electrons at the Fermi energy. For very small metal particles this results in a broadening of the plasmon resonance transition but not in a shift of the resonance frequency. For still smaller particles (see Section 1.5.2) a quantum mechanical approach is necessary.

For a nanoparticle with radius R close to the exciting wavelength λ, the plasmon resonance will be shifted to longer wavelengths. The related plasmon dielectric function is given by Kreibig and Vollmer [1995] according to

$$\begin{aligned}\varepsilon(\omega,R) = \varepsilon_{\text{bulk}}(\omega) + \omega_{\text{p}}^2(1((\omega^2+\gamma_0^2)^{-1} - 1(\omega^2+\gamma(R)^2)^{-1}) + \\ i\omega_{\text{p}}^2\omega^{-1}(\gamma(R)(\omega^2+\gamma(R)^2)^{-1} - \gamma_0(\omega^2+\gamma_0^2)^{-1})\end{aligned} \quad . \quad (1.10)$$

The imaginary part describes the absorption of particles in the intermediate range of sizes. Typical experimental absorption spectra are shown for gold particles in Fig. 1.21 for various crystal diameters.

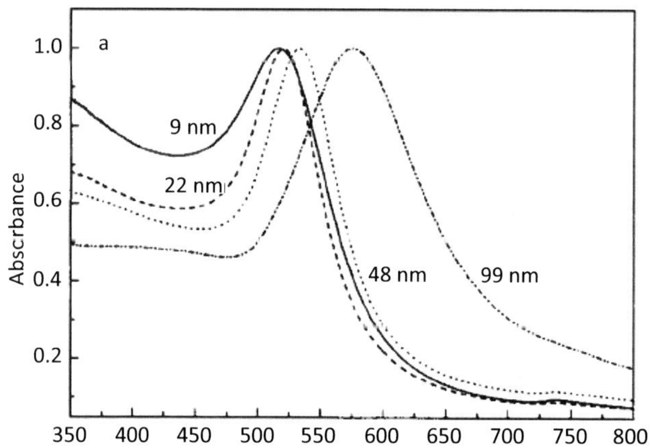

Figure 1.21 Absorption spectra of Au particles in water for different diameters. Reprinted with permission from Link, S., and El-Sayed, M. A. (1999). Size and temperature dependence of the plasmon absorption of colloidal gold nanoparticles, *J. Phys. Chem. B*, **103**, 4212–4217. Copyright (1999) American Chemical Society.

So far, we have only discussed spherical nanoparticles. However, the shape and symmetry of a metal particle will also influence the dielectric function. Without going into detail, experimental examples from a recent review [Crut et al., 2014] are shown in the case of single-particle detection in Fig. 1.22. A comprehensive review of single-metal particle detection has been published by Zijlstra and Orrit (2011), including applications such as second harmonic generation.

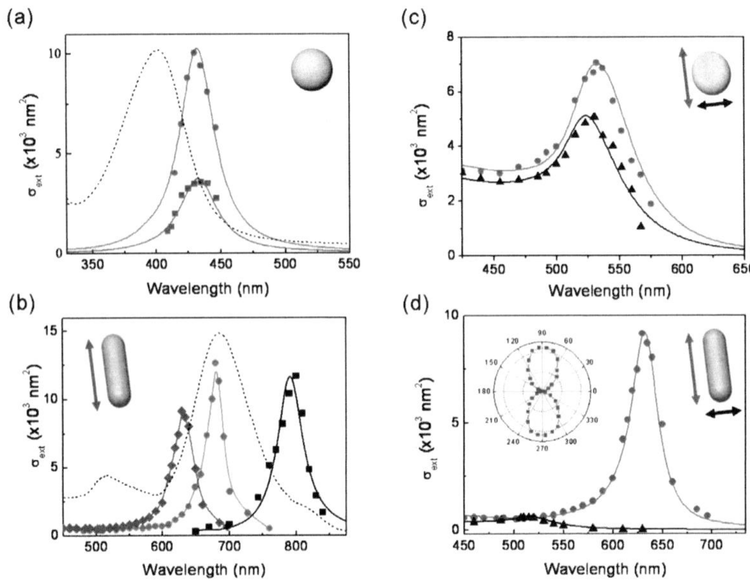

Figure 1.22 Absolute extinction cross-sectional (σ_{ext}) spectra of different single-metal nanoparticles measured by spatial modulation spectroscopy. (a) Silver nanospheres with diameters of 30 nm (circles) and 21 nm (squares). (b) Gold nanorods with aspect ratios of 2.5, 3.0, and 4.0 (with increasing SPR wavelength) for light polarized along their main axis (the dotted line shows the absorption spectrum of the initial solution). (c) A quasi-spherical gold nanoparticle for incident light polarized along two perpendicular directions (the wavelength and amplitude changes reflect its slight ellipticity with dimensions of 50 and 46 nm for the long and the short axis, respectively). (d) A single gold nanorod for light polarized along its long and short axes (long axis of 20.5 nm and aspect ratio $Z = 2$). Inset: σ_{ext} polarization dependence measured at 630 nm. The full lines are fits by assuming a Lorentzian shape in (a) and (b), an ellipsoidal model in (c), or finite element modeling in (d). Reproduced from Crut, A., Maioli, P., Del Fatti, N., and Vallée, F. (2014). Optical absorption and scattering spectroscopies of single nano-objects, *Chem. Soc. Rev.*, **43**, 3921–3956 with permission of The Royal Society of Chemistry.

1.5.2 Metal Clusters

Plasmonic metal nanoparticles do not fluorescence. The apparent bright color is related to strong absorption in a relatively narrow wavelength range. This is different for very small clusters. Approaching metal clusters from the atomic side it is possible to form selectively small clusters of metals with N atoms, for example, in the gas phase and depositing them into a rare gas matrix [Harbig et al., 1990]. Such clusters show upon excitation several discrete and strong optical transitions related to the number of constituting atoms and geometry of the (molecular) cluster. Structures and optical energies can be calculated by adapted quantum mechanical models for small clusters, see the example [Sieber et al., 2004].

The preparation of very small metal clusters by other means than by gas-phase techniques is especially important with respect to fluorescence-based applications. Already in the 1980s, particles were shown to exhibit a pronounced fluorescence, namely metal nanoclusters, especially Ag clusters [Ozin and Hugues, 1983; Treguer et al., 2005]. In experiments with Ag clusters formed by only a few atoms embedded in rare gas matrices, it turned out that the absorption and emission characteristics of such clusters strongly depend on the cluster size [Ozin and Hugues, 1983; Harb et al., 2008; Shang et al., 2011]. A change by a single atom or even a change in the atomic arrangement can lead to strong modifications of optically allowed transitions [Ozin and Hugues, 1983; Harb et al., 2008; Bonačić Koutecký et al., 2001].

The optical behavior of small clusters has been reviewed, especially for Au clusters, by Dickson and coworkers [Zheng et al., 2007]. Experiments on metal nanoclusters in liquids proved a strong dependence of the optical emission energy E_{em} on N number of atoms forming these clusters and the Fermi energy E_{Fermi} of the respective metal according to [Zheng et al., 2004]

$$E_{em} \approx E_{Fermi}/N^{1/3}. \tag{1.11}$$

Figure 1.23 shows the quantitative correlation of the emission energy with N number of atoms per cluster.

Figure 1.24 shows schematically optimal potentials adapted for the corresponding cluster sizes reaching from a spherical potential to the particle in the box model.

In the case of very small clusters, most or even all atoms are at the surface of the cluster and are therefore strongly influenced by environmental changes or surface-attached ligands which are necessary to stabilize the clusters [Zheng et al., 2003; Oemrasingh et al., 2012]. Similarly as for colloidal semiconductor QDs, metal clusters in solution have to be protected by ligands or shells from further aggregation or chemical reactions.

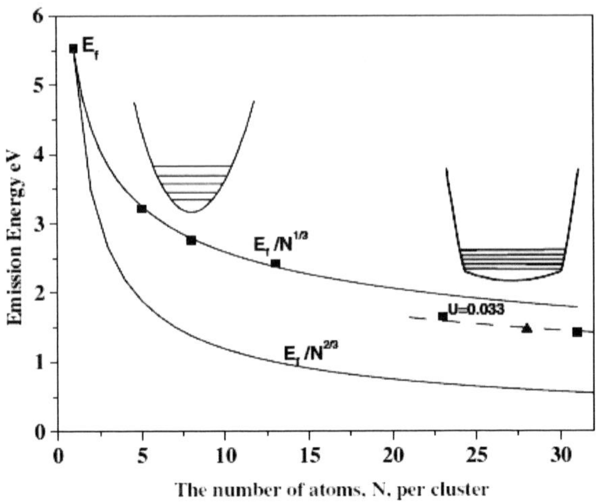

Figure 1.23 Correlation of the number of N atoms per cluster with emission energy. The correlation is quantitatively fitted with $E_{Fermi}/N^{1/3}$, as predicted by the jellium model [Kreibig and Vollmer, 1995; Haberland, 1994]. Emission energies of Au_{23} and Au_{31} exhibit slight deviations from the scaling. Consistent with the narrow excitation and emission spectra, the potential confining the free electrons flattens slightly for Au_{23} and Au_{31} with an anharmonicity of $U = 0.033$. The black triangle characterizes Au_{28}. The square-box model ($6/5E_f/N^{2/3}$) [Kreibig and Vollmer, 1995], which is also shown in the figure, and Kubo's predicted model E_f/N [Kubo, 1962] cannot accurately fit the emission energy scaling of the Au clusters. Reprinted (figure) with permission from [Zheng, J., Zhang, C., and Dickson, R. M. (2004). Highly fluorescent, water-soluble, size-tunable gold quantum dots, *Phys. Rev. Lett.*, **93**, 077402.] Copyright (2004) by the American Physical Society.

Small Au and Ag clusters have been used as fluorescence labels in molecular biology [Zheng et al., 2007]. Even the luminescence of single-metal clusters has been detected, as outlined in the short review in Chapter 5. Single-metal cluster detection extends the range of using metal clusters as probes, for example, in single-particle-tracking experiments [Krause et al., 2013]. A comparison of ensemble Ag cluster fluorescence spectra and single ones on a SiO_2 substrate is shown in Fig. 1.25. Clusters have been preselected in diameter via synthesis in zeolithes [Kahle et al., 2014]. Additional size selection occurs via selective laser excitation [Krause et al., 2013].

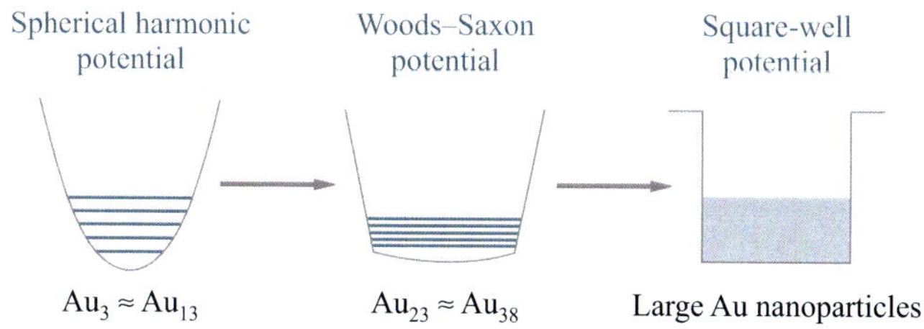

Figure 1.24 Schematic of size-dependent surface potentials of Au clusters on different size scales. For the smallest Au clusters (Au_3 to Au_{13}), cluster emission energies can be well fit with the energy scaling law $E_{Fermi}/N^{1/3}$, where N is the number of atoms in each cluster, indicating that electronic structure transitions of these small Au clusters are well described by a spherical harmonic potential. With increasing size, small anharmonicities distort the potential well, which at larger sizes gradually distorts into a Woods–Saxon potential surface [Klemenger, 1985] and eventually becomes a square-well potential, characteristic of electrons in large metal nanoparticles [Kreibig and Vollmer, 1995; Zheng et al., 2007].

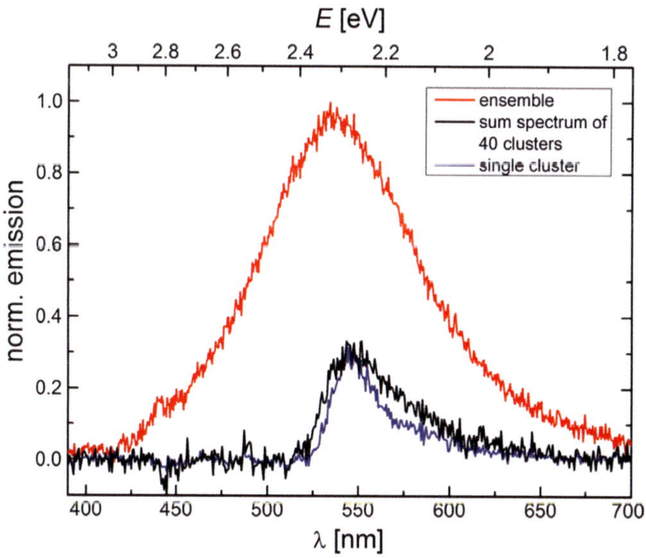

Figure 1.25 Comparison of fluorescence spectra of an ensemble (red, broad line; excitation 275 nm), a single Ag cluster (blue, slightly structured line; excitation 488 nm), and the sum of 40 single Ag cluster spectra (black line) at 293 K. From Krause, S., Heidernätsch, M., Hartmann, M., and Borczyskowski, C. von (2014). Optical investigation of diffusion of single Ag markers in confined water films, J. Nanopart. Res., 16, 2700. With permission from Springer.

1.6 Conclusion

This short introduction into optical features of nanoparticles shows that one parameter, namely crystal size, controls both for metals and semiconductor particles the tunability of physical parameters. The basic concept is the cross-over from atomic-like structures to bulk solid material, which is partly described by a particle-in-a-box model (quantum confinement) and also by molecular structures in the case of clusters containing less than 100 atoms. The small size in the range of 0.5 nm (Ag clusters) to 5 nm (semiconductor QDs) and finally to 50 nm (plasmonic metal particles) spans a scale of 2 orders of magnitude with properties all in the visible range. Additionally, systems are quite stable under moderate conditions, offering a wide field for applications in various areas of sensoric applications, labeling in life sciences, and photovoltaics. Moreover, all the discussed entities are potential single-particle emitters or absorbers, making them usable, if properly detected, in liquid solutions, including water, which has already stimulated tremendous applications in life sciences. Surfaces have on the nanoscale a considerable impact, which is an additional advantage in offering a nearly unlimited range for tailored functionalization.

References

Akimov, I. A., Hundt, A., Flissikowski, T., Kratzert, P., and Henneberger, F. (2003). Energy spectrum of negatively charged single quantum dot: trion and charged biexciton states, *Physica E*, **17**, 31–34.

Baker, D. R., and Kamat, P. V. (2010). Tuning the emission of CdSe quantum dots by controlled trap enhancement, *Langmuir*, **26**, 11272–11276.

Blaudeck, T. (2007). *Self-Assembly of Functionalized Porphyrin Molecules on Semiconductor Nanocrystal Surfaces*, PhD Thesis, TU Chemnitz, Germany.

Blaudeck, T., Zenkevich, E. I., Cichos, F., and Borczyskowski, C. von (2008). Probing wave functions at semiconductor quantum-dot surfaces by non-FRET photoluminescence quenching, *J. Phys. Chem. C*, **112**, 20251–20257.

Boldt, K., Schwarz, K. N., Kirkwood, N., Smith, T. A., and Mulvaney, P. (2014). Electronic structure engineering in ZnSe/CdS type-II nanoparticles by interface alloying, *J. Phys. Chem. C*, **118**, 13276–13284.

Bonaĉić-Koutecký, V., Veyret, V., and Mitrić, R. (2001). Ab initio study of the absorption spectra of Ag_n (n = 5–8) clusters, *J. Chem. Phys.*, **115**, 10450.

Borczyskowski, C. von and Zenkevich, E. (2014). Formation principles and exciton relaxation in semiconductor quantum dot–dye nanoassemblies, in *Quantum Dot Molecules, Lecture Notes in Nanoscale Science and Technology* (Wu, J., and Wang, Z. M., eds.), Springer Science+Business Media, New York, Vol. 14, pp. 77–148.

Brown, P. R., Kim, D., Lunt, R. R., Zhao, N., Bawendi, M. G., Grossman, J. C., and Bulović, V. (2014). Energy level modification in lead sulfide quantum dot thin films through ligand exchange, *ACS Nano*, **8**, 5863–5872.

Brus, L. E. (1984). Electron-electron and electron-hole interactions in small semiconductor crystallites: the size dependence of the lowest excited electronic state, *J. Chem. Phys.*, **80**, 4403.

Bube, R. H. (1987). *Electrons in Solids*, Academic Press, San Diego.

Burdett, J. A. (1984). From bonds to bands and molecules to solids, *Prog. Solid State Chem.*, **15**, 173–255.

Califano, M., Franceschetti, A., and Zunger, A. (2007). Lifetime and polarization of the radiative decay of excitons, biexcitons, and trions in CdSe nanocrystal quantum dots, *Phys. Rev. B*, **75**, 115401.

Chamarro, M., Gourdon, C., Lavallard, P., Lublinskaya, O., and Ekimov, A. I. (1996). Enhancement of electron-hole exchange interaction in CdSe nanocrystals: a quantum confinement effect, *Phys. Rev. B*, **53**, 1336.

Cheatnoy, N., Harris, T. D., Hull, R., and Brus, L. E. (1986). Luminescence and photophysics of CdS semiconductor clusters: the nature of the emitting electronic state, *J. Phys. Chem.*, **90**, 3393.

Chen, O., Zhao, J., Chauhan, V. P., Cui, J., Wong, C. W., Harris, D. K., Wei, H., Han, H.-S., Fukumura, D., Jain, R. K., and Bawendi, M. G. (2013). Compact high-quality CdSe–CdS core–shell nanocrystals with narrow emission linewidths and suppressed blinking, *Nat. Mater.*, **12**, 445–451.

Chung, I., and Bawendi, M. G. (2004). Relationship between single quantum-dot intermittency and fluorescence intensity decays from collections of dots, *Phys. Rev. B*, **70**, 165304.

Clemenger, K. (1985). Ellipsoidal shell structure in free-electron metal-clusters, *Phys. Rev. B*, **32**, 1359.

Cichos, F., Martin, J., and Borczyskowski, C. von (2004). Emission intermittency in silicon nanocrystals, *Phys. Rev. B*, **70**, 1153141.

Cichos, F., Borczyskowski, C. von, and Orrit, M. (2007). Power-law intermittency of single emitters, *Curr. Opin. Colloid Interface Sci.*, **12**, 272–284.

Crut, A., Maioli, P., Del Fatti, N., and Vallée, F. (2014). Optical absorption and scattering spectroscopies of single nano-objects, *Chem. Soc. Rev.*, **43**, 3921–3956.

De Mello Donega, C., Bode, M., and Meijerink, A. (2006). Size- and temperature dependence of exciton lifetimes in CdSe quantum dots, *Phys. Rev. B*, **74**, 085321–085329.

Dabbousi, B. O., Rodriguez-Viejo, J., Mikulec, F. V, Heine, J. R., Mattoussi, H., Ober, R., Jensen, K. F., and Bawendi, M. G. (1997). (CdSe)ZnS core-shell quantum dots: synthesis and characterization of a size series of highly luminescent nanocrystallites, *Phys. Chem. B*, **101**, 9463.

Efros, A. L., Rosen, M., Kuno, M., Nirmal, M., Norris, D. J., and Bawendi, M. (1996). Band-edge exciton in quantum dots of semiconductors with a degenerate valence band: dark and bright exciton states, *Phys. Rev. B*, **54**, 4843.

Einevoll, G. T. (1992). Confinement of excitons in quantum dots, *Phys. Rev. B*, **45**, 3410.

Ekimov, A. I., Hache, F., Schanne-Klein, M. C., Ricard, D., Flytzanis, C., Kudryavtsev, I. A., Yazeva, T. V., Rodina, A. V., and Efros, Al. L. (1993). Absorption and intensity-dependent photoluminescence measurements of cdse quantum dots: assignment of the first electronic transitions, *J. Opt. Soc. Am. B*, **10**, 100.

Eychmüller, A. (2000). Structure and photophysics of semiconductor nanocrystals, *J. Phys. Chem. B*, **104**, 6514.

Fischer, S. A., Crotty, A. M., Kilina, S. V., Ivanov, S. A., and Tretiak, S. (2012). Passivating ligand and solvent contributions to the electronic properties of semiconductor nanocrystals nanoscale, *Nanoscale*, **4**, 904–914.

Franceschetti, A., and Zunger, A. (1997). Direct pseudopotential calculation of exciton coulomb and exchange energies in semiconductor quantum dots, *Phys. Rev. Lett.*, **78**, 915.

Franceschetti, A., Fu, H., Wang, L. W., and Zunger, A. (1999). Many-body pseudopotential theory of excitons in InP and CdSe quantum dots, *Phys. Rev. B*, **60**, 1819.

Frenzel, J., Joswig, J. O., and Seifert, G. (2007). Optical excitations in cadmium sulfide nanoparticles, *J. Phys. Chem. C*, **111**, 10761–10770.

Frenzel, J., Thieme, S., Seifert, G., and Joswig, J.-O. (2011). Optical excitations in CdSe/CdS core–shell nanoparticles, *J. Phys. Chem. C*, **115**, 10338–10344.

Gaponenko, S. V. (1998). *Optical Properties of Semiconductor Nanocrystals, Cambridge Studies in Modern Optics*, Cambridge University Press, Cambridge.

Gaponenko, S. V. (2010). *Introduction to Nanophotonics*, Cambridge University Press, Cambridge.

Gómez, D. E., Califano, M., and Mulvaney, P. (2006). Optical properties of single semiconductor nanocrystals, *Phys. Chem. Chem. Phys.*, **8**, 4989.

Gómez-Campo, F. M., and Califano, M. (2012). Hole surface trapping in CdSe nanocrystals: dynamics, rate fluctuations, and implications for blinking, *Nano Lett.*, **12**, 4508–4517.

Gavrilenko, V. I. (2011). *Optics of Nanomaterials*, Pan Stanford Publishing, Singapore.

Hache, F., Klein, M. C., Ricard, D., and Flytzanis, C. (1991). Photoluminescence study of Schott commercial and experimental CdSe-doped glasses: observation of surface states, *J. Opt. Soc. Am. B*, **8**, 1802.

Haberland, H. (1994). *Clusters of Atoms and Molecules: Theory Experiment, and Clusters of Atoms*, Springer-Verlag, Berlin.

Harb, M., Rabilloud, F., Simon, D., Rydlo, A., Lecoultre, S., Conus, F., Rodrigues, V., and Félix, C. (2008). Optical absorption of small silver clusters: Ag_n, (n = 4–22), *J. Chem. Phys.*, **129**, 194108.

Harbich, W., Fedrigo, S., Meyer, F., Lindsay, D. M., Lignieres, J., Rivoal, J. C., and Kreisle, D. (1990). Deposition of mass selected silver clusters in rare gas matrices, *J. Chem. Phys.*, **93**, 8535.

Issac, A., Borczyskowski, C. von, and Cichos, F. (2005). Correlation between photoluminescence intermittency of CdSe quantum dots and self-trapped states in dielectric media, *Phys. Rev. B*, **71**, 161302.

Issac, A. (2006). *Photoluminescence Intermittency of Semiconductor Quantum Dots in Dielectric Environments*, PhD Thesis, TU Chemnitz, Germany.

Ji, X., Copenhaver, D., Sichmeller, C., and Peng, X. (2008). Ligand bonding and dynamics on colloidal nanocrystals at room temperature: the case of alkylamines on CdSe nanocrystals, *J. Am. Chem. Soc.*, **130**, 5726–5735.

Kahle, I., Krause, S., Krasselt, C., Jakob, A., Oehlke, A., Georgi, C., Schulze, S., Lang, L., Hietschold, M., Spange, S., and Borczyskowski, C. von (2014). Assemblies from metallic and semiconducting nanocrystals, *Appl. Phys. A*, **115**, 617–625.

Kaschke, M., Ernsting, N. P., Müller, U., and Weller, H. (1990). Ultrafast electron ejection and trapping in semiconductor colloids after multiple photon absorption, *Chem. Phys. Lett.*, **168**, 543.

Kelly, L., Coronado, E., Zhao, L. L., and Schatz, G. C. (2003). The optical properties of metal nanoparticles: the influence of size, shape, and dielectric environment, *J. Phys. Chem. B*, **107**, 668–677.

Kilina, S., Ivanov, S., and Tretiak, S. (2009). Effect of surface ligands on optical and electronic spectra of semiconductor nanoclusters, *J. Am. Chem. Soc.*, **131**, 7717–7726.

Klimov, V. I. (2007). Spectral and dynamical properties of multiexicons in semiconductor nanocrystals, *Ann. Rev. Phys. Chem.*, **58**, 635.

Knowles, K. E., Frederick, M. T., Tice, D. B., Morris-Cohen, A. J., and Weiss, E. A. (2012). Colloidal quantum dots: think outside the (particle-in-a-)box, *J. Phys. Chem. Lett.*, **3**, 18–26.

Kovalenko, M. V., Manna, L., Cabot, A., Hens, Z., Talapin, D. V., Kaga, C. R., Klimov, V. I. Rogach, A. L., Reiss, P., Milliron, D. J., Guyot-Sionnnest, P., Konstantatos, G., Parak, W. J., Hyeon, T., Korgel, B. A., Murray, C. B., and Heiss, W. (2015). Prospects of nanoscience with nanocrystals, *ACS Nano*, **9**, 1012–1057.

Kowerko, D., Schuster, J., Amecke, N., Abdel-Mottaleb, M., Dobrawa, R., Würthner, F., and Borczyskowski, C. von (2010). FRET and ligand related non-FRET processes in single quantum dot-perylene bisimide assemblies, *Phys. Chem. Chem. Phys.*, **12**, 4112.

Krause, S., Hartmann, M., Kahle, I., Neuman, M., Heidernätsch, M., Spange, S., and Borczyskowski, C. von (2013). Optical tracking of single Ag clusters in nanostructured water films, *J. Phys. Chem. C*, **117**, 24822–24829.

Krause, S., Heidernätsch, M., Hartmann, M., and Borczyskowski, C. von (2014). Optical investigation of diffusion of single Ag markers in confined water films, *J. Nanopart. Res.*, **16**, 2700.

Kreibig, U., and Vollmer M. (1995). *Optical Properties of Metal Clusters*, Springer, Berlin.

Kubo, R. (1962). Electronic properties of metallic fine particles, *J. Phys. Soc. Jpn.*, **17**, 975.

Leatherdale, C. A., and Bawendi, M. G. (2001). Observation of solvatochromism in CdSe colloidal quantum dot, *Phys. Rev. B*, **63**, 165315.

Maier, S. A. (2007). *Plasmonics: Fundamentals and Applications*, Springer Verlag, Berlin.

Murray, C. B., Norris, D. J., and Bawendi, M. G. (1993). Synthesis and characterization of nearly monodisperse CdE (E = sulfur, selenium, and tellurium) semiconductor nanocrystallites, *J. Am. Chem. Soc.*, **115**, 8706.

Nirmal, M., Norris, D. J., Kuno, M., Bawendi, M. G., Efros, Al. L., and Rosen, M. (1995). Observation of the "dark exciton" in CdSe quantum dots, *Phys. Rev. Lett.*, **75**, 3728.

Nirmal, M., Dabbousi, B. O., Bawendi, M. G., Macklin, J. J., Trautman, J. K., Harris, T. D., and Brus L. E. (1996). Fluorescence intermittency in single cadmium selenide nanocrystals, *Nature*, **383**, 802.

Link, S., and El-Sayed, M. A. (1999). Size and temperature dependence of the plasmon absorption of colloidal gold nanoparticles, *J. Phys. Chem. B*, **103**, 4212–4217.

Lueth H. (2001). *Solid Surfaces, Interfaces and Thin Film*, Springer-Verlag, Berlin.

MacDonald, R. L., and Lawandy, N. M. (1993). Observation of charge screening in semiconductor nanocrystals, *Phys. Rev. B*, **47**, 1961.

Manna, L. Scher, E. C., and Alivisatos, A. P. (2002). Shape control of colloidal semiconductor nanocrystals, *J. Cluster Sci.*, **13**, 521.

Němec, P., and Maly, P. (2000). Temperature study of trap-related photoluminescence decay in CdS_xSe_{1-x} nanocrystals in glass, *J. Appl. Phys.*, **87**, 3342.

Oemrawsingh, S. S. R., Markešević, N., Gwinn, E. G., Eliel, E. R., and Bouwmeester, D. (2012). Spectral properties of individual DNA-hosted silver nanoclusters at low temperatures, *J. Phys. Chem. C*, **116**, 25568–25575.

Ozin, G. A., and Hugues, F. (1983). Silver atoms and small silver clusters stabilized in zeolite Y: optical spectroscopy, *J. Phys. Chem.*, **87**, 94–97.

Peng, X., Manna, L., Yang, W., Wickham, J., Scher, E., Kadavanich, A., and Alivisatos, A. P. (2000). Shape control of CdSe semiconductor nanocrystals, *Nature*, **404**, 59.

Peterson, J. J., and Nesbitt, D. J. (2009). Modified power law behavior in quantum dot blinking: a novel role for biexcitons and Auger ionization, *Nano Lett.*, **9**, 338–345.

Prezhdo, O. V. (2008). Multiple excitons and the electron-phonon bottleneck in semiconductor quantum dots: an *ab initio* perspective, *Chem. Phys. Lett.*, **460**, 1.

Prezhdo, O. V. (2009). Photoinduced dynamics in semiconductor quantum dots: insights from time-domain ab initio studies, *Acc. Chem. Res.*, **42**, 2005–2016.

Puzder, A., Williamson, A. J., Zaitseva, N., Galli, G., Manna, L., and Alivisatos., A. P. (2004). The effect of organic-ligand binding on the growth of CdSe nanoparticles probed by ab-initio calculations, *Nano Lett.*, **4**, 236.

Qin, H., Niu, Y., Yeng, R., Lin, X., Lai, R., Fang W., and Peng, X. (2014). Single-dot spectroscopy of zinc-blende CdSe/CdS core/shell nanocrystals: nonblinking and correlation with ensemble measurements, *J. Am. Chem. Soc.*, **136**, 179–187.

Rabani, E., Hetényi, B., Berne, B. J., and Brus, L. E. (1999). Electronic properties of CdSe nanocrystals in the absence and presence of a dielectric medium, *J. Chem. Phys.*, **110**, 5355.

Rogach, A. L. (2008). *Semiconductor Nanocrystal Quantum Dots*, Springer, Wien, New York.

Roussignol, P., Ricard, D., Lukasik, J., and Flytzanis, C. (1987). New results on optical phase conjugation in semiconductor-doped glasses, *J. Opt. Soc. Am. B*, **4**, 5.

Schapotschnikow, P., Hommersom, B., and Vlugt, T. J. H. (2002). Adsorption and binding of ligands to CdSe nanocrystals, *J. Phys. Chem. C*, **113**, 12690–12698.

Schlegel, G., Bohnenberger, J., Potapova, I., and Mews, A. (2002). Fluorescence decay time of single semiconductor nanocrystals, *Phys. Rev. Lett.*, **88**, 1374011.

Shang, L., Dong, S., and Nienhaus, G. U. (2011). Ultra-small fluorescent metal nanoclusters: synthesis and biological applications, *Nano Today*, **6**, 401–418.

Sieber, C., Buttet, J., Harbich, W., Mitrić, C. F. R., and Bonačić-Koutecký, V. (2004). Isomer-specific spectroscopy of metal clusters trapped in a matrix: Ag_9, *Phys. Rev. A*, **70**, 041201(R).

Smith, A. M., Mohs, A. M., and Nie, S. (2009). Tuning the optical and electronic properties of colloidal nanocrystals by lattice strain, *Nat. Nanotechnol.*, **4**, 56–63.

Thoai D. B. T., Hu, Y. Z., and Koch, S. W. (1990). Influence of the confinement potential on the electron-hole-pair states in semiconductor microcrystallites, *Phys. Rev. B*, **42**, 261–266.

Tian, Y., Wu, C., and Fendler, J. H. (1994). Fluorescence activation and surface state reactions of size-quantized cadmium sulfide particles in Langmuir-Blodgett films, *J. Phys. Chem.*, **98**, 4913.

Treguer, M., Rocco, F., Lelong, G., Nestour, A. L., Cardinal, T., Maali, A., and Lounis, B. (2005). Fluorescent silver oligomeric clusters and colloidal particles, *Solid State Sci.*, **7**, 812–818.

Vela, J., Htoon, H., Chen, Y., Park, Y.-S., Ghosh, Y., Goodwin, P. M., Werner, J. H., Wells, N. P., Casson, J. L., and Hollingsworth, J. A. (2010). Effect of shell thickness and composition on blinking suppression and the blinking mechanism in 'giant' CdSe/CdS nanocrystal quantum dots, *J. Biophotonics*, **3**, 706–717.

Voznyy, O. (2011). Mobile surface traps in CdSe nanocrystals with carboxylic acid ligands, *J. Phys. Chem. C*, **115**, 15927–15932.

Wang Y. (1991). Local field effect in small semiconductor clusters and particles, *J. Phys. Chem.*, **95**, 1119.

Wang, Y., Suna, A., Mahler, W., and Kasowski, R. (1987). PbS in polymers. From molecules to bulk solids, *J. Chem. Phys.*, **87**, 7315.

Wang, L.-W., and Zunger, A. (1998). High-energy excitonic transitions in CdSe quantum dots, *J. Phys. Chem. B*, **102**, 6449.

Wang, X., Qu, L., Zhang, J., Peng, X., and Xiao, M. (2003). Surface-related emission in highly luminescent CdSe quantum dots, *Nano Lett.*, **3**, 1103–1106.

Wei, S. H., Zhang, S. B., and Zunger, A. (2000). First-principles calculation of band offsets, optical bowings, and defects in CdS, CdSe, CdTe, and their alloys, *J. Appl. Phys.*, **87**, 1304–1311.

Woggon, U., Rückmann, I., Kornack, J., Müller, M., Cesnulevicius, J., Petrauskas, M., and Kolenda, J. (1992). Interface transfer processes and spectral hole burning in CdSe quantum confined cluster, *J. Cryst. Growth*, **117**, 608.

Woggon, U. (1997). *Optical Properties of Semiconductor Quantum Dots.* Springer Tracts in Modern Physics, vol. 136, Springer, Berlin.

Xie, R., U. Kolb, U., Li, J. U., Basché, T., and Mews, A. (2005). Synthesis and characterization of highly luminescent CdSe-Core CdS/Zn0.5Cd0.5S/ ZnS multishell nanocrystals, *J. Am. Chem. Soc.*, **127**, 7480–7488.

Yoffe, A. D. (2002). Low-dimensional systems: quantum size effects and electronic properties of semiconductor microcrystallites (zero-dimensional systems) and quasi-two-dimensional systems, *Adv. Phys.*, **51**, 799.

Zenkevich, E. I., Stupak, A., Goehler, C., Krasselt, C., and Borczyskowski, C. von (2015). Tuning electronic states of a CdSe/ZnS quantum dot by only one functional dye molecule, *ACS Nano*, **9**, 2886–2907.

Zenkevich, E. I., and Borczyskowski, C. von, eds. (2016). *Self-Assembled Organic-Inorganic Nanoassemblies,* Pan Stanford.

Zhang, J. Z., O'Neil, R. H., Roberti, T. W., McGowen, J. L., and Evans, J. E. (1994). Femtosecond studies of trapped electrons at the liquid–solid interface of aqueous CdS colloids, *Chem. Phys. Lett.*, **218**, 479.

Zhao, J., Nair, G., Fisher, B. R., and Bawendi, M. G. (2010). Challenge to the charging model of semiconductor-nanocrystal fluorescence intermittency from off-state quantum yields and multiexciton blinking, *Phys. Rev. Lett.*, **104**, 157403.

Zheng, J., Petty, J. T., and Dickson, R. M. (2003). High quantum yield blue emission from water-soluble Au_8 nanodots, *J. Am. Chem. Soc.*, **125**, 7780–7781.

Zheng, J., Zhang, C., and Dickson, R. M. (2004). Highly fluorescent, water-soluble, size-tunable gold quantum dots, *Phys. Rev. Lett.*, **93**, 077402.

Zheng, J., Nicovich, P. R., and Dickson, R. M. (2007). Highly fluorescent noble-metal quantum dots, *Annu. Rev. Phys. Chem.*, **58**, 409–431.

Zijlstra, P., and Orrit, M. (2011). Single metal nanoparticles: optical detection, spectroscopy and applications, *Rep. Prog. Phys.*, **74**, 106401–106456.

Chapter 2

New Insights: Photophysics of CdSe Quantum Dots. A Temperature-Related Approach

Eduard Zenkevich,[a] Alexandre Stupak,[b] and Christian von Borczyskowski[c]

[a]*Department of Information Technologies and Robototechnique, National Technical University of Belarus, Prospect Nezavisimosti 65, 220013 Minsk, Belarus*

[b]*B.I. Stepanov Institute of Physics, National Academy of Sciences of Belarus, Prospect Nezavisimosti 70, 220072 Minsk, Belarus*

[c]*Institute of Physics, Center for Nanostructured Materials and Analytics, Technische Universität Chemnitz, Reichenhainer Str. 70, D-09107 Chemnitz, Germany*

borczyskowski@physik.tu-chemnitz.de

Science and technology of nanoparticle-based materials such as semiconductor quantum dots (QDs) involve surface and interfacial phenomena that may be tuned by tailoring the surface state energy and by varying the specific electronic interactions with chemical entities attached to such surfaces and interfaces. The main obstacle on the path to developing efficient QD-based energy materials is our limited understanding of QD surfaces, their

Tuning Semiconducting and Metallic Quantum Dots: Spectroscopy and Dynamics
Edited by Christian von Borczyskowski and Eduard Zenkevich

ISBN 978-981-4745-24-6 (Hardcover), 978-981-4745-25-3 (eBook)
www.panstanford.com

interaction with surface-ligated molecules, and their impact on the nature and properties of QD excited states, photoluminescence (PL) parameters, and charge or energy transfer between QDs and attached molecules. Optical properties might be tuned specifically via the interaction of semiconductor nanoparticles with functional organic molecules. Of special interest is the basic understanding of what happens with the excited states of QDs of a one-molecule-to-one-QD basic nanoassembly. If successfully applied, the large variety of functional organic molecules at hand allows for a broad scenario for modification of optical QD properties.

In this chapter, we present a reinvestigation of spectral-kinetic properties (including absorption, PL, PL excitation spectra, PL decay data) for CdSe/ZnS QDs in solvents being obtained in a wide temperature range. We present new insights into optical features which have not been published or recognized. They became now obvious due to a comprehensive study bringing together subtle details which have possibly been overlooked previously. We would like to underline several principal aspects which should be taken into account upon analysis of PL characteristics for QDs which are planned to be part of heterogeneous nanoassemblies with functional organic molecules. This approach seems to be useful for further development of defined multicomponent structures for exploitation as artificial light-harvesting complexes, electro- and photochemical devices, or nanosensors.

2.1 Introduction

Colloidal semiconductor nanocrystals (quantum dots [QDs]) are increasingly important for applications such as sensorics and photovoltaics [De Mello Donega et al., 2011; Huang et al., 2012; McBride et al., 2013; Sharon et al., 2011; Tanne et al., 2011]. Due to the inherent size-dependent quantum confinement the color of QDs can be tuned across a wide optical range, which opens a path for tailored optical features [Klimov, 2000; Woggon et al., 2006; Klimov, 2007; Rogach, 2008; Gaponenko, 2010; Talapin et al., 2010; Smith and Nie, 2010]. Additionally, formation of nanoassemblies between colloidal nanocrystals and functional dye molecules adds a new class of nanodevices [Schmelz et al., 2004; Clapp et al., 2004; Zenkevich et al., 2005; Dayal et al., 2006; Boulesbaa et al., 2007; Ren et al., 2008; McDowell et al., 2010; Dworak et al., 2014; Zenkevich et al., 2011; Kowerko et al., 2010; Tagliazucchi et al., 2011; Lemon et al., 2013; Adegoke et al., 2013; Tripathi et al., 2015; Hadar et al., 2015; Aldeek et al., 2015]. Since the surface/volume ratio for nanoscaled materials is large surface properties are of central importance for a basic understanding of both optical properties of nanocrystals and envisaged applications. Some applications rely on as stable as possible surface properties; others, such as sensoric applications, might profit from high sensitivity of electronic conditions at the surface.

Though colloidal materials have been intensively studied during the past decades essential properties of their surface and their response to environmental conditions are not understood despite many detailed experiments already in existence. Here we attempt to discuss new insights into surface-related properties of colloidal CdS/ZnS nanocrystals in relation to their intrinsic optical properties. With respect to these properties, various theoretical approaches have been set up, including the fine structure of the band-edge exciton [Efros et al., 1996; Norris and Bawendi, 1996; Kuno et al., 1997; Klimov et al., 2000; Klimov, 2007; Salman et al., 2007; Salman et al., 2009] and exciton–phonon coupling [Takagahara, 1993; Salvador et al., 2006; Liptay and Ram 2006; Liptay et al., 2007; Gautham and Bawendi, 2007; Sagar et al., 2008; Dzhagan et al., 2009a; Dzhagan et al., 2009b; Kelley, 2010] to describe the physical mechanisms responsible for QD absorption and photoluminescence (PL) and their temperature dependence, including quantum confinement, electronic fine structure, Stokes shifts, spectral line widths, and trap states (both for bulk and for single-QD detection).

While QD absorption spectra rely on the overall crystal structure and size as well as on electronic hole and electron states, the light-emitting properties of QDs can only be understood by taking into account—besides the fine structure of the band-edge 1S(e)-$1S_{3/2}$(h) transitions—various surface states, trap states in the energy gap between valence and conduction bands [Frenzel et al., 2007; Kilina et al., 2013], the existence of ligand-mediated states [Kilina et al.,

2009; Kilina et al., 2012; Blaudeck et al., 2012], and possible phase transitions of surfactant layers in liquid solutions upon temperature lowering [Wuister et al., 2004a; Wuister et al., 2004b; Zenkevich et al., 2012].

Even single quantum object detection often reveals broad distributions and related variations of optical properties during the experimental time. This implies that one has to question ergodicity when comparing ensemble and time-averaged experiments on single entities. Moreover, it is often reported for CdSe/ZnS QDs of various sizes that the optical absorption and PL may differ significantly among individual QDs in one and the same ensemble [Empedocles and Bawendi, 1999; Dayal and Burda, 2007; Koberling et al., 2002; Ji et al., 2008] or according to different preparation and detection conditions [Quagliano, 2004; McBride et al., 2004; Funston et al., 2008; Munro et al., 2007; Schreuder et al., 2009]. This also includes the dependence of optical properties on temperature [Crooker et al., 2003; Labeau et al., 2003; Wuister et al., 2004a; Wuister et al., 2004b; Califano et al., 2005; De Mello Donega et al., 2006; Zenkevich et al., 2012], ligands [Kaluzhny and Murray, 2005; Munro and Ginger, 2008; Blaudeck et al., 2012], and solvent properties [Issac et al., 2012; Gerlach et al., 2013] (polarity, ligand exchange via solubility changes, etc.), as well as on possible photodegradation effects.

In this chapter we will concentrate on ensemble experiments to unravel as many as possible subtle optical properties of core and surface-related electronic states, which will give new insights into several optical observables both with respect to a basic understanding of surface properties and with envisaged applications.

2.2 Experiments

For ensemble and single-particle experiments colloidal core/shell CdSe/ZnS QDs (d_{CdSe} = 3.0 nm; two ZnS capping monolayers) passivated by trioctylphosphine oxide (TOPO) were obtained from Evident Technologies, Inc, Troy, NY, USA. The absorbance of the QDs in the initial toluene solution was adjusted to be lower than 0.1 OD in order to avoid nonlinear absorption and reabsorption effects. The concentrations varied in the range of $(1–10) \times 10^{-7}$ M. In ensemble experiments stability and purity of the QD solutions were checked by measuring the quantum yield (QY) stability at least for over 3 h after preparation.

We have studied the temperature dependence of absorption and PL properties (from 77 K up to 300 K) for CdSe/ZnS and QD-porphyrin nanoassemblies [Zenkevich et al., 2012; Zenkevich et al., 2015] in toluene or a methylcyclohexane/toluene (6:1) mixture. The latter forms an optical transparent glass at low temperatures. The respective glass transition temperatures are the following: 146.7 K for methylcyclohexane, 180 K for toluene [Murov et al., 1993], and 151.6 K for methylcyclohexane/toluene (6:1) mixtures [Marcus, 2002].

In ensemble experiments, absorption spectra were recorded with a Shimadzu 3001 UV/Vis or Cary-500 M Varian spectrometer. Emission spectra were measured with a SFL-1211A (Solar, Belarus) or Shimadzu RF-5001PC spectrofluorimeter. Temperature experiments were carried out using a home-built cryostat (Solar, Belarus) equipped with a temperature controller (ΔT = 0.5 K). Single-particle experiments at ambient conditions were performed in a home-built laser scanning confocal microscope described earlier [Kowerko et al., 2010].

2.3 Results

The obtained temperature dependencies for integrated optical band intensities, energetic positions of the corresponding energetic positions, and spectral full-width-at-half-maximum (FWHM) have been extracted after spectral deconvolution into corresponding Gaussian components across a temperature range from 90 K to 395 K. Additionally, we have recorded at 290 K photoluminescence excitation (PLE) spectra and PL emission at various excitations. This extensive spectroscopic investigation has been complemented by time-resolved detection of PL emission at 77 K and 300 K. To maintain track of the complex data set we will discuss the above-mentioned observables step by step and will relate them with each other in Section 2.3.3 before we finally discuss the physical interpretation in Section 2.4.

2.3.1 Absorption Spectra

As can be seen from Fig. 2.1A the absorption of CdSe/ZnS QDs shifts to the red with temperature increase. The optical density is nearly temperature independent (±2.5 %) for wavelengths below 490 nm. This ensures that excitation always occurs at the same conditions independent of temperature. Figure 2.1B shows the temperature dependence of the energetic position E_{max} of the maximum of the lowest absorption band based on spectral deconvolution by one Gaussian component, which takes both an experimental background and the onset of energetically higher absorption bands into account. An example for the fit quality is presented as an inset at T = 243 K. The details of the fitting results have been published recently by us [Zenkevich et al., 2015]. It is obvious that the first absorption band maximum is nearly continuously shifted to the red upon temperature increase, though an already reported [Zenkevich et al., 2012] small deviation is observed close to T_{crit} = 210 K. The continuous red shift of the lowest absorption band (E_{max}) upon temperature increase is due to temperature increasing exciton–phonon coupling [Htoon et al., 2004; Norris et al., 1996; Labeau et al., 2003; Mello et al., 2006; Klimov 2000a] and/or a thermally driven band-gap widening [Valerini et al., 2005] due to thermal expansion of the QD. Figure 2.1C shows that the line width (FWHM) also increases with temperature increase. However, we observe a sudden decrease (followed by a subsequent further increase) of FWHM values at nearly the same temperature (220 K) at which we detect a sudden (small) jump of absorption energy (see Fig. 2.1B). The abrupt change in line width at 220 K has an impact on the absorption band intensity, as marked by a dashed arrow in Fig. 2.1A, as can be seen in Fig. 2.1D.

Quantitatively, the observed overall temperature tendencies for E_{max} and FWHM data coincide—besides the discontinuity at $T_{crit} \approx$ 220 K—with the corresponding results reported for CdSe/ZnS QDs of various sizes using an one-component Gaussian deconvolution [Valerini et al., 2005; Liptay et al., 2007]. Figure 2.1D shows both for the optical density (2) and for the integrated absorption (1) a slight increase with decreasing temperature, which might be explained by moderate changes of the excitonic parameters such as fine structure and oscillator strengths. Noticeably, small discontinuities are observed at T_{crit}, indicating that surface-related states contribute to overall absorption.

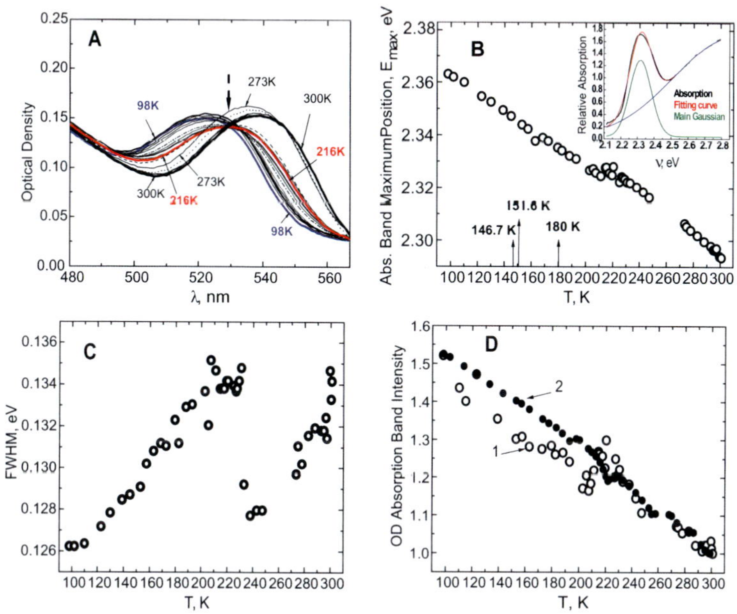

Figure 2.1 Temperature dependence of optical properties for CdSe/ZnS QDs in a methylcyclohexane/toluene (6:1) mixture. Dependencies shown in all parts of Fig. 2.1 were obtained upon temperature lowering; those obtained upon temperature increase were practically the same. (A) Spectral optical density changes in the range of the two energetically lowest transitions 1S(e)-1$S_{3/2}$(h) and 1S(e)-2$S_{3/2}$(h); the corresponding nomenclature for states and transitions was taken from Efros et al. (1996), Kuno et al. (1997), Klimov (2007), and Gaponenko (1998). Variations of the base line and stray light have been taken into account. (B) Absorption energy (E_{max}) of the lowest 1S(e)-1$S_{3/2}$(h) absorption band fitted by one Gaussian component as a function of temperature. The insert presents an example of spectral deconvolution at T = 243 K. The broad component at short absorption wavelengths corresponds to a transition to a higher excitonic state(s); for details of the fitting procedure see Zenkevich et al. (2015). Arrows indicate the glass transition temperatures of methylcyclohexane (146.7 K), a methylcyclo-hexane/toluene (6:1) mixture (151.6 K), and toluene (180 K), respectively [Murov et al., 1993; Marcus, 2002]. (C) Temperature dependence of the absorption line width at half-maximum (FWHM). (D) Temperature dependence of the normalized integrated absorption (1) and normalized optical density (2) for the first excitonic band (fitted by one Gaussian component). Both curves were normalized to the corresponding values at T = 300 K.

Since we observe deviations from smooth temperature behavior and a quantitatively not satisfying single Gaussian fitting procedure, a two-component Gaussian deconvolution (including background correction) has been applied. Figure 2.2A shows that deconvolution gives an energy separation $\Delta E_{H,L} \approx 75–80$ meV between a high-energy (HE) and a low-energy (LE) absorption band. Figure 2.2A also shows that both components, HE and LE, shift monotonously to lower energies upon temperature increase. Interestingly, the energy separation $\Delta E_{H,L}$ remains almost constant without any deviation (kink at T_{crit} marked by a dashed ellipse) from a monotonous behavior, as has on the contrary been observed in the case of a single Gaussian deconvolution in the range of 210–220 K.

However, the mean weighted presentation MW (for explanation see Zenkevich et al. [2015]) of the two-component fit shows such a kink similar to the ones observed in Fig. 2.1B. Additionally, the nearly linearly temperature-dependent slopes dE_{max}/dt are for all components different below compared to those above T_{crit}.

It follows from Fig. 2.2B that the line widths (FWHM) for the HE component are broader compared to those for the LE component. The FWHM increases nearly linearly with temperature increase (but differently comparing the two components), both below and above T_{crit}. The difference in the FWHM for the two components becomes larger above T_{crit}. Importantly, fitting of integrated absorption intensities shows in Fig. 2.2C that the relative ratio of the absorption intensities (measured as the area under the corresponding Gaussian line at each temperature) becomes abruptly larger in favor of the HE component above T_{crit}. Taking the change in relative intensities of these two components into account, the superposition of the two explains the sudden changes of E_{max} and the related FWHM values at T_{crit} in the case of the one Gaussian component presentation.

The two most important experimental findings with respect to the temperature-dependent absorption are as follows: (i) the absorption usually related to the lowest $1S(e)$-$1S_{3/2}(h)$ transition of CdSe/ZnS QDs [Efros et al.,1996; Klimov, 2007; Gaponenko, 1998] is split into two components about 80 meV apart and (ii) the apparent deviations (of a type which one would expect in the presence of a phase transition) from a monotonous temperature behavior (typical for electron–phonon coupling) is merely due to a relative change of the absorption probabilities of the two involved components HE and

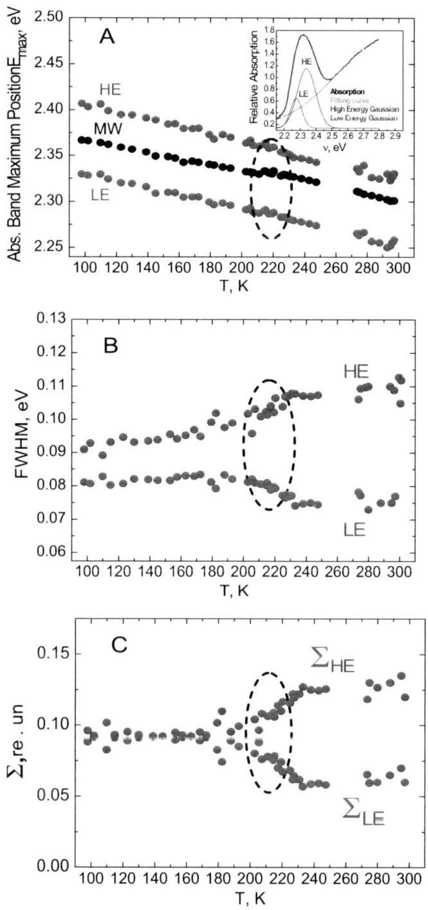

Figure 2.2 Temperature dependence of absorption parameters for CdSe/ZnS QDs fitted by two Gaussian components for the lowest excitonic absorption band. The dashed ellipsoid marks the range of a nonmonotonous temperature behavior at T_{crit}. (A) Absorption energies E_{max} of the deconvoluted absorption maxima for the low-energy (LE) and high-energy (HE) Gaussian components as a function of temperature. MW corresponds to the temperature dependence of the lowest absorption energy, resulting from the two mean weighted Gaussian components: $E = E_{HE} \times [\Sigma_{HE}/(\Sigma_{HE} + \Sigma_{LE})] + E_{LE} \times [(\Sigma_{LE}/(\Sigma_{HE} + \Sigma_{LE})]$. The inset presents an example of the deconvolution by two Gaussian components at $T = 243$ K. (A third component takes corrections caused by stray light and baseline variations into account.) (B) Temperature dependence of line widths (FWHM) for the LE and HE Gaussian components. (C) Temperature dependence of integrated absorption intensities Σ for the LE and HE Gaussian components (measured as an integral of the corresponding Gaussian lines).

LE, respectively. Whether similar features are observed for the PL will be discussed in the following section.

2.3.2 Photoluminescence Properties

2.3.2.1 PL spectra

PL spectra of QDs have been obtained following excitation at $\lambda_{exc} \leq 450$ nm that is energetically high into the QD absorption range to avoid size selection due to the remaining size distribution [Gaponenko, 1998; Petrov et al., 2006]. The PL emission can be approximated nearly by a Gaussian profile, though systematic deviations can be identified, as will be discussed later in detail. E_{max} of the PL band shifts upon temperature increase from 77 K to 300 K continuously by about 64 meV to lower energies under constant excitation conditions (λ_{exc} = 450 nm), while the FWHM value increases by about 19 meV. The integrated intensity decreases by about a factor of 2. These experimental findings are shown in Fig. 2.3.

Similar results are obtained for λ_{exc} = 410 nm. Changes in all three observables (E_{max}, FWHM, integrated intensity) follow nearly continuously the variation of temperature. However, a small but clearly noticeable discontinuity is observed close to T_{crit} = 220 K (see broken arrows in Fig. 2.3). This finding is similar to the absorption features and has been reported earlier [Zenkevich et al., 2012]. It will be discussed in more detail later. As observed in many other reported experiments line widths are smaller for PL bands compared to those of absorption bands belonging to the long-wavelength excitonic transition. We will explain in the discussion that this is mainly due to the fact that the HE absorption band does not result in a following-up PL.

Since the PL band shape does not obey a simple Gaussian profile it can be approximated more accurately, assuming a superposition of at least two Gaussian bands (see Fig. 2.4), as has also been reported recently [Zenkevich et al., 2012; Jakubek et al., 2008]. The two related bands differ considerably in line widths, which we have marked by the labels N (narrow) and B (broad). Although the separation in energy is only a few millielectron volts, Fig. 2.4A shows that each component behaves noticeably differently upon temperature variation.

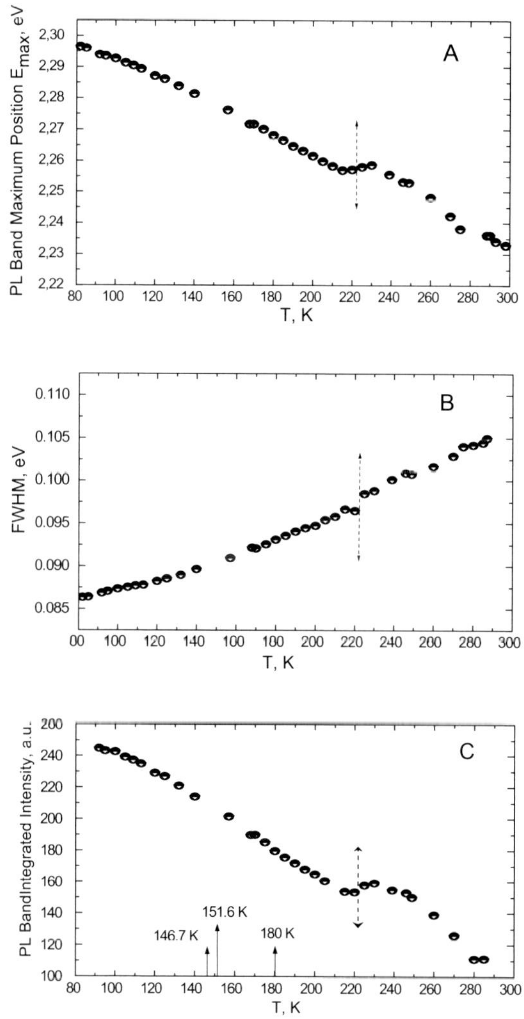

Figure 2.3 Temperature dependence of PL spectra for CdSe/ZnS QDs in a methylcyclohexane/toluene (6:1) mixture at λ_{exc} = 450 nm (fitted by one Gaussian component). Arrows indicate glass transition temperatures for methylcyclohexane (146.7 K), a methylcyclohexane/toluene (6:1) mixture (151.6 K), and toluene (180 K), respectively. A discontinuity of the temperature dependence (kink) is indicated by a broken arrow. (A) Energy band maximum E_{max}. (B) FWHM. (C) Integrated intensity.

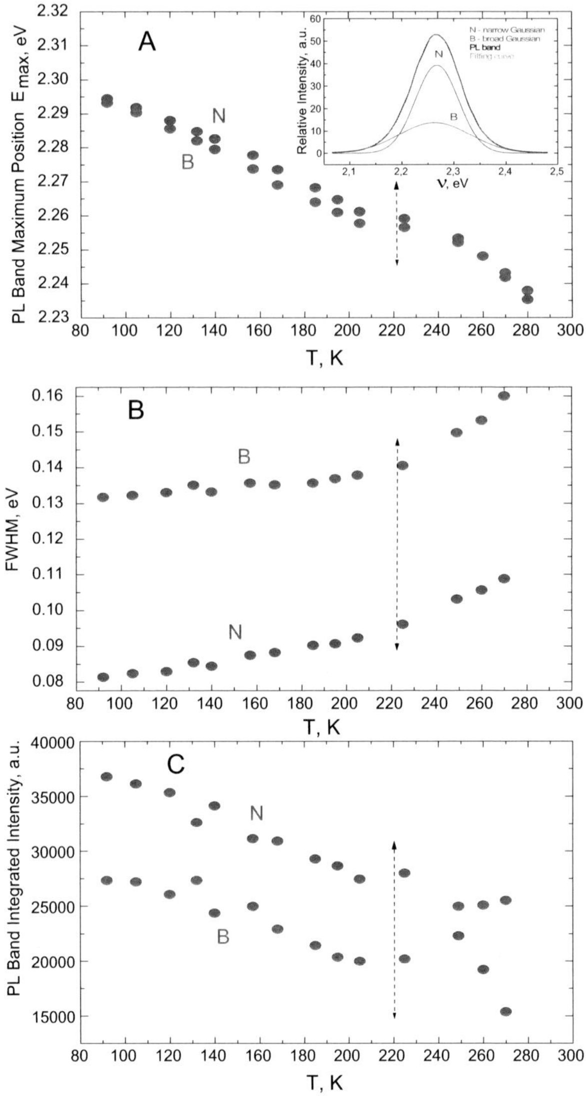

Figure 2.4 Temperature dependence of PL spectral parameters for CdSe/ZnS at λ_{exc} = 450 nm fitted by two Gaussian components, B (broad) and N (narrow), respectively. The temperature of a kind of phase transition at T_{crit} ≈ 220 K is indicated by a dashed arrow. (A) PL band maximum E_{max} for N and B components. The inset presents an example of the deconvolution of the QD PL band by two Gaussian components at T = 185 K. (B) FWHM for N and B components. (C) Integrated intensities for N and B components.

Above $T_{crit} \approx 220$ K the energy separation nearly vanishes and is also very small at the lowest temperatures. The line width of both components increases continuously with temperature. The ratio FWHM (B)/FWHM (N) decreases up to 225 K and becomes constant thereon (see Fig. 2.4B). Both (integrated) intensities initially decrease with temperature, but at 225 K there is an abrupt increase in the intensity for band N, accompanied by an abrupt decrease in the intensity for band B, as shown in Fig. 2.4C. Taking all these observations into account sudden (phase transition like) changes occur close to $T_{crit} \approx 220$ K, which also explains the discontinuities of the PL properties at this temperature, evident in case of an approximation by only one Gaussian component (see Fig. 2.3). For amine (AM)-capped CdSe/ZnS QDs the deconvolution of the PL band into two Gaussian components gives at 295 K an energy separation between the N and B components of $\Delta E \approx 18.5$ meV, which decreases slightly to 13.6 meV within an observation time of 350 min [Zenkevich et al., 2012].

2.3.2.2 Time-resolved photoluminescence

It is known from many experiments (e.g., Petrov et al., 2006; Salman et al., 2009; Kern et al., 2011; Knowles et al., 2011) that the PL of QDs (including the one from CdSe/ZnS) decays in a multiexponential manner, the details of which depend on the concrete properties (e.g., QD size, capping shell, details of synthesis, solvent, ligands, sample preparation). PL decay times are in the range of a few hundred picoseconds to some hundred nanoseconds. Much shorter times have also been reported and have been related to intraband relaxation of hot electrons or holes [Klimov, 2000a]. In the context of the present result we do not take these very short processes into account but concentrate on the PL from the band-edge states. Our special emphasis is to map the distribution of PL decay times onto the range of PL energies (restricting to temperatures between 77 K and 295 K).

Figure 2.5A shows PL decay traces at 77 K for the two detection wavelengths, $\lambda_{em} = 515$ nm and 550 nm, in the respective wings of the PL band, while Fig. 2.5B shows at 295 K the decay close to the centre of the PL band at $\lambda_{em} = 555$ nm. It can be clearly seen that the decay kinetics depend both on temperature and λ_{em}. Additionally, no built-up components are observed within the experimental time resolution of ~0.1 ns. This implies that the individual PL components decay independently, prevailing that the nonexponentiality is not caused, for example, by an energy cascade from energetically high to energetically low QD states.

We have analyzed the PL decay at various detection wavelengths λ_{em} within the PL band both at high and low temperatures. A deconvolution of the decay by three decay components was sufficient for the present results on CdSe/ZnS QDs in a methylcyclohexane/toluene (6:1) mixture, though more components have been reported for other but closely related systems [Petrov et al., 2006; Salman et al., 2009; Zenkevich et al., 2015]. We determined the average decay time $<\tau>$ according to

$$<\tau> = (\Sigma A_i \tau_i^2)/(\Sigma A_i \tau_i). \tag{2.1}$$

The results are shown in Fig. 2.5C,D in relation to the respective normalized steady-state PL emission band. It is seen (Fig. 2.5C) that the PL decay time $<\tau>$ is increasing toward longer wavelengths at 77 K. In contrast, at 295 K (Fig. 2.5D) $<\tau>$ is nearly constant across the PL band but with a small but noticeable minimum of $<\tau>$ close to the centre of the PL band. Additionally, the PL decay at long wavelengths is considerably shortened at 295 K compared to 77 K (from 18–18.5 ns down to 14–15 ns). Fitting results are shown in Table 2.1 in more detail. A comparison with single-QD data shows similar results. From this we conclude that the multiexponential decay is predominantly not due to an inhomogeneous distribution of individual QDs but a sequential occupation of different electronic states, as is learned from PL blinking experiments [Schmidt et al., 2014; Zenkevich et al., 2015].

We have concluded from a comparison of single- and ensemble-QD PL decay data (see Table 2.1) that QDs in liquid solutions (e.g., ensemble data at 290 K) also show PL blinking phenomena, which implies that they are charged (or at least differently polarized) states [Cichos et al., 2007; Zenkevich et al., 2015]. In such a case we expect a pronounced sensitivity upon the dielectric properties of the environment [Isaac et al., 2005 ; El Kork et al., 2011; Krasselt et al., 2011 ; Schmidt et al., 2012; Isaac et al., 2012; Schmidt et al., 2014]. For this reason we have performed time-resolved experiments for the same QDs in different solvents.

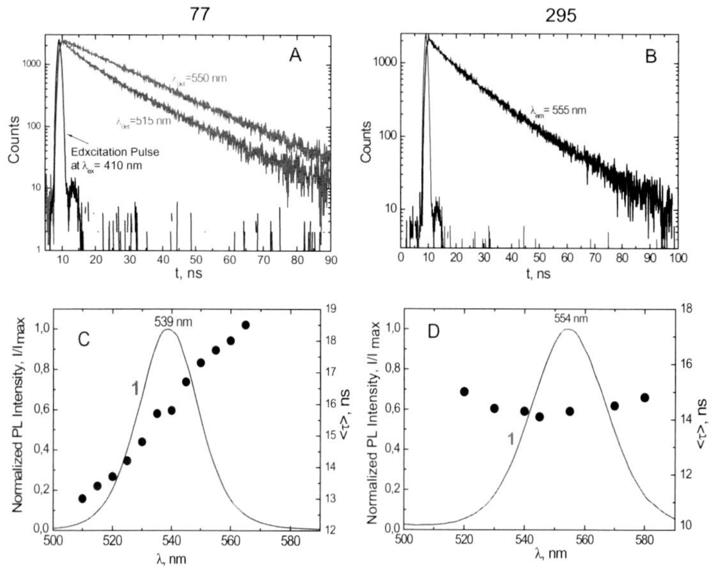

Figure 2.5 Normalized PL bands (1), PL decay traces (A, B), and (black solid circles) PL decay times $<\tau>$ (C, D) at various detection wavelengths λ_{em} for CdSe/ZnS at 77 K (A, C) and 295 K (B, D), respectively. Excitation is at λ_{exc} = 410 nm. $<\tau>$ corresponds to $<\tau> = (\Sigma A_i \tau_i^2)/(\Sigma A_i \tau_i)$.

Figure 2.6 shows a comparison of time-tagged PL spectra for CdSe/ZnS QDs in a methylcyclohexane/toluene (6:1) mixture (A) and in toluene (B), respectively, at 77 K for delay times between 0.2 and 30 ns after the laser pulse. It can be seen in Fig. 2.6A that upon increasing delay time the short-wavelength component of the PL band shifts evidently to lower energies by about 5 meV, which is similar to the PL energy band separation observed for the two Gaussian components in the deconvoluted PL spectrum (see Fig. 2.4A). Notably, in pure toluene solutions at 77 K, the corresponding shift for the same CdSe/ZnS QDs is much more pronounced upon increasing delay time (Fig. 2.6B). We would like to note that toluene is slightly more polar than the methylcyclohexane/toluene mixture. In contrast to experiments at 77 K, no pronounced spectral shifts are observed at 295 K for either of the two solvents. However, previous experiments on slightly larger CdSe/ZnS QDs have shown time-related spectral shifts in toluene also at room temperature [Petrov et al., 2006].

Table 2.1 PL decay times τ_i, normalized amplitudes A_i, and intensities I_i for CdSe/ZnS QDs at T = 77 K and T = 290 K

	τ_i (ns)	Single QD[a] CdSe/ZnS		Ensemble experiments[b,c]	
T (K)	λ/nm	High	Low	High	Low
77 K	τ_1			0.5	
	A_1			0.2	
	I_1			1	
	τ_2			4	3.5
	A_2			0.28	0.2
	I_2			14	5
	τ_3			14.5	16
	A_3			0.52	0.8
	I_3			85	95
	$<\tau>$			14.5	17.5
290 K	τ_1	2	0.8	0.8	0.8
	A_1	0.13	0.44	0.37	0.33
	I_1			4	3
	τ_2	9	8	7.7	7
	A_2	0.31	0.27	0.36	0.36
	I_2			35	30
	τ_3	21	18	18	18
	A_3	0.56	0.29	0.27	0.31
	I_3			61	67
	$<\tau>$	13.9	14.4	14.5	14.5

[a]Single-QD data according to Zenkevich et al. (2015).
[b]Parameter ranges correspond to the variation of the QD PL detection at short and long wavelengths at the FWHM values of the PL band. Typical errors for τ_i are 0.2–0.5 ns. Typical errors for A_i are 5%.
[c]λ_{exc} = 410 nm in the methylcyclohexane/toluene (6:1) mixture.

An inhomogeneous distribution of PL decay times might stem from the remaining size distribution of QDs in an ensemble. However, according to previous experiments [Petrov et al., 2006], we expect across the respective emission range a much smaller variation of

QD PL decay times by at most 20%. Therefore we conclude that the individual QDs in the ensemble subsequently go through a series of different electronic states, which themselves have different PL energies and decay times (and QYs). This conjecture matches closely the observed spectral diffusion processes observed for single QDs [Neuhauser et al., 2000; Martin et al., 2012; Schmidt et al., 2014]. Additionally, it has to be mentioned that continuously recorded PL spectra will contain a reduced component of short-lived states since these add only to a minor degree to the integrated PL intensity (see also Table 2.1 for the determined PL intensities I_i compared to amplitudes A_i).

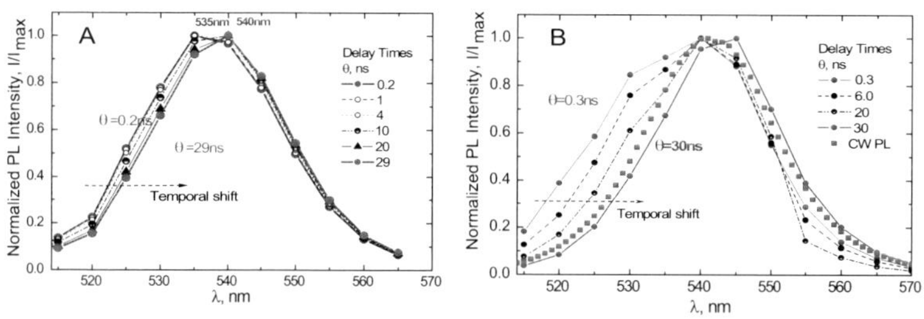

Figure 2.6 PL spectra at 77 K for CdSe/ZnS QDs at various delay times θ after a laser pulse with λ_{exc} = 410 nm in a methylcyclohexane/toluene (6:1) mixture (A) and toluene (B). Black squares in (B) correspond to the continuously recorded PL spectrum. The PL maximum is at 540 nm in a steady-state experiment. At 295 K spectra are nearly the same at various delay times.

Shorter decay times would have been expected in the short-wavelength range in case that those energetically higher states populate lower-lying states, especially at low temperatures, with an effective Boltzmann population across the PL band. In that case the decay times of energetically high states would correlate with PL built-up times of lower-lying states. This is clearly not observed in the present and earlier reported [Petrov et al., 2006] experiments. For this reason we conclude that we observe at least three different (not subsequently populated) classes of PL decay channels reflecting the presence of at least three different but energetically close-lying electronic states belonging to the same individual QD.

Comparing the results shown in Figs. 2.5 and 2.6 we conclude that the short-lived state will not contribute significantly to the steady-state PL spectrum and that the two relatively long-lived states are energetically separated by a few millielectron volts at 77 K in the solvent mixture. However, they are at about the same energy at 295 K, which goes in line with the deconvolution of the PL spectra (see Fig. 2.4A). Additionally, the small time-dependent spectral shift to lower energies at 77 K and the small minimum for $<\tau>$ observed in the PL band centre at 295 K suggest that the decay component A_2 (higher energy at 90 K) is at 290 K narrower in spectral distribution than the A_3 component. A_2 and A_3 are, with respect to the (continuous-wave) PL spectra, by far the dominant components.

All these experimental findings related to PL decay dynamics (energetic position, line width, temperature dependence, PL intensity) completely agree with the observations following the deconvolution of the (steady state) PL bands. Therefore we assign at 77 K the shorter decay time of the two long-lived decay components A_2 and A_3 (see Table 1) to the band N and the longer one to the band B.

2.3.2.3 PL excitation spectra

One experimental observation remains to be discussed. At PL decay times longer than $\tau \approx 30$ ns (see, for example, Fig. 2.6B) also the LE part of the PL band starts to shift to longer wavelengths (though PL intensities are already extremely low, which is hidden in the figure according to normalization of the PL intensity). A similar observation has been observed in previously reported experiments in toluene solutions of CdSe/ZnS [Petrov et al., 2006]. In both cases we find decay times up to 100 ns but with very low PL intensities. We assign this emission to energetically low–lying QD surface and/or trap states. In the present experiments we did not follow the decay kinetics in detail (since they turned out to be very similar to those already reported [Petrov et al., 2006]). However, our analysis concentrates on the following in PLE spectra under various conditions. Such experiments have to the best of our knowledge not yet been reported.

Figure 2.7A shows the investigated range of PLE spectra at 295 K. It is clearly seen that the PL excitation band broadens and shifts noncontinuously as a function of the PL detection wavelength rather than linearly, as one would expect according to size selection across the PL band.

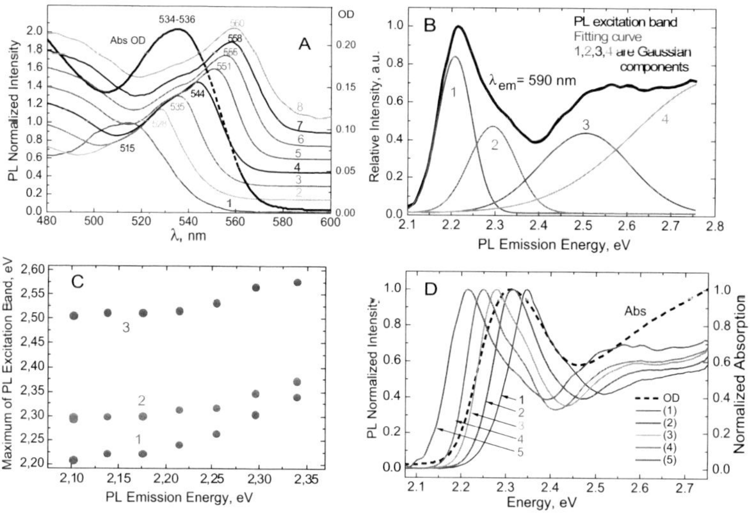

Figure 2.7 PL excitation (PLE) spectra of CdSe/ZnS at 295 K. (A) Dependence of PLE spectra on emission (detection) wavelength. The maximum of the PL emission upon high energetic excitation is at 555 nm (2.234 eV): λ_{em} = 520 nm (1), λ_{em} = 530 nm (2), λ_{em} = 540 nm (3), λ_{em} = 550 nm (4), λ_{em} = 560 nm (5), λ_{em} = 570 nm (6), λ_{em} = 580 nm (7), and λ_{em} = 590 nm (8). Maxima of every PL excitation spectrum (in wavelengths) are depicted for each line. The absorption spectrum is shown by a (partly dashed) black line. (B) Typical spectral deconvolution of a QD PLE spectrum by four Gaussian components. PL emission is 590 nm. (C) Dependence of the three deconvoluted energetically lowest Gaussian PL excitation components 1, 2, and 3 on detection (emission) PL energy. (D) Comparison of normalized PL excitation spectra with the normalized absorption spectrum (dashed line: OD) for five detection wavelengths: λ_{em} = 530 nm (1), λ_{em} = 540 nm (2), λ_{em} = 550 nm (3), λ_{em} = 560 nm (4), and λ_{em} = 590 nm (5).

Figure 2.7B presents the results of Gaussian fits of the PLE spectrum for λ_{em} = 590 nm at the very-low-energy side of the PL band. At least four excitation bands are needed to reproduce the excitation spectra adequately. The PLE energies for the three energetically lowest bands (1), (2), and (3) are shown in Fig. 2.7C as a function of the detection energy (corresponding to λ_{em} within the PL band excited at very high excitation energies). Inspection of Fig. 2.7C shows that the two higher energies (2) and (3) are energetically split by about 220 meV in parallel to each other over the total PL detection range, while the lowest excitation energy (1) is at variance with this behavior. An energy separation of ~220 meV corresponds to the energy separation between the $1S_{3/2}1s$ and $2S_{3/2}1s$ excitonic states for CdSe typical for QDs of 3 nm core diameter [Efros et al., 1997; Klimov et al., 2007].

It follows from the results obtained at 2.22 eV emission energy (λ_{em} = 560 nm, see Fig. 2.7) that the maximum of the lowest PL excitation band $^1S_{3/2}1s$ exhibits an energy splitting of 86 meV, which is comparable to the splitting of the absorption band into the HE and LE bands. The separation between PLE bands (1) and (2) becomes larger with increasing λ_{em} (decreasing PL emission energy). We assign this (detection wavelength dependent) shift to lower energies of excitation bands (and the related very long-lived PL) to intraband states. Obviously these states are not effectively populated via QD band states since they hardly show up in PL spectra upon high-energy excitation.

On the other hand, at the emission energy 2.34 eV (λ_{em} = 530 nm) the corresponding energy separation between the lowest (deconvoluted) excitation bands is only 33 meV, becoming still smaller with further decreasing λ_{em}. In addition, it is seen from Fig. 2.7D that beginning at the emission energy of 2.34 eV (λ_{em} = 530 nm), the maximum of the corresponding PLE spectrum is shifted to higher energies with respect to the maximum of the lowest absorption band.

The relative increase in the slope of the correlation between E_{exc} and E_{em} (Fig. 2.7C) above about 2.225 eV is related to the overlap of the states $^1S_{3/2}1s$ and $^2S_{3/2}1s$ accompanied by a selection of both small and large QDs within the respective states.

Additionally, comparing PLE spectra in Fig. 2.7D normalized with respect to the lowest absorption maximum, it is immediately evident that the PL QY is reduced upon excitation into QD states higher than $E_{exc} \geq 2.45$ eV with respect to excitation to the energetically lowest QD states. This implies that nonradiative decay channels become very effective at high excitation energies. Similar observations on increased nonradiative relaxation channels at high excitation energies have been reported in recent single-QD experiments [Schmidt et al., 2014; Galland et al., 2012] and for QD bulk solutions [Salman et al., 2009]. Moreover, the relative contribution of the PLE band (3) to the QY is decreasing with increasing PLE energy.

Comparing excitation spectra with the absorption spectra it is evident that the lowest PL excitation band is always considerably narrower than the corresponding absorption band, which is partly related to an effective size selection and is termed "fluorescence line narrowing" (FLN) [Nirmal et al., 1994]. Tentatively we deduce from the comparison of line widths the relative contribution of the QD size distribution to be at maximum 50 meV, that is, 30% of the overall line width of absorption [Zenkev et al., 2015]. Upon excitation into energetically high absorption bands this site selection is lost and we will observe a PL emission without any size selection. Thus the size-related contribution to the inhomogeneous PL emission is relatively small, which allows us to deconvolute the contributions of several electronic states to the PL of QDs, as has been performed [Zenkevich et al., 2015] and is shown in Fig. 2.4.

2.3.3 Summary of Experimental Findings

According to the complex and careful investigation of spectroscopic data the following scheme emerges (see Fig. 2.15). The energetically lowest absorption band is (independent of the temperature) split into at least two bands energetically separated by 70–80 meV. Absorption reveals a phase-transition-type behavior close to 220 K. At this temperature the relative contribution of the two absorption bands changes considerably (see Fig. 2.2C), which indicates a change in relative oscillator strengths of the absorption bands.

The PL band is split into a broad (B) and narrow (N) component separated (at 77 K) by about only 10 meV, which is considerably smaller than the nondeconvoluted PL line width (80 meV at 90 K and 105 meV at 290 K). Band N is characterized by an at least two times shorter PL decay time τ_i compared to band B. It has to be emphasized that the related states do not communicate with each other, which implies that they are very distinct but detectable—at least for TOPO-capped CdSe/ZnS QDs—in all experiments, even under various conditions, though the details such as energy separation and absolute decay times vary.

Upon temperature increase the PL energy shifts to lower energies, line widths become broader, PL intensities decrease, and PL decay times increase. While the N band PL energy follows monotonously the temperature increase, the B band energy shows a phase-transition-type behavior, resulting in a jump of energy at about 220 K. The transition complicates a quantitative analysis of temperature-dependent PL observables in the frame of electron–phonon coupling.

Energetically below these states we find from PLE a broad distribution of weakly absorbing and emitting states with PL decay times as long as 100 ns.

2.4 Discussion

The number of papers that deal with colloidal II/VI QDs is numerous, even when only concentrating on basic aspects. What can under such circumstances—as we announce in the title—new insights be? Before we turn to the discussion of our results we have to clarify (i) what new insights we will concentrate and (ii) what the theoretical frame for such a discussion is:

(i) New insights we like to refer to in the case of TOPO-capped CdSe/ZnS QDS are related to (1) the temperature-dependent Stokes shift, (2) the temperature dependence of the fine structure, (3) a ligand-induced phase transition, (4) the identification of several emissive excitonic band-edge states, and (5) the assignment of deep photoluminescent trap states. In some cases we will draw close relations to single-QD data, which helps analyze dynamics due to changes of electronic properties in the course of the experimental observation time that is on long times, say, longer than 1 ms. Such times are much longer than any intrinsic dynamic of isolated QDs. In Chapter 3 of this book we will tightly focus on electron–phonon coupling processes. Our interpretation is based on the spectral deconvolution of absorption, PL, and PLE spectra in a combination with PL time-resolved data over a broad temperature range from 77 K to 300 K.

(ii) The most elementary description of electronic states and the related optical oscillator strengths is as we like to term it the "solid state or quantum mechanical" model. Such an approach is well documented for CdSe QDs [Efros et al., 1996; Klimov, 2007] (see also many textbooks). It takes hole and electron states, including spin–orbit coupling, into account. Accordingly, the lowest excitonic state is eightfold degenerated for spherically symmetric QDs and is split into five levels (fine structure). Energies and transition probabilities depend on the crystal symmetry, the intrinsic crystal field (for hexagonal lattice structures), and the electron–hole exchange interaction. Size-dependent quantum confinement is taken into account within a multiband effective mass approximation. The relative ordering of the fine structure levels, their energetic separations, and oscillator strengths depend critically on crystal size and symmetry. Two of the five states, including the ground state, are optically not active (dark excitons), while the other three are optically allowed (bright excitons). Analysis of the size dependence of the band-edge exciton splitting (fine structure) for hexagonal CdSe QDs of different shapes (presented, for example, in Efros et al. [1996] and Nirmal et al. [1995]) shows that the energy separation between optically active exciton sublevels ($\pm1^L$ and $\pm1^0$, 0^0) for QDs which have been studied by us (CdSe core radius a = 1.5 nm) may vary between 30 meV (spherical symmetry) and 70 meV (oblate symmetry). However, it is generally agreed that CdSe/ZnS QDs are close to a spherical symmetry [Liptay et al., 2007; Htoon et al., 2004]. We present in Fig. 2.8B adapted from the literature [Efros et al., 1996; Klimov, 2007] fine structure energies as they are related to the crystal structure. The corresponding energy differences are in the range of the experimentally observed absorption and PL line widths despite the line broadening due to unavoidable inhomogeneities such as the remaining size distribution, variations in crystal symmetry among the individual nanocrystals, and the physicochemical nature of the surface or interfaces interacting with surrounding solvent molecules or substrates. They can partly be overcome by single-particle detection (see Chapter 4), which, however, is generally limited to PL emission. However, a complete description of QDs has to be based on absorption and emission, which we intend to follow with our experiments.

(iii) In this elementary (solid state) approach it is difficult to handle separately crystal imperfections, surface-related states, capping layers, and the influence of ligands. This can be (partly) overcome by numerical models (molecular approach) within an effective medium approach, for example, Frenzel et al. (2007), Kilina et al. (2009), Kilin et al. (2007), Voznyy and Sargent (2014), and Gomez-Campos and Califano (2012). A broad distribution of discrete electron and hole state energies and related oscillator strengths can be calculated via this approach. Though the applicability is presently limited to relatively small QDs, detailed qualitative insights are provided into the general nature of surface-related states, also for larger QDs (e.g., influence of dangling bonds [Frenzel et al., 2007], ligands [Kilina et al., 2009], or intraband states [traps] [Frenzel et al., 2007; Kilina et al., 2009; Kilin et al., 2007]). In fact, such an approach makes the discrimination between exciton, surface, intraband, or trap states in some respect obsolete, since all features naturally emerge from the calculation directly. In the discussion of our results we will make use of both terminologies depending on which of them explains most adequately our findings.

An additional degree of complexity is added by the observation that QDs are often charged (see Chapter 4) or subject to oxidation, ligand exchange [Ji et al., 2008], or even changes of the surface and interface morphology, for example, on temperature variation [Mello Donega et al., 2006]. Investigations of single QDs reveal PL intermittency (blinking) [Cichos et al., 2007; Riley et al., 2012], which has been explained either by photoinduced charging (see Chapters 4 and 6) or a multiple relaxation center model [Frantzusov et al., 2013; Schmidt et al., 2014]. It has been shown that such effects are even present in liquid solutions. Additionally, it has been reported that all these effects are also present in ensemble experiments in the form of reversible photobleaching but are heavily masked by ensemble averaging [Chung et al., 2006; Cichos et al., 2004; Schuster et al., 2007]. Moreover, due to weak ergodicity breaking QD ensemble and time-averaging single-QD data cannot uniquely but qualitatively be related to each other. All the above effects will to a certain extent modify electronic energies and oscillator strengths [Zenkevich et al., 2015].

Temperature variation in the range from 77 K to 300 K—as performed in the present experiments—will further illuminate the complex distribution of QD state energies. In a similar work on CdSe [Liptay et al., 2007; Salman et al., 2007] and CdSe/ZnS [Valerini et al., 2005] temperatures have been varied between 5 K and 300 K but without detailed spectral analysis. Contrary to the work of Liptay et al. (2007) the present investigation deals with core/shell CdSe/ZnS QDs instead of CdSe core QDs.

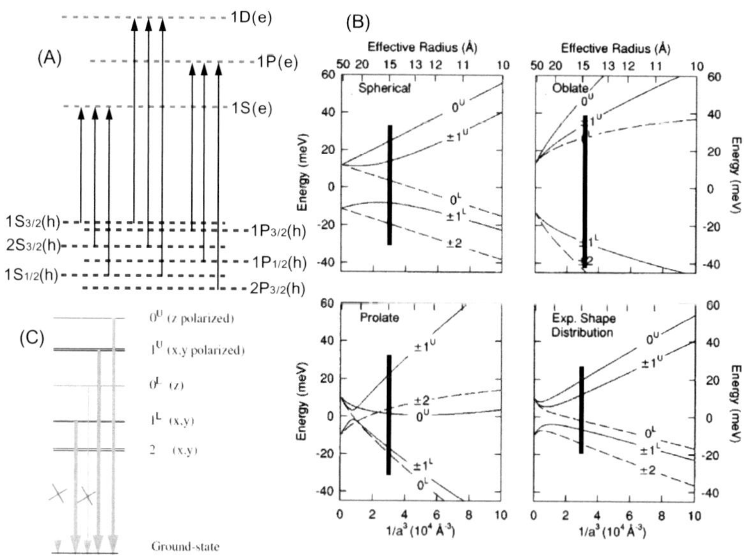

Figure 2.8 (A) Schematic presentation of allowed interband optical absorption for CdSe QDs in condition of quantum confinement and mixing between different valence subbands (Klimov, 2007). (B) Size dependence of the exciton fine structure in hexagonal CdSe QDs with various ellipticity μ for spherical particles ($\mu = 0$, top left), oblate particles ($\mu = -0.28$, top, right), prolate particles (μ – 0.28, bottom left), and QDs having a size-dependent ellipticity as determined from SAXS and TEM measurements (bottom right). Black solid and dashed curves correspond to states related to optically allowed and forbidden transitions, respectively. Black solid, vertical bars indicate the size and fine structure for CdSe QDs with radius a = 1.5 nm studied and discussed in this chapter. Reprinted (figure) with permission from [Efros, A. L., Rosen, M., Kuno, M., Nirmal, M., Norris, D. J., and Bawendi, M. (1996). Band-edge exciton in quantum dots of semiconductors with a degenerate valence band: dark and bright exciton states, *Phys. Rev. B*, **54**, 4843–4856.] Copyright (1996) by the American Physical Society. (C) Schematic presentation of the fine structure splitting of the PL band-gap transitions of CdSe QDs, as given by the Efros model depicted in (B). Reprinted from Salman, A. Al, Tortschanoff, A., van der Zwan, G., van Mourik, F., and Chergui, M. (2009). A model for the multi-exponential excited-state decay of CdSe nanocrystals, *Chem. Phys.*, **357**, 96–101, Copyright (2009), with permission from Elsevier.

Here we will summarize shortly the basic implications of the influence of temperature on PL spectral-kinetic parameters for core/shell CdSe/ZnS QDs. In both of the above theoretical approaches it was supposed that a Boltzmann equilibrium will be established among the near-band-edge (NBE) states (or fine structure states), resulting in temperature-dependent PL energies, Stokes shifts, and (multiexponential) PL lifetimes, as described in various publications [Mello et al., 2006; Labeau et al. 2003; Crooker et al., 2003; Salman et al., 2009; Schmidt et al. 2014]. Since energy differences between the lowest dark and bright fine structure states are low, Boltzmann equilibrium among these states is established already well below 80 K [Labeau et al. 2003; Liptay et al., 2007]. Our temperature data start from 77 K upward and are therefore not influenced by thermalization of the lowest dark state.

Also exciton-acoustic phonon scattering is temperature dependent and will influence absorption, PL spectra, related spectral widths, and PL decay rates due to the involved dephasing [Liptay and Ram, 2006; Liptay et al., 2007]. Moreover, the coupling to optical phonons should also be taken into account [Valerini et al. 2005; Dzhagan et al., 2009a]. Additionally, the influence of temperature on coupling to optical phonons mainly stems from thermal excitation to a higher electronic QD states [Liptay and Ram, 2006]. The closest energetically higher one is the $2S_{3/2}$ state, approximately 200 meV higher in energy than the $1S_{3/2}$ state.

Finally, with respect to temperature-induced changes of the semirigid ligand layer the ZnS shell may influence both absorption and PL. Such variations are probably most effectively described, though not yet reported, by a density functional theory (DFT) approach. Within the quantum mechanical model an appropriate description appears as a modification of the fine structure. A related model was recently proposed [Mourik et al., 2008; Salman et al., 2009] for CdSe QDs which is based on both electric linear dichroism data and results of two-photon excitation within the band gap. The interpretation of their results shows that a (externally induced) permanent ground-state dipole moment may cause a strong perturbation of intrinsic excitonic (fine structure) states. This results in considerable changes of oscillator strengths and the multiexponential PL decay kinetics, while fine structure energies are basically unaffected. Correspondingly, it is probable that the stochastic nature of such a dipole moment will be governed not only by modifications of the crystal structure of the QD but also by temperature-dependent fluctuations of the ligand and/or ZnS layers, thus leading to a complex temperature dependence of the first excitonic band parameters both in absorption and PL properties as well as in related PL decay kinetics.

2.4.1 Temperature-Dependent Stokes Shift and Fine Structure

All our experiments are above 80 K where the influence of the dark exciton is negligible [Liptay et al., 2007; Eilers et al., 2014; Mello et al., 2006; Labeau et al., 2003]. In Fig. 2.9A we compare the temperature dependencies of spectral characteristics for deconvoluted Gaussian components, including the LE and HE components of the first excitonic absorption band and both narrow N and broad B components of the PL emission. While we observe for CdSe/ZnS QDs a very small splitting of the PL bands ($\Delta E_{B,N} \leq 10$ meV), the splitting of absorption bands is considerable ($\Delta E_{HE,LE} \approx 75 \div 80$ meV). To the best of our knowledge a spectral deconvolution of the lowest absorption band has not yet been performed across such a wide temperature range. However, several attempts have been reported to unravel the energies of the lowest band-edge absorption at low temperatures, for example, via FLN and PLE of CdSe QDs [Norris et al., 1996] and by PLE of single CdSe/ZnS QDs [Htoon et al., 2004]. Both reports take the coupling to LO phonons explicitly into account. Norris et al. find for similar QDs, as presently reported, (size-dependent) energy splitting for the lowest absorption band of about 50–65 meV, which they assign to fine structure splitting. However, according to a priori calculations [Efros et al., 1996], these values are by about a factor of 2 too large [Liptay et al., 2007].

Htoon et al. (2004) report PLE data on single CdSe/ZnS QDs (which are not obscured by ensemble averaging), a somewhat different result, which is also valid for related ensemble experiments. According to their results they find a relatively small fine structure splitting in agreement with a priori calculations. Interestingly, at 10 K they find besides narrow excitation bands (with a 3÷4 meV line width) at the band edge about a 50 meV wide absorption gap followed by a broad absorption band (quasi-continuum) centered at ~90 meV above the centre of the band-edge transitions. They assign this quasi continuum to a high density of hole states with a related ultrafast (temperature independent) intraband carrier relaxation. This energy separation is very similar to the 80 meV splitting of bands, which we identified in the present absorption analysis. We are apt to believe that the splitting between HE and LE components is related to the reported hole state absorption since 80 meV are neither in agreement with the calculated (reasonable) fine structure energies [Efros et al., 1996] nor in agreement with the second excitonic absorption band ~200 meV apart (evaluated from the analysis of PLE spectra deconvolution; see Fig. 2.7B,C). This assignment is supported by the observation that on temperature increase the FWHM of the HE band is broader than the one of the LE band and changes from 90 meV to 110 meV (see Fig. 2.8C). On the contrary, the FWHM of the LE band is sensitive to the phase transition varying by only 2 meV between 90 K and 200 K but becomes narrower by 10 meV at T_{crit}. The HE band does not show a significant change in FWHM at T_{crit}. We would like to note that the fine structure remains buried also in the present experiments within the LE absorption band. However, the jump in FWHM at T_{crit} might indicate that the contribution of the fine structure decreases by about 10 meV or that the contribution of the lowest and highest components decreases symmetrically in oscillator strength.

We observe temperature dependence of the energy separation between the PL band and the mean weighted absorption band (MW), as shown in Fig. 2.9. The shift changes from about (80 meV) at 90 K to 70 meV at 280 K. The value of 80 meV is similar to the shift of 90 meV reported by Liptay et al. (2007). They deduced from the observed temperature dependence a fine structure scheme, which, however, is neither in agreement with calculated values nor does it reproduce the shape of the absorption band. We suggest that these failures of the fine structure identification from the temperature dependence of the apparent Stokes shift is closely related to the observation of a split absorption band both by Htoon et al. (2004) and in the present experiments. Accordingly, we calculate the Stokes shift as the energy separation ΔE_S of N or B components of the PL band with respect to the LE component of the first excitonic absorption band, as shown in Fig. 2.9B. The Stokes shift ΔE_S changes from $\Delta E_S \approx 39$ meV at 90 K to $\Delta E_S \approx 22$ meV at 280 K. Not surprisingly, ΔE_S decreases at T_{crit} for both PL energies by a sudden jump of about 13 meV and 10 meV, respectively, and remains nearly constant thereafter. The two deconvoluted PL components show slightly different temperature dependence for ΔE_S below T_{crit}. ΔE_S decreases more slowly with temperature for the PL component B, while it decreases faster for component N. Since the two deconvoluted absorption components do not change in energy separation over the total temperature range, the change in ΔE_S with respect to temperature is not specifically related to one of these absorption components.

As has been suggested by Liptay et al. (2007) a decrease in ΔE_S on temperature increase implies that PL emission is shifted to higher energies due to Boltzmann related population of the higher PL level(s), which are assumed to be related to the fine structure. Since we also observe a temperature-dependent decrease of the Stokes shift at low temperatures, we have to discuss which energy level might be responsible for such a thermal population followed by a decrease of ΔE_S. Since we find that the PL is related to both NBE and surface trap (ST) states [Zenkevich et al., 2015], an alternative approach is needed. We assign above T_{crit} the nearly temperature-independent Stokes shift ΔE_S = 22 meV to one of the NBE states [Zenkevich et al., 2015]. We therefore attempt to extrapolate this NBE-related Stokes shift down to 90 K with the assumption that the related Stokes shift does not depend on temperature. However, Liptay et al. (2007) observed in the temperature range from 290 K to 90 K an increase in the Stokes shift by 6 meV. Extrapolating ΔE_S = 22 meV (or 28 meV, taking the observation of Liptay into account) results in a PL energy of ~2.308 eV (2.302 eV) at 90 K, that is, by 14 meV (8 meV) above the PL energy of probably ST states at 2.294 eV (see Fig. 2.4A). Part of the observed temperature dependence of the PL emission below T_{crit} can now be explained by the thermal population of the NBE states according to the energy separation of $\Delta E_{NBE\text{-}ST}$ = 14 meV (8 meV). Depending on the QYs Φ_i of the thermally mixed (NBE; ST) states, this will result in a temperature-dependent apparent energy shift of the ST states to higher energies, which corresponds to a reduction of the (apparent) Stokes shifts, as can be clearly seen in Fig. 2.9B. We observe between 90 K and 180 K (210 K) a reduction of the Stokes shifts by ~7 meV (5 meV) for N (B) components of the QD PL bands. To rationalize these shifts we have to assume that the ratio of the QYs behaves like $\Phi_{NBE} \approx 2.3\ \Phi_{ST}$. The reduction of the Stokes shift is somewhat less pronounced for the PL band B compared to the PL band N, which is reasonable since it is slightly lower in energy and considerably broader.

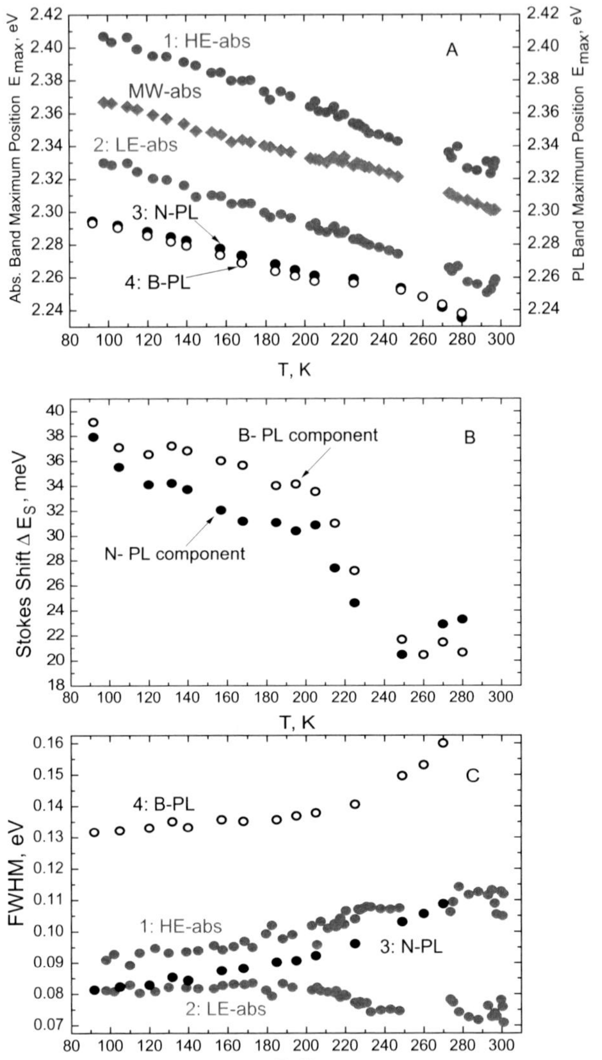

Figure 2.9 Comparison of temperature-dependent properties for the two spectrally deconvoluted Gaussian components HE (1) and LE (2) in absorption as well as for N (3) and B (4) components of PL. (A) Band maximum energy positions $E_{\max,i}$. MW corresponds to the mean weighted absorption. (B) Stokes shift ΔE_S for the two PL components N and B with respect to the lower absorption band LE (2). (C) FWHM values for deconvoluted absorption and PL bands.

Three aspects have to be summarized, as also illustrated in Fig. 2.10:

- According to the outlined procedure, we obtain above T_{crit} an estimation of the energy separation between ST and NBE states, which amounts to 14 meV (B) and 8 meV (N).
- The NBE emission is at least 22 (÷2) meV Stokes-shifted from the lowest absorption band LE, which is in reasonable agreement with Stokes shifts of about 17 meV reported by PLE experiments [Norris et al., 1996] for a similar type of QD. This implies that the fine structure of the presently investigated CdSe/ZnS QD is in reasonable agreement with literature values.
- In fact, the obvious jump of ΔE_S at T_{crit} on temperature increase is due to a jump of ST state energies above those of the NBE states.

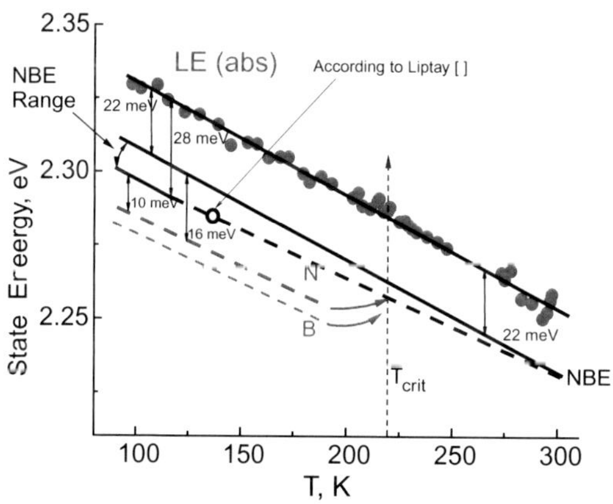

Figure 2.10 Schematic presentation of PL energies of ST (N and B) and NBE states energetically below the lowest absorption band LE as a function of temperature. The broken NBE line corresponds to a finding of Liptay et al. (2007). Black solid circles correspond to experimental absorption data following spectral deconvolution into two bands, HE and LE.

2.4.2 Temperature Dependence of Spectral Properties

2.4.2.1 Spectral width and oscillator strength

It is generally assumed that the temperature dependence of the spectral width of QD absorption or PL bands either relates to (inhomogeneous) broadening due to a (Boltzmann equilibrated) distribution of emitting PL states and a remaining size distribution or relates to (homogeneous) broadening caused by electron–phonon coupling. We compare the absorption line widths (FWHM) in Fig. 2.1C with the ones of PL in Fig. 2.3B. It can be clearly seen that they are at all temperatures considerably broader for absorption than for PL emission. This can be easily explained by the fact that absorption is split by two equally (below T_{crit}) weighted spectral bands separated by about 80 meV, while the energy separation of the two PL bands is <10 meV. However, PL may in principle originate from a distribution of exciton or trap states which emit according to the respective Boltzmann equilibrium, whereas absorption does not depend on Boltzmann equilibration. This might artificially reduce the PL line width. That this possibility is not sufficient, the differences in FWHM will be discussed with respect to PLE in Section 2.4.5. More detailed information is obtained for FWHMs when comparing the temperature dependencies of the deconvoluted spectral bands in Figs. 2.2B and 2.4B.

A compilation of the corresponding data is presented in Fig. 2.9C, which shows the temperature dependence of the spectral widths for the two absorption and two PL band components. Four observations immediately emerge from a comparison of FWHMs:

- The PL band B is much broader than the three other bands, which are of comparable line width. We assign this band to a broad distribution of (isolated) trap states ST different from NBE states.
- The temperature dependence of the PL bands is (below T_{crit} = 220K) nearly the same as for the absorption band HE.
- A pronounced change of the slope d(FWHM)/dT is observed for PL bands at T_{crit}.
- Additionally, the FWHM decreases at T_{crit} for the LE absorption by −10 meV, while the respective slope d(FWHM)/dT is nearly the same below and above T_{crit}. The difference in FWHM of −10 meV has to be compared to −6 meV for line narrowing at T_{crit} observed for the total absorption width shown in Fig. 2.1C. For the HE absorption band there is hardly any sudden change of the FWHM at T_{crit} and the temperature dependence can be

approximated nearly linearly across the total temperature range.

Now we discuss in more detail the influence of temperature on PL line widths. We know from QD-dye nanoassembly formation reported recently [Zenkevich et al., 2015] that PL stems below T_{crit} from specific STs. Above T_{crit}, PL is more closely related to NBE states. Obviously, the related temperature dependence (above T_{crit}) is stronger for NBE states in comparison to ST states. The latter dependence is nearly linear across the temperature range below T_{crit}, while it becomes steeper for NBE states above T_{crit}.

Changes in the line width of PL are probably not related to temperature-dependent fine-structure-level populations (see the previous section) but are more likely due to exciton–phonon coupling, as has been reported for CdSe QDs [Liptay et al., 2007] and as is discussed in Chapter 3. When comparing the slope d(FWHM)/dT (~0.115 meV/K) estimated in a temperature range of 80–300 K for the PL band of CdSe QDs with d_{CdSe} = 3.0 nm [Liptay et al., 2007] with the ones observed for TOPO-caped CdSe/ZnS QDs (d_{CdSe} = 3.0 nm) in the present experiment, we find the following:

- The dependence of the FWHM is small in the case of the nondeconvoluted absorption band (see Fig. 2.1C). This is due to the fact that absorption is (inhomogeneously) broadened for two independent reasons, namely the presence of the two absorption bands HE and LE and the size distribution. In this context additional (homogeneous) temperature-dependent contributions will not broaden the line effectively. In fact, the influence of temperature on the FWHM of the LE band is (besides at T_{crit}) negligible.
- Below T_{crit} (220K) slopes d(FWHM)/dT calculated by a linear approximation are found to be ~0.056 meV/K and ~0.112 meV/K for the B and N components, respectively. The N component behaves very similar to the HE band.
- Above T_{crit} the slopes for N and B bands become much larger and are probably not linear any more. The FWHM of the HE band shows a slope of ~0.112 across the total temperature range. Thus, a quantitative agreement between Liptay's data and our results is seen only for the HE band. Alternatively, thermal population of LO phonons has to be discussed. Details

of electron–phonon coupling will be discussed in Chapter 3 of this book.

Concluding, we tentatively assign the variation of FWHMs for N and B bands above T_{crit} and probably also below T_{crit} to a thermal equilibration of a distribution of NBE and ST states, as has been suggested by recent work of Jones et al. [2009a, 2009b].

Next, we explain the relative change in oscillator strength of the two absorption bands HE and LE (the latter is a superposition of all fine structure energies with related oscillator strengths) at the phase transition (see Fig. 2.2C). As has been shown and argued recently [Mourik et al., 2008] changes in the ground-state dipole moment, for example, by crystal distortions or nonspherical surface modifications, might result in considerable mixing of fine structure levels combined with strong variations of apparent oscillator strengths. Since we do not observe any energy jump of either of the two absorption bands HE and LE, we suggest that a decrease (see Fig. 2.2C) of oscillator strength within the LE (fine structure) states takes place above the phase transition [Zenkevich et al., 2015], which results in a sudden decrease of the optical density (see Fig. 2.1D) above T_{crit}. Simultaneously the FWHM of the LE band is suddenly reduced above T_{crit}. Both effects result in an apparent shift of the nondeconvoluted absorption band to higher energies (Fig. 2.1B). Following theoretical arguments [Liptay et al., 2007; Norris et al., 1996] oscillator strengths, especially for small QDs, are compared to fine structure splitting much more sensitive, for example, to the size and/or symmetry of QDs. On the contrary, the HE band is somewhat less influenced by the phase transition (compare Figs. 2.2B and 2.2C), which seems to be reasonable since the hole wavefunction is more confined to the QD core compared to the electron wavefunction [Efros et al., 1996]. Concluding, the temperature dependence of absorption is sensitive to the phase transition, though to a much less extent than the PL. Nevertheless this finding indicates that absorption not only is related to CdSe core properties but also feels the influence of the ZnS shell.

It is obvious that the overall temperature effects observed for the first excitonic absorption band LE of CdSe/ZnS QDs cannot be explained merely by exciton–phonon coupling. Excluding the regime close to T_{crit} (broken ellipsoids in Fig. 2.2B) we observe a continuous increase of FWHM of HE and LE energies but differently with respect to the two bands. Generally, temperature effects on line widths are explained as being due to electron–phonon coupling either by LO optical [Valerini et al., 2005] or acoustic phonons [Fernee et al., 2008; Liptay et al., 2007]. Without going into detail (see Chapter 3 in this book), scattering with acoustic phonons will increase the line width linearly with increasing T [Liptay et al., 2007], while (LO) phonons contribute—depending on the electron–phonon coupling and LO frequency—exponentially on temperature increase [Valerini et al., 2005].

Line broadening of absorption is independent of the specific band nearly linear with T over the complete temperature range (besides the change at T_{crit}). This implies that predominantly acoustic phonon scattering takes place. Figures 2.2B and 2.8C show that the amount of temperature dependence of the FWHM of the two absorption bands itself depends on the specific band. The LE component is obviously less temperature dependent compared to the HE component, which suggests a difference in electron–phonon coupling. The reason of the different temperature behavior for the LE and HE components may be explained by the presence of an energetically higher electronic state (see the energy-level scheme in Fig. 2.8). In this case, electron–phonon coupling for the HE component feels more strongly the thermal mixing with this more closely separated state.

2.4.2.2 Absorption and PL energies

The deconvoluted absorption energies (Fig. 2.2A) decrease within experimental accuracy linearly with increasing temperature. For high temperatures ($T > 250$ K) it has been shown that the experimental value of dE/dT is independent of the size of CdSe/ZnS QDs and agrees with the average bulk value of $dE_{bulk}/dT = -0.36$ meV/K [Liptay et al., 2006]. The results depicted for absorption of TOPO-capped CdSe/ZnS QDs in Fig. 2.2A give in a linear approximation $dE/dT = -0.42$ meV/K for the HE component, −0.40 meV/K for the LE component, and −0.33 meV/K for the MW data. From this we conclude that absorption roughly reflects the properties of CdSe/ZnS.

In comparison, Fig. 2.3A reveals that there are two ranges of the temperature dependence of the lowest PL energy E_{max}. The temperature dependence gives $dE/dT = -0.42$ meV/K above T_{crit}, which is in agreement with data from absorption, indicating that we are dealing with QD states of the same nature. Assuming a linear relationship, we obtain −0.32 meV/K below T_{crit}. Linearity is, however, evidently not fulfilled: the increment is lower below about 150 K compared to the temperature range from 150 K to 220 K. The temperature dependence of PL has been explained, for example, by Valerini et al. (2005) as being due to band-gap shrinking (due to thermal expansion) with increasing temperature. This should also apply to absorption. However, absorption and PL show comparable temperature dependency only above T_{crit}. We therefore conclude in analogy to the aforementioned temperature-dependent Stokes shifts that the PL below T_{crit} stems from QD states of a different nature than the one from above T_{crit}.

2.4.3 Ligand-Induced Phase Transition

In contrast to experimental results presented and discussed recently [Liptay et al., 2007], some specific properties in temperature dependencies for E_{max} and the FWHM in absorption and PL emission are observed in the present case. As can be seen in Figs. 2.1–2.4, these dependencies (upon fitting the first excitonic band in absorption by Gaussian components) change not monotonously with temperature. These nonmonotonous behaviors for E_{max} and the FWHM (which we termed for simplicity "kink" [Zenkevich et al., 2012]) are observed between 220 and 230 K, far from the glass transition temperatures of methylcyclohexane (146.7 K) and toluene (180 K), respectively [Zenkevich et al., 2015; Murov et al. 1993], as well as for a methylcyclohexane/toluene (6:1) mixture (151.6 K) [Marcus, 2002]. We have shown recently [Zenkevich et al., 2012] that such a kink is not caused by changes of the morphology of the matrix upon temperature lowering but can be related to a freezing (phase transition) of the TOPO ligand layer in the temperature range of $T_{crit} \approx 200–230$ K. Notably, such a phase transition of the ligand layer causes strain-induced deformations of the ZnS shell. Previously detailed investigations of such a phase transition have been reported [Mourik et al., 2008]. Obviously this results in new PL transitions which are below T_{crit} lower in energy compared to NBE states of CdSe/ZnS. From PL experiments it cannot be decided whether these new ST states are electron or hole trap states. Additionally, the concentration is very low, which implies that they cannot be detected in absorption. However, we find that the phase transition shows up in absorption via a change of optical oscillator strength and FWHM (Fig. 2.2C) of the fine structure–related LE absorption band. Following arguments posed recently [Mourik et al., 2008] strain may induce an effective electric dipole moment which correspondingly mixes fine structure levels. Obviously this effect is more strongly felt by the LE absorption band than by the HE band, which is related to hole states [Htoon et al., 2004].

How can this finding be explained? According to Peng et al. (1997), Dabbousi et al. (1997), and Cao and Banin (2000), absorption (and PL) of core-shell CdSe/ZnS QDs are red-shifted with respect to merely core CdSe QDs with identical core radius because in the former case, the excitonic wavefunction tunnels into the ZnS shell (and even beyond), increasing the delocalization of the electron-hole pair, thus lowering the confinement energy and consequently widening the energy gap. Correspondingly, due to different thermal expansion coefficients for the semirigid surfactant layer and the ZnS shell as well as ligand-induced strain, slight changes of electron delocalization may take place in the range of the ligand phase transition. In its turn, temperature-induced deformations of the semirigid ligand layer and the ZnS shell will have influence on the splitting of CdSe fine structure levels as well as on the oscillator strengths of the corresponding excitonic transitions of the CdSe QD core. Obviously, the latter will result in changes of the LE absorption band with respect to the oscillator strength and FWHM. Taking this conjecture further, experimental findings indicate that the overall fine structure splitting is above T_{crit} reduced by about 10 meV (as concluded from the change in the FWHM) and the overall oscillator strength is decreased by roughly a factor of ~1.6 (see Fig. 2.2C). In this respect it is interesting to note that for CdSe QDs with $d_{CdSe} = 3.0$ nm, according to Efros et al. (1996) calculated oscillator transition strengths for interband optically active transitions from the ground state (fine structure model) may be decreased by ~1.25–2.0 times when comparing spherical with oblate QDs.

With respect to the HE band we observe a noticeable increase only for the oscillator strength above T_{crit}. This might indicate that the number of hole states is increased by a factor of 1.3. Comparing changes of optical properties in the vicinity of T_{crit} reveals that the related range of temperature in which noticeable variations become visible is slightly different for absorption and PL. In the case of absorption the range is 210–230 K (Fig. 2.2), while it is 215–225 K in PL (Fig. 2.3). This might be related to the fact that absorption monitors "true" exciton states, while PL corresponds to a composition of the ST and NBE states.

Finally, it should be noted that such a kink, reflecting a phase transition of the capping ligand layer upon temperature lowering, is also observed for PL properties of AM-capped CdSe/ZnS QDs in a methylcyclohexane/toluene (6:1) mixture at $T_{crit} \approx 240$ K, which is about 20 K higher than for TOPO ligands [Zenkevich et al., 2012]. Contrary to samples in glass-forming solutions, for nanoassemblies based on TOPO-capped CdSe/ZnS QDs and pyridyl-substituted Cu-porphyrins (1:1) spin-coated onto a quartz substrate, no kink is observed in a temperature range of 80–295 K [Zenkevich et al., 2015]. This is probably due to the fact that the TOPO capping layer is (almost) completely removed or that the remaining ligands can freely reorient due the absence of an embedding solvent.

2.4.4 Correlation of Kinetics and Spectra of CdSe/ZnS Photoluminescence

Here we will in a first step correlate in close analogy to a recent publication [Zenkevich et al., 2015] the results from spectral deconvolution of PL (Fig. 2.4A) with those for the PL decay (Table 2.1 and Fig. 2.5). To the best of our knowledge a spectral deconvolution into two spectral components has not been achieved before across such a broad temperature range. In all our experiments deconvolution into two Gaussian components reduces considerably the fitting error in comparison to only one component. We like to emphasize that the assumption of only two components is somewhat arbitrary but has to be taken as an indication that the distribution of states has to be approximated by at least two components.

At first glance the identification of only two spectral components is in contrast to the finding of at least three PL decay times. However, inspection of Table 2.1 reveals that the integrated intensities I_i of the fastest-decaying component ($\tau_1 < 1$ ns) are less than 5% of the total intensity both at 77 K and 295 K and will therefore neither contribute noticeably to $<\tau>$ (see Eq. 2.1) nor to the QD PL spectra. We have to discuss our results for the two temperatures separately.

The normalized amplitudes A_2 (A_3) of the PL decay components decrease (increase) continuously from short to long wavelengths for CdSe/ZnS QDs at 77 K (see Table 2.1 and Fig. 2.5C). However, the integrated intensity of the (short) component A_2 is relatively small (16% at high and 5% at low energies), whereas the spectral deconvolution results in almost equal integrated intensities (I_N:I_B $\approx$ 1.3). This implies that both deconvoluted spectral components decay roughly within 15 ns, with a preference that component N decays somewhat faster than component B (see Fig. 2.5). The PL decay component A_2 (with $\tau \approx 4$ ns) is energetically above the N and B energies by about 10 meV, as can be concluded from Fig. 2.6A. The third very short decay component with amplitude A_1 is still further in the wing of the PL band and is thus at higher energy than the related energies of A_3 or A_2 components. According to Fig. 2.6B we speculate that the A_1 related energy is by about another 10 meV above the one of the A_2 component. Since the contributions of the two short-lived components A_1 and A_2 depend strongly on the solvent polarity, we conclude that some charge transfer character might be involved. The assignment of PL decay times to spectrally deconvoluted PL energies is included in the scheme in Fig. 2.11.

In case the distribution of energies identified at 77 K is temperature independent the overall line broadening of about 20 meV when passing to 290 K will wipe out all time-correlated spectral information and PL decay times cannot be assigned anymore to PL energies, as is observed experimentally (see Table 2.1). Consequently, only relatively small variations of $<\tau>$ are observed across the PL band at 290 K (see Fig. 2.5D). This implies that all deconvoluted spectral components are determined mainly by I_2 and I_3 at 290 K (above T_{crit}), while $<\tau>$ is almost constant across the FWHM of the PL band for CdSe/ZnS QDs with a slight preference of A_3 at low energies.

PL decay times τ_2 and τ_3 are shorter at 77 K compared to 290 K. This is quite unusual but can be explained by the fact that the nature of states below and above T_{crit} is different (ST versus NBE states). Additionally, a quite striking observation is that the contribution of the short decay components A_1 and A_2 is relatively increased at 290 K compared to 77 K. Whether this can be related to the different nature of emitting states at these two different temperatures remains presently an open question.

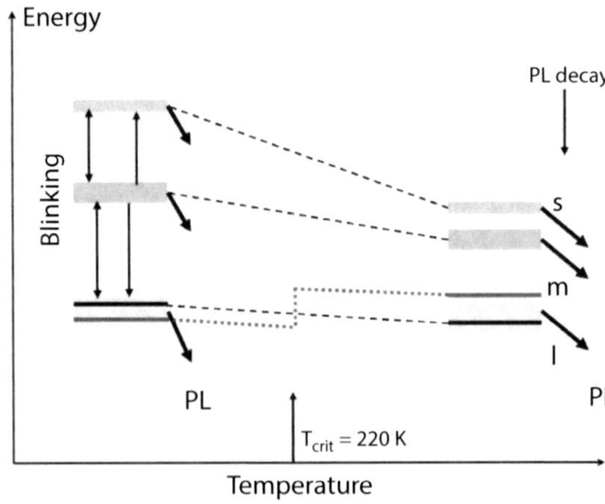

Figure 2.11 Qualitative presentation of PL energies at low (left) and high (right) temperature, derived from the combined analysis of experimental PL decays (τ_1, τ_2, τ_3) and deconvoluted Gaussian components of the PL band. T_{crit} indicates the phase transition. At least three principally different states are indicated at each temperature according to their PL decay times: low (l), medium (m), and short (s). According to blinking analysis (see Chapter 6) of single QDs [Schmidt et al., 2014] they are interrelated (time-resolved single-QD data show that a short-lived component of type τ_1 is predominantly detected at the red edge of the PL at 290 K). The general shift to low energies with increasing temperature is due to exciton–phonon coupling (see Chapter 3). At the phase transition temperature T_{crit}, specific band-edge emissive states are shifted up in energy on temperature increase [Zenkevich et al., 2015].

The following discussion is closely related to recently presented results [Zenkevich et al., 2015]. We did not observe build-up components of the PL decay longer than the intrinsic time resolution of about 0.2 ns [Zenkevich et al., 2005; Petrov et al., 2006]. Therefore internal conversion processes such as energy transfer among three individual energy levels of a single QD, resulting in shortening of energetically higher states, can be safely excluded. This consequently implies that either individual QDs (of the ensemble) have different decay dynamics or—in close relation to blinking and spectral diffusion processes observed for single QDs [Cichos et al., 2007; Riley et al., 2012; Cordones and Leone, 2013]—the electronic states of QDs change subsequently on (slow) timescales between various electronic configurations, such as charged/uncharged QDs or different trap states are switched on and off during observation time. The latter behavior is suggested by blinking related multiple multiple-relaxation-center models [Frantzusov et al., 2013; Schmidt et al., 2014]. We have recently shown that this conjecture holds at least qualitatively [Schmidt et al., 2014; Zenkevich et al., 2015].

The above finding proves that the energetically identified state is, for a once-excited QD, at a given time the only emissive one. Such a finding can easily be explained by experiments on blinking single QDs. We suggest that blinking and spectral diffusion (corroborated by the observation of various emissive states in closely related ensemble experiments) is due to subsequently created different ligand-/dye-related surface states with varying PL energies and decay dynamics. Such fluctuations are probably due to temperature-independent tunneling of ligand binding between neighboring but differently coordinated Cd and/or Se surface atoms. The mobility of surface ligands or atoms has recently been calculated [Voznyy, 2011; Vozzny and Sargent, 2014] or even experimentally been observed [Pennycook et al., 2012]. Additionally, there is ample evidence that the PL decay of even a single QD is multiexponential in nature [Schlegel et al., 2002; Fisher et al., 2004; Biebricher et al., 2006; Zhang et al., 2006].

The identification of at least three PL decay times is in agreement with many previous results on CdSe/ZnS QDs [Petrov et al., 2006; Schmidt et al., 2014; Blaudeck et al., 2012; Salman et al., 2009; Knowles et al., 2011; Kern et al., 2011; Schlegel et al., 2002; Fisher et al., 2004; Biebricher et al., 2006]. Depending on QD size, preparation, and experimental conditions the time ranges are typically 0.5–2 ns, 3–10 ns, and 15–25 ns. Recent careful investigations have shown that the times are broadly distributed [Schmidt et al., 2014; Kowerko et al., 2010; Borczyskowski et al., 2014; Petrov et al., 2006] and are dependent on detection wavelength [Petrov et al., 2006], PL intensity [Salman et al., 2009; Schlegel et al., 2002; Fisher et al., 2004; Biebricher et al., 2006], and temperature [Salman et al., 2009]. Naturally, such decays could also be approximated by a stretched exponential decay [Petrov et al., 2006; Schlegel et al., 2002; Fisher et al., 2004; Biebricher et al., 2006]. However, we prefer a description by individual components since we intend to relate PL decay times to well-defined electronic states.

With respect to time-resolved experiments we finally conclude that there are at least three basically different types of emissive states for each individual QD (see Table 2.1). In addition, upon temperature variation the ordering and energetic position at least of some of these states are changed at the phase transition of the capping ligand shell at $T_{crit} \approx 220$ K.

Recent calculations have shown that the energies of band-edge states of QDs depend on ligand attachment (saturation of dangling bonds) [Frenzel et al., 2007; Kilina et al., 2009], solvent [Fischer et al., 2012], and dye attachment [Kilin et al., 2007]. We have repeatedly shown that such dye attachments will result in QD PL quenching (and modification of blinking), which is neither related to Förster resonance energy transfer (FRET) nor related to charge transfer [Zenkevich et al., 2005; Kowerko et al., 2010; Blaudeck et al., 2012]. It is only a matter of semantics to assign such states either to band-edge states or shallow traps. In both cases, they will have properties closely related to those of QDs themselves.

Using DFT and time-dependent DFT quantum mechanical methods it has recently been shown [Kilina et al., 2009; Kilina et al., 2012] that ligands like $OPMe_3$ (TOPO-like), NH_2Me (AM-like), and pyridine introduce new states to the QD band-edge states of both the conduction and valence bands. The energies of these new states depend on the number and kind of ligands as well as on the position of the ligands with respect to the QD crystal structure [Kilina et al., 2009]. Depending on the relative coverage of ligands calculated (lowest) absorption energies shift to lower (high coverage) or higher energies (low coverage) in the range of +/− 200 meV [Kilina et al., 2009]. Due to steric ligand-related reasons only about 30% of the dangling bonds of the investigated CdSe/ZnS QDs are covered by TOPO ligands [Bullen and Mulvaney 2006; Ji et al., 2008; Kowerko et al., 2010]. From inspection of the calculations, it becomes immediately evident that already minor variations in the ligand coverage (number, position, chemical nature) will create new or energetically shifted surface related states across an energy range of about 30 meV, as we have experimentally observed.

We have discussed our results deduced from spectral deconvolution of ensemble data and compared them with calculations and experiments in the spirit of single-QD moieties. However, ensemble data are naturally superimposed by the size distribution of the QDs and abundant inhomogeneities of surface morphology and crystal structure. Whether inhomogeneities of energies and decay dynamics are the governing mechanism in ensemble experiments is of crucial importance. Respective experiments in two different matrices show [Schmidt et al., 2014; Göhler, 2014] that also single QDs prevail a nonvanishing inhomogeneity in PL energies and dynamics, which depends strongly on the observation time being closely related to blinking and spectral diffusion. Moreover, PL energies at a given PL intensity change in the course of blinking time traces. Such effects are known as spectral diffusion [Fernee et al., 2012; Beyler et al., 2013]. The manifestation of spectral diffusion reflects in line with the previous discussion of ligand–dye interaction with the QD surface that QD surface states are by no means a constant set of energies but undergo permanent fluctuations on timescales longer than milliseconds [Fernee et al., 2012; Beyler et al., 2013; Müller et al., 2005].

A further indication that the energetically lowest PL emissive states are of a different nature below and above T_{crit} is related to the influence of solvents on the PL decay at the respective temperature regimes. Figure 2.6 shows a pronounced and short-lived blue-shifted PL component at temperatures below T_{crit} for CdSe/ZnS in toluene and to a much smaller extent in (the less polar) methylcyclohexane/toluene (6:1) mixture. We emphasize that the short-lived PL can be also identified from Table 2.1 for the mixture of solvents. This means that it is not newly created in a specific solvent but only its contribution becomes larger in toluene. Obviously this manifestation depends on the polarity of the solvent from which we conclude that this short-lived PL stems from a charged or at least polar state. At 290 K (and thus above T_{crit}), also observed short-lived PL is energetically much less separated from the two other PL decay components since the detected spectral shifts are much smaller. Small time-dependent spectral shifts have recently been reported for a slightly different QD [Petrov et al., 2006] in toluene at 290 K.

2.4.5 Near-Band-Edge and Deep-Level Trap States

More information on electronic states close to the band edge can be obtained by PLE spectra. PLE reduces the influence of size distribution on optical densities to a certain extent, resulting in an increased spectral resolution of a quasi-absorption. Additionally, PLE is selective with respect to a specifically identified (even weak)

PL. Figures 2.7A and 2.7D show PLE spectra shifting with the detecting emission wavelength λ_{em}. In comparison to the also shown absorption spectrum, PLE spectra are narrower in the spectral width. Both effects are commonly explained by size selection. It is expected that the second-lowest electronic transition $^2S_{3/2}$ (besides the fine structure splitting) is about 200 meV above the lowest one ($^1S_{3/2}$) [Petrov et al. 2006; Valerini et al., 2005]. For this reason PLE spectra are broadened at certain λ_{em} (e.g., 515 nm in Fig. 2.7A) since small QDs are excited via the absorption band $^2S_{3/2}$ and larger QDs get excited via the 200 meV energetically lower $^1S_{3/2}$ absorption band [Petrov et al., 2006; Valerini et al., 2005]. From the overall behavior one can deduce the excitonic absorption energies and size distribution [Blaudeck et al., 2012].

Figure 2.7B shows the spectral deconvolution of the lowest PLE bands by four Gaussian lines for a typical PL emission at λ_{em} = 530 nm. Note that the first absorption band is approximated by the two Gaussian components (1) and (2). Excitation energies (1), (2), and (3) depend on λ_{em}, as shown in Fig. 2.7C. Moreover, PL excitation spectra show in Fig. 2.7D that the PL quantum efficiency (PLE intensity) depends on λ_{em} and is relatively reduced to what one would expect from absorption when exciting into states energetically above $^1S_{3/2}1s$. The relevant spectral indication is the relative ratio of PLE and absorption intensity at high compared to low excitation energies. This phenomenon has been discussed recently in close relation to blinking phenomena of single QDs [Zenkevich et al., 2015] and points toward increased nonradiative relaxation channels upon high-energy excitation.

Complementary results from the spectral deconvolution of PLE spectra are given in Fig. 2.12. We show in Fig. 2.12A that the relative energy separation of PL excitation band (1) from PL excitation band (2) depends according to Figs. 2.7C and 2.12A strongly on the detection wavelength, while the energy separation between the excitation bands (2) and (3) is independent of the PL detection wavelength. The latter is expected in case that size distribution is the origin of the distribution of both PL and absorption. Upon excitation into band (4) size selection is obviously lost. The energy separation between bands (2) and (3) is close to 200 meV, as is expected for the two lowest absorption bands of CdSe/ZnS with the given size [Valerini et al., 2005].

Remarkably, when changing the emission (detection) energy below about 2.225 eV, the absolute energies of excitation bands (2) and (3) do not change with E_{em}, opposite to what is expected in case of continued size selection. This might indicate that an increasing part of PL lower than 2.225 eV stems from more or less size/independent deep-level trap (DLT) states which become populated via bands (2) and (3). The energy separation of bands (1) and (2) is 100 meV at 2.105 eV, from there on decreasing to 80 meV at 2.2.200 eV, but starts to vanish rapidly for detection above the maximum of the PL band (2.233 eV) at higher detection energies.

We observe for the lowest deconvoluted PLE bands a kind of "turnover" for all fitting parameters below detection energy of about 2.225 eV, which is about 25 meV below the LE absorption band at 2.25 eV (see Fig. 2.2A). This separation is close to the Stokes shift of 27 meV, as discussed in Section 2.4.2. We take this as an indication that upon excitation in the vicinity of this energy range, the electronic character of PL states starts to change from NBE to DLT states. Below 2.225 eV we probably observe a mixture of DLT, ST, and NBE emission. The relative contribution of PLE band (3) and energetically still higher excitation bands to the PL emission is steadily increasing with decreasing emission energy (see Fig. 2.12C) contrary to the one of band (2). This tendency is also directly evident from Fig. 2.7D. According to this observation we conclude that DLT and ST states are besides direct excitation more effectively populated via QD states energetically above the lowest (LE) absorption band. From inspection of the FWHM given in Fig. 2.12B we find that the FWHM of less than 70 meV is not increasing with E_{em} but decreases, while the ones related to bands (2) and (3) increase in parallel. This indicates that the latter are related to size selection of intrinsic QD states. The lowest energy of PLE band (2) is close to 2.28 eV, which is close to the LE energy of 2.26 eV. On the contrary, above 2.225 eV the intrinsic PL of NBE states of CdSe/ZnS is dominating. We do not find an indication of LE–HE splitting. It is also not buried in the FWHM since the FWHM of band (2) is as narrow as 90 meV. From this we safely conclude that excitation of the HE band does not result in PL emission.

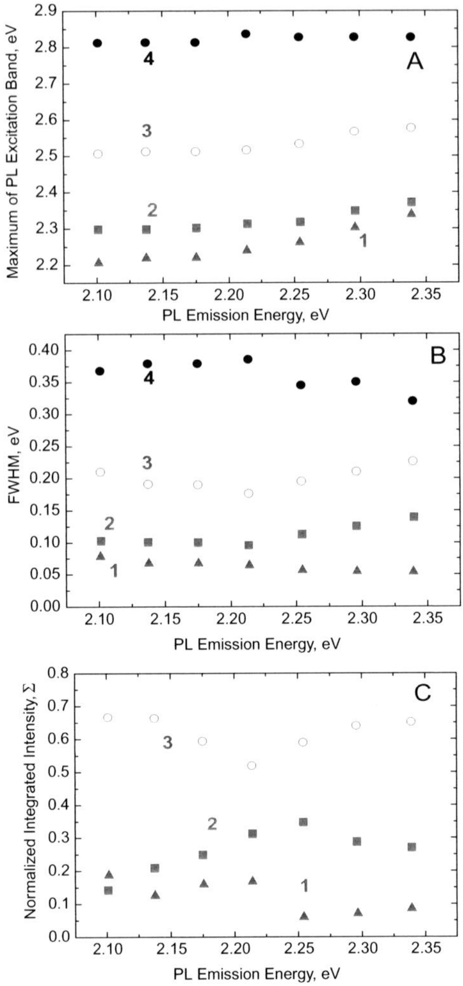

Figure 2.12 Deconvolution of PLE spectra of CdSe/ZnS at 295 K as a function of PL emission (detection) energy E_{em}. The data for four Gaussian components in the region of the first excitonic transition (see Fig. 2.7B) are as follows: 1 (triangles), lowest-energy component; 2 (squares), medium-energy component; 3 (open circles), highest-energy component; and 4 (solid circles), component presenting transition to high excitonic states. (A) Maximum of PLE bands. (B) FWHM values of deconvoluted PLE bands. (C) Normalized integrated intensities of three PLE bands in the region of the first excitonic transition.

Complementary to PLE we have recorded (spectrally deconvoluted) PL as a function of excitation energy, as shown in Fig. 2.13. At excitation energies above 2.6 eV the PL is split into two PL bands separated by about 20 meV. This is somewhat larger than observed in Fig. 2.2, which is most likely due to the fact that experiments have been performed on two different but nominally identical QDs which are obviously slightly smaller comparing PL energies.

As expected, no size selection is effective upon excitation into high-energy bands of QDs. An intermix of size and electronic state selection (see the discussion above and Petrov et al. [2006]) occurs in the excitation energy range from 2.3 to 2.5 eV, which causes PL bands to shift and broaden. In this range, spectral deconvolution into two bands becomes meaningless to a certain extent, since it would be necessary to assign at least four PL transitions. The noticeable inset of intermixed spectral and size selection at 2.5 eV (Fig. 2.13A) is accompanied by an increased energy separation (Fig. 2.13A) and line broadening (Fig. 2.13B) of the formerly two bands. Between 2.3 and 2.45 eV only one of the two bands dominates the PL (Fig. 2.13C).

Excitation at 2.25 eV, which is below the maximum of the $^1S_{3/2}$ absorption (at 2.3 eV), results in only one PL band (or at most two bands with a very small splitting; see Fig. 2.13A). This is a remarkable result which can be explained as follows: Excitation above 2.5 eV occurs for all QD sizes into the second electronic absorption band. Excitation below 2.3 eV happens for all sizes only into the lowest excitation band. Obviously, this latter excitation results in only one PL band. This implies that the now missing band needs a higher-energy excitation. This might be explained by the fact that following high-energy excitation the excess energy might be accompanied by a creation of hot electrons, as has been reported repeatedly [Klimov, 2000a]. Here, hot electrons might effectively contribute to the formation of charged or polarized QDs, reducing the PL QY, as can be seen from Fig. 2.7D. PL emission follows below 2.23 eV linearly the excitation energy according to size selection rules. Remarkably, excitation at 2.2 eV results in PL emission at 2.175 meV, which corresponds to a Stokes shift of ~25 meV, very similar to 27 meV, as has been identified from the analysis in Section 2.4.1. PLE at detection energy E_{em} = 2.2 eV results in a lowest excitation energy of 2.225 eV, which implies a shift of also 25 meV.

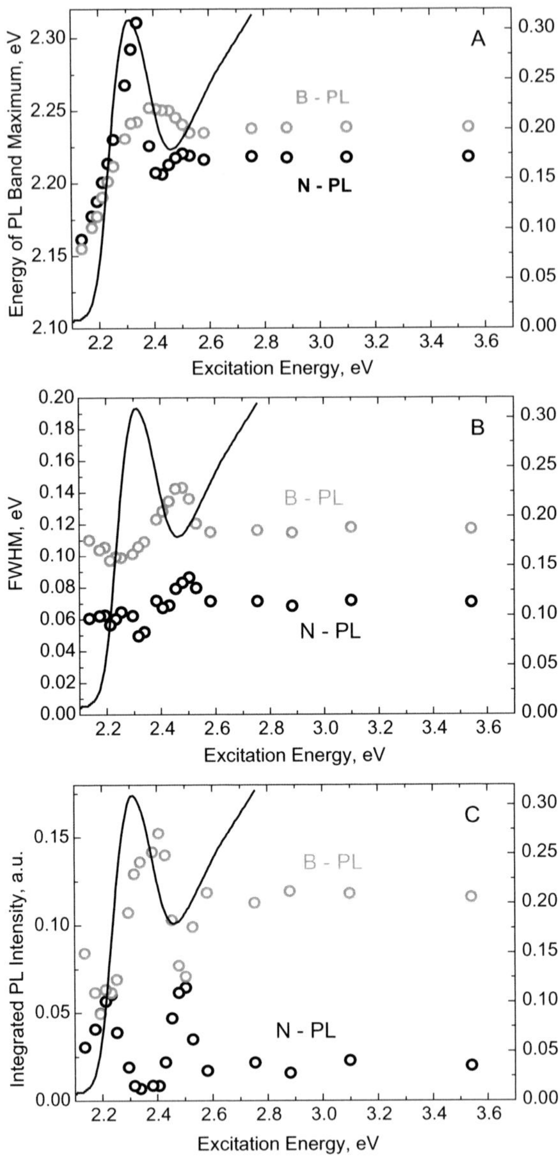

Figure 2.13 Properties of two deconvoluted PL bands for CdSe/ZnS as a function of excitation energy at 291 K. The absorption is also included as a black line (optical density is shown on the right scale). (A) Energy of the PL band maxima. (B) Spectral line widths (FWHM) of the PL bands. (C) Integrated PL intensities.

We take these observations as an indication that we are in fact close to the absorption band edge of intrinsic exciton states for the largest QDs. In that case, we will no longer observe (due to size selection) a decrease in excitation energy on decrease of PL detection energy. We propose that we observe the contribution of emission from deep traps which can now be excited more effectively in comparison to band-edge states.

To get even more insight into this question we inspect in Fig. 2.14A the PL emission at very low excitation energy (i.e., PL was monitored for λ_{exc} in the long-wavelength tail of the first excitonic absorption band). It can be clearly seen that some kind of shoulders appear at the low energy of the PL shifted by about 200 meV from the maximum of the PL band.

Time-resolved experiments shown in Figs. 2.5C, 2.6A, and 2.6B indicate that PL at very low energy shows much longer decay times compared to those of NBE states. Similar results have been reported earlier by us [Petrov et al., 2006] resulting in PL decay times of up to 100 ns accompanied by energy shifts ~50–200 meV below the band edge. The (integrated) intensity of such trap states is far below 3% of the maximal PL intensity at excitation above 3.0 eV. However, it is not clear whether deep traps are populated via bands (1) and (2) or whether NBE states and trap states spectrally overlap, thus contributing to the same PL energy, as might be concluded from the only smoothly varying shift of excitation energy with increasing PL energy. DLT states have a very low QY [Frenzel et al., 2007]. Similar observations have been previously reported by us [Petrov et al., 2006].

Additional information is obtained from PLE spectra at the lowest investigated PL energies 2.138 (580 nm) and 2.102 eV (590 nm) shown in Fig. 2.14B. Evidently, for detection at 2.102 eV the excitation band is broadened compared to a 2.138 eV detection energy, which is opposite to what one would expect according to size selection. Additionally, the inset shows an excitation band at ~2.171 eV when subtracting the PLE spectra obtained for 2.138 or 2.102 eV. The combined information obtained from PLE and PL emission at very low excitation energy strongly indicates that deep trap states can be excited directly. All the above findings are collected in Fig. 2.15.

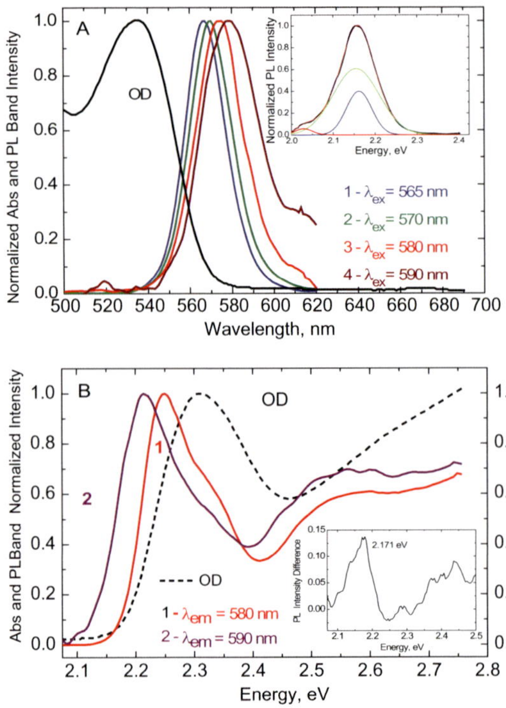

Figure 2.14 Spectral properties of PL emission for CdSe/ZnS QDs at 291 K. (A) Dependence of normalized PL spectra on λ_{exc}, especially at the long-wavelength tail of the first excitonic absorption band. The inset shows the deconvoluted PL emission with excitation at 580 nm (2.138 eV). (B) Normalized PLE spectra at detection wavelengths 590 nm (2.102 eV) and 580 nm (2.138 eV) with the manifestation of weak emission from deep traps (see inset). The dotted line corresponds to the normalized absorption. The inset shows the PLE intensity difference upon subtracting the PLE spectra obtained for 2.138 or 2.102 eV and superimposed at the same maximum position. The presence of an additional red-shifted PLE band at ~2.171 eV is evidently noticeable.

2.5 Conclusion

In fact, here we presented and discussed the results of comprehensive and comparative experimental investigation of spectral-kinetic properties (including absorption, PL, PLE, and PL decay data) for CdSe/ZnS QDs in solvents being obtained in a wide temperature

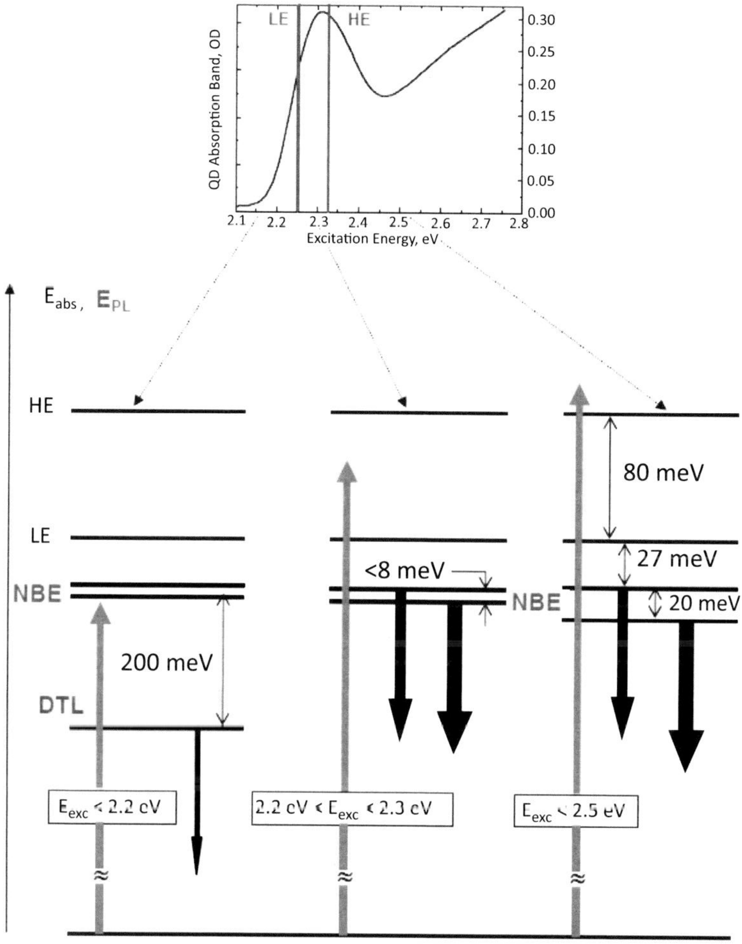

Figure 2.15 Scheme of NBE (middle and right) and deep-level traps DLT (left) PL emission as a function of excitation energy. Absorption features and PL features are shown in gray and black color, respectively. The width of black arrows reflects qualitatively the change of the PL efficiency from the corresponding NBE and DLT states. Top: QD absorption spectrum in the region of the first excitonic band; energy positions E_{max} of the deconvoluted absorption maxima for the low-energy (LE) and high-energy (HE) Gaussian component at 293 K are shown by red and blue lines, correspondingly; dotted arrows show the corresponding energy excitation ranges resulting in variation of PL properties.

range. To our knowledge, such a wide spectroscopic approach has been realized for the first time. The main conclusions evaluated from temperature experiments with ensemble QDs and comparison with single QDs are as follows:

- Spectral deconvolution of the absorption reveals two absorption bands separated by about 80 meV. The energetically higher one is attributed to a distribution of hole states. Exciting these hole states does not result in PL emission.
- The temperature dependence of the PL is related to optical phonons (see also Chapter 3).
- Ligands induce a type of phase transition which couples both to the PL from NBE states and to the absorption. Additionally, the phase transition changes the fine structure dipole transition strength considerably (observed in absorption). Moreover, below the phase transition temperature, STs carry the dominant part of the PL, and the above NBE states are responsible for the PL.
- PL from traps deep in the band gap are observed, which can be also excited directly, resulting in weak PL with decay times up to 100 ns.
- Multiexponential PL decay is explained by random switching among states with different polarity or charge transfer character. In that sense it reflects blinking phenomena observed in single-QD detection.
- PL spectra have to be deconvoluted into at least two spectral bands at all temperatures. The energy separation, line width, PL intensity, and decay time depend critically on the embedding environment, the ligand shell, and dye molecule attachment. This indicates that PL is a complex interplay of NBE, surface, and deep trap states.

Our elaborate spectroscopic investigation demonstrates how conventional spectroscopy and the interdependence of different approaches by absorption, PL emission, and time resolution, all as a function of temperature, is able to unravel detailed information on the lowest optical active exciton state of CdSe/ZnS QDs. Such detailed information, on the one hand, clarifies several open or controversial explanations or observations in QD spectroscopy and, on the other hand, allows for tailored tuning of optical properties of QDs. Moreover, recent improvements of calculations of optical observables of QDs can now be compared with detailed experimental information. Finally we state that we have established a promising link between ensemble and single-QD spectroscopy.

References

Beyler, A. P., Marshall, L. F., Cui, J., Brokmann, X., and Bawendi, M. G. (2013). Direct observation of rapid discrete spectral dynamics in single colloidal CdSe-CdS core shell quantum dots, *Phys. Rev. Lett.*, **111**, 177401.

Blaudeck, T., Zenkevich, E. I., Abdel-Mottaleb, M., Szwaykowska, K., Kowerko, D., Cichos, F., and Borczyskowski, C. von (2012). Formation principles and ligand dynamics of nanoassemblies of CdSe quantum dots and functionalized dye molecules, *ChemPhysChem*, **13**, 959–972.

Biebricher, A., Sauer, M., and Tinnefeld, P. (2006). Radiative and nonradiative rate fluctuations of single colloidal semiconductor nanocrystals, *J. Phys. Chem. B*, **110**, 5174–5178.

Boulesbaa A., Issac, A. Stockwell, D., Huang, Z., Huang, J., Guo, J., and Lian, T. (2007). Ultrafast charge separation at CdS quantum dot/rhodamine B molecule interface, *J. Am. Chem. Soc.*, **129**, 15132.

Borczyskowski, C. von, Zenkevich, E. (2014). Formation principles and exciton relaxation in semiconductor quantum dot–dye nanoassemblies, in *Quantum Dot Molecules, Lecture Notes in Nanoscale Science and Technology* (Wu, J., and Wang, Z. M., eds.), Springer, Dordrecht, Heidelberg, New York, London, Vol. 14, pp. 77–148.

Bullen, C., and Mulvaney, P. (2006). The effects of chemisorption on the luminescence of CdSe quantum dot, *Langmuir*, **22**, 3007–3013.

Califano M., Franceschetti, A., and Zunger, A. (2005). Temperature dependence of excitonic radiative decay in CdSe quantum dots: the role of surface hole traps, *Nano Lett.*, **5**, 2360–2364.

Cao, Y. W., and Banin, U. (2000), Growth and properties of semiconductor core/shell nanocrystals with InAs cores, *J. Am. Chem. Soc.*, **122**, 9692–9702.

Chung, I., Witkoskie, J. B., Cao, J., and Bawendi, M. G. (2006). Description of the fluorescence intensity time trace of collections of CdSe nanocrystal quantum dots based on single quantum dot fluorescence blinking statistics, *Phys. Rev. E*, **73**, 011106.

Cichos, F., Martin, J., and Borczyskowski, C. von (2004). Emission intermittency in silicon nanocrystals, *Phys. Rev. B*, **70**, 1153141.

Cichos, F., Borczyskowski, C. von, and Orrit, M. (2007). Power-law intermittency of single emitters, *Curr. Opin. Colloid Interface Sci.*, **12**, 272–284.

Clapp, A. R., Medintz, R. I. L. J., Mauro, M. Fisher B. R., Bawendi M. G., and Mattoussi, H. (2004). Fluorescence resonance energy transfer between quantum dot donors and dye-labeled protein acceptors, *J. Am. Chem. Soc.*, **126**, 301–310.

Cordones, A. A., and Leone, S. R. (2013). Mechanisms for charge trapping in single semiconductor nanocrystals probed by fluorescence blinking, *Chem. Soc. Rev.*, **42**, 3209–3221.

Crooker S. A., Barrick, T. Hollinthworth, J. A., and Klimov, V. I. (2003). Multiple temperature regimes of radiative decay in CdSe nanocrystal quantum dots: intrinsic limits to the dark-exciton lifetime, *Appl. Phys. Lett.*, **82**, 2793–2795.

Dabbousi, B. O., Redriguez-Vejo, J., Mikulec, F. V., Heine, J. R., Mattousi, H., Ober, R., Jensen, K. F., and Bawendi, M. G. (1997). (CdSe)ZnS core–shell quantum dots: synthesis and characterization of a size series of highly luminescent nanocrystallites, *J. Phys. Chem. B*, **101**, 9463–9475.

Dayal, S., Lou, Y., Samis, A. C. S., Berlin, J. C., Kenney, M. E., and Burda, C. (2006). Observation of non-Förster-type energy-transfer behavior in quantum dot-phthalocyanine conjugates, *J. Am. Chem. Soc.*, **128**, 13974–13975.

Dayal, S., and Burda, C. (2007). Surface effects on quantum dot-based energy transfer, *J. Am. Chem. Soc.*, **129**, 7977–7981.

Dworak, L., Matylitzky, V. V., Ren, T., Basche, T., and Wachtveitl, J. (2014). Acceptor concentration dependence of förster resonance energy transfer dynamics in dye-quantum dot complexes, *J. Phys. Chem. C*, **118**, 4396–4402.

Dzhagan, V. M., Valakh, M. Y., Raevska, O. E., Stroyuk, O. L., Kuchmiy, S. Y., and Zahn, D. R. T. (2009a). The influence of shell parameters on phonons in core-shell nanoparticles: a resonant Raman study, *Nanotechnology*, **20**, 365704.

Dzhagan V., Rayevska, O. Stroyuk, O., Kuchmiy, S., and Zahn, D. R. T. (2009b). Resonant Raman spectroscopy of confined and surface phonons in CdSe-caped CdS nanoparticles, *Phys. Stat. Sol. C*, **6**, 2043–2046.

Eilers, J., Hest J. Van, Meijerink, A., and Donega, C. De M. (2014). Unravelling the size and temperature dependence of exciton lifetimes in colloidal ZnSe quantum dots, *J. Phys. Chem. C*, **118**, 23313–23319.

Efros, A. L., Rosen, M., Kuno, M., Nirmal, M., Norris, D. J., and Bawendi, M. (1996). Band-edge exciton in quantum dots of semiconductors with a degenerate valence band: dark and bright exciton states, *Phys. Rev. B*, **54**, 4843–4856.

El-Kork, N., Huisken, F., and Borczyskowski, C. von (2011). Dielectric effects on the optical properties of single silicon nanocrystals, *J. Appl. Phys.*, **110**, 074312.

Empedocles, S. A., and Bawendi, M. G. (1999). Influence of spectral diffusion on the line shapes of single CdSe nanocrystallite quantum dots, *J. Phys. Chem. B*, **103**, 1826–1830.

Fadi, A., Hawkins, D., Safi, M., Palui, G., Dawson, P. E., Alabugin, I. V., and Mattoussi, H. (2015). UV and sunlight driven photoligation of quantum dots: understanding the photochemical transformation of the ligands, *J. Am. Chem. Soc.*, **137**, 2886–2903.

Fernee, M. J., Plahotnik, T., Louyer, Y., Littleton, B. N., Potzner, C., Tamarat, P., Mulvaney, P., and Lounis, B. (2012). Spontaneous spectral diffusion in CdSe quantum dots, *J. Phys. Chem. Lett.*, **3**, 1716–1729.

Fischer, S. A., Crotty, A. M., Kilina, S. V., Ivanov, S. A., and Tretiak, S. (2012). Passivating ligand and solvent contributions to the electronic properties of semiconductor nanocrystals, *Nanoscale*, **4**, 904–914.

Fisher, B. R., Eisler, H. J., Stott, N., and Bawendi, M. G. (2004). Emission intensity dependence and single-exponential behaviour in single colloidal quantum dot fluorescence lifetimes, *J. Phys. Chem. B*, **108**, 143–148.

Frenzel, J., Joswig, J. O., and Seifert, G. (2007). Optical excitations in cadmium sulfide nanoparticles, *J. Phys. Chem. C*, **111**, 10761–10770.

Frantsuzov, P. A., Volkán-Kacscó, S., and Jankó, B. (2013). Universality of the fluorescence intermittency in nanoscale systems: experiment and theory, *Nano Lett.*, **13**, 402–408.

Funston, A. M., Jasieniak, J. J., and Mulvaney, P. (2008). Complete quenching of CdSe nanocrystal photoluminescence by single dye molecules, *Adv. Mater.*, **20**, 4274–4280.

Galland, C., Ghosh, Y., Steinbrück, A., Sykora, M., Hollingsworth, J. A., V. I., and Htoon, H. (2011). Two types of luminescence blinking revealed by spectroelectrochemistry of single quantum dots, *Nature*, **479**, 203–207.

Gaponenko, S. V. (1998). *Introduction to Nanophotonics*, Cambridge University Press, Cambridge.

Goupalov, S. V., Suris, R. A., Lavallard, P., and Citrin, D. S. (2001). Homogeneous broadening of the zero-optical-phonon spectral line in semiconductor quantum dots, *Nanotechnology*, **12**, 518.

Gerlach, F., Täuber, D., and Borczyskowski, C. von (2013). Correlated blinking via time dependent energy transfer in single CdSe quantum dot-dye nanoassemblies, *Chem. Phys. Lett.*, **572**, 90–95.

Göhler, C. (2014). *Fast Spectral Analysis of Blinking Processes in Semiconductor Nanocrystals*, Master Thesis, TU Chemnitz, Germany.

Gomez-Campos, F. M., and Califano, M. (2012). Hole surface trapping in CdSe nanocrystals: dynamics, rate fluctuations, and implications for blinking, *Nano Lett.*, **12**, 4508–4517.

Hada, I., Halivni, S. Even-Dar N., Faust, A., and Banin, U. (2015). Dimensionality effects on fluorescence resonance energy transfer between single semiconductor nanocrystals and multiple dye acceptors, *J. Phys. Chem. C*, **119**, 3849–3856.

Htoon, H., Cox, P. J., and Klimov, V. I. (2004). Structure of excited-State transitions of individual semiconductor nanocrystals probed by photoluminescence excitation spectroscopy, *Phys. Rev. Lett.*, **93**, 187402.

Huang, J., Mulfort, K. L., Du, P., and Chen, L. X. (2012). Photodriven charge separation dynamics in CdSe/ZnS core/shell quantum dot/cobaloxime hybrid for efficient hydrogen production, *J. Am. Chem. Soc.*, **134**, 16472–16475.

Issac, A., Borczyskowski, C. von, and Cichos, F. (2005). Correlation between photoluminescence intermittency of CdSe quantum dots and self-trapped states in dielectric media, *Phys. Rev. B*, **71**, 161302 (R).

Issac, A., Krasselt, C., Cichos, F., and Borczyskowski, C. von (2012). Influence of the dielectric environment on the photoluminescence intermittency of CdSe quantum dots, *ChemPhysChem*, **13**, 3223–3230.

Jakubek, Z. J., deVries, J., Lin, S., Ripmeester, J., and Yu, K. (2008). Exciton recombination and unconverted photoluminescence in colloidal CdSe quantum dots, *J. Phys. Chem. C*, **112**, 8153–8158.

Ji, X., Copenhaver, D., Sichmeller, C., and Peng, X. (2008). Ligand bonding and dynamics on colloidal nanocrystals at room temperature: the case of alkylamines on CdSe nanocrystals, *J. Am. Chem. Soc.*, **130**, 5726–5735.

Jones, M., Lo, S. S., and Scholes, G. D. (2009a). Quantitative modeling of the role of surface traps in CdSe/CdS/ZnS nanocrystal photoluminescence decay dynamics, *Proc. Natl. Acad. Sci. U.S.A.*, **106**, 3011–3016.

Jones, M., Lo, S. S., and Scholes, G. D. (2009b). Signitures of exciton dynamics and carrier trapping in the time-resolved photoluminescence of colloidal CdSe nanocrystals, *J. Phys. Chem. C*, **113**, 18632–18642.

Kaluzhny, G., and Murray, R. W. (2005). Ligand effects on optical properties of CdSe nanocrystals, *J. Phys. Chem. B*, **109**, 7012–7021.

Kelley, A. M. (2010). Electron-phonon coupling in CdSe nanocrystals, *J. Phys. Chem. Lett.*, **1**, 1296–1300.

Kern, S. J., Sahu, K., and Berg, M. A. (2011). Heterogeneity of the electron-trapping kinetics in CdSe nanoparticles, *Nano Lett.*, **11**, 3493–3498.

Kilin, D. S., Tsemekham, K., Zenkevich, E. I., Prezho, O. V., and Borczyskowski, C. von (2007). Ab initio study of transfer dynamics from a core-shell semiconductor quantum dot to a porphyrin sensitizer, *J. Photochem. Photobiol. A*, **190**, 342–351.

Kilina, S. V., Ivanov, S., and Tretiak, S. (2009). Effect of surface ligands on optical and electronic spectra of semiconductor nanoclusters, *J. Am. Chem. Soc.*, **131**, 7717–7726.

Kilina, S., Velizhanin, K. A., Ivanov, S., Prezhdo, O. V., and Tretiak, S. (2012). Surface ligands increase photoexcitation rates in CdSe quantum dots, *ACS Nano*, **6**, 6515–6524.

Kilina, S. V., Neukirch, A. J., Habenicht, B. F., Kilin, D. S., and Prezhdo, O. V. (2013). Quantum zeno effect rationalizes the phonon bottleneck in semiconductor quantum dots, *Phys. Rev. Lett.*, **110**, 180404.

Klimov, V. I. (2000a). *Handbook of Nanostructured Materials and Nanotechnology* (Nalwa, H. S., ed.), Academic Press, USA, Vol. 4, pp. 451–527.

Klimov, V. I., Mikhailovsky, A. A., McBranch, D., Leatherdale, C. A., and Bawendi, M. G. (2000b). Mechanisms for intraband energy relaxation in semiconductor quantum dots: the role of electron-hole interactions, *Phys. Rev. B*, **61**, R13349– R13352.

Klimov, V. I. (2007). Spectral and dynamical properties of multiexcitons in semiconductor nanocrystals, *Ann. Rev. Phys. Chem.*, **58**, 635–673.

Knowles, K. E., Mc Arthur, E. A., and Weiss, E. A. (2011). A multi-timescale map of radiative and nonradiative decay pathways for excitons in CdSe quantum dots, *ACS Nano*, **5**, 2026–2035.

Koberling, F., Mews, A., Philipp, G., Kolb, U., Potapova, I., Burghard, M., and Basche, T. (2002). Fluorescence spectroscopy and transmission electron microscopy of the same isolated semiconductor nanocrystals, *Appl. Phys. Lett.*, **81**, 1116–1118.

Kowerko, D., Schuster, J., Amecke, N., Abdel-Mottaleb, M., Dobrawa, R., Wuerthner, F., and Borczyskowski, C. von (2010). FRET and ligand related non-FRET processes in single quantum dot–perylene bisimide assemblies, *Phys. Chem. Chem. Phys.*, **12**, 4112.

Krasselt, C., Schuster, J., and Borczyskowski, C. von (2011). Photoinduced hole trapping in single semiconductor quantum dots at specific sites at silicon oxide interfaces, *Phys. Chem. Chem. Phys.*, **13**, 17084–17092.

Kuno, M., Lee, J. K., Dabbousi, B. O., Mikulec, F. V., and Bawendi, M. G. (1997). The band edge luminescence of surface modified CdSe nanocrystallites: probing the luminescing state, *J. Chem. Phys.*, **106**, 9869–9882.

Labeau, O., Tamarat, P., and Lounis, B. (2003). Temperature dependence of the luminescence lifetime of single CdSe/ZnS quantum dots, *Phys. Rev. Lett.*, **90**, 257404.

Lemon, C. M., Karnas, E., Bawendi, M. G., and Nocera, D. G. (2013). Two-photon oxygen sensing with quantum dot-porphyrin conjugates, *Inorg. Chem.*, **52**, 10394–10406.

Liptay, T. J., and Ram, R. J. (2006). Temperature dependence of the exciton transition in semiconductor quantum dots, *Appl. Phys. Lett.*, **89**, 223132.

Liptay, T. J., Marshall, L. F., Rao, P. S., Ram, R. J., and Bawendi, M. G. (2007). Anomalous Stokes shift in CdSe nanocrystals, *Phys. Rev. B*, **76**, 155314.

Marcus, Y. (2002). *Solvent Mixtures*, Marcel Dekker, New York.

Martin, J., Cichos, F., and Borczyskowski, C. von (2012). Spectral diffusion of quasi localized excitons in single silicon nanocrystals, *J. Lumin.*, **132**, 2161–2165.

McBride, J. R., Kippeny, T. C., Pennycook, S. J., and Rosenthal, S. J. (2004). Aberration-corrected Z-contrast scanning transmission electron microscopy of CdSe nanocrystals, *Nano Lett.*, **4**, 1279–1283.

McBride, J. R., Pennycook, T. J., Pennycook, S. J., and Rosenthal, S. J. (2013). The possibility and implications of dynamic nanoparticle surfaces, *ACS Nano*, **7**, 8358–8365.

McDowell, M., Wright, A. E., and Hammer, N. I. (2010). Semiconductor nanocrystals hybridized with functional ligands: new composite materials with tunable properties, *Materials*, **3**, 614–637.

Mello Donega, C. de, Bode, M., and Meijerink, A. (2006). Size-and temperature-dependence of exciton lifetimes in CdSe quantum dots, *Phys. Rev. B*, **74**, 085320.

Mello Donega, C. de (2011). Synthesis and properties of colloidal heteronanocrystals, *Chem. Soc. Rev.*, **40**, 1512–1546.

Mourik, F. van, Giraud, G., Tonti, D., Chergui, M., and van der Zwan, G. (2008). Linear dichroism of CdSe nanodots: large anisotropy of the band-gap absorption induced by ground-state dipole moments, *Phys. Rev. B*, **77**, 165303.

Müller, J., Lupton, J. M., Rogach, A. L., Feldmann, J., Talapin, D. V., and Weller, H. (2005). Monitoring surface charge migration in the spectral dynamics of single CdSe/CdS nanodot/nanorod heterostructures, *Phys. Rev. B*, **72**, 205339.

Munro, A. M., Jen-La Plante, I., Ng, M. S., and Ginger, D. S. (2007). Quantitative study of the effects of surface ligand concentration on CdSe nanocrystal photoluminescence, *J. Phys. Chem. C*, **111**, 6220–6227.

Munro, A. M., and Ginger, D. S. (2008). Photoluminescence quenching of single CdSe nanocrystals by ligand adsorption, *Nano Lett.*, **8**, 2585–2590.

Murov, S., Carmichael, I., and Hug, G. (1993). *Handbook of Photochemistry*, Marcel Dekker, New York.

Nair, G., and Bawendi, M. G. (2007). Carrier multiplication yields of CdSe and CdTe nanocrystals by transient photoluminescence spectroscopy, *Phys. Rev. B*, **76**, 081304.

Neuhauser, R. G., Shimizu, K. T., Woo, W. K., Empedocles, A. A., and Bawendi, M. G. (2000). Correlation between fluorescence intermittency and spectral diffusion in single semiconductor quantum dots, *Phys. Rev. Lett.*, **85**, 3301–3304.

Nirmal, M., Murray, C. B., and Bawendi, M. G. (1994). Fluorescence line narrowing in CdSe quantum dots: surface localisattion of the photegenerated exciton, *Phys. Rev. B*, **50**, 2293–2300.

Nirmal, M., Norris, D. J., Kuno, M., Bawendi, M. G., Efros, A. L., and Rosen, M. (1995). Observation of the "dark exciton" in CdSe quantum dots, *Phys. Rev. Lett.*, **75**, 3728.

Norris, D. J., and Bawendi, M. G. (1996). Measurement and assignment of the size-dependent optical spectrum in CdSe quantum dots, *Phys. Rev. B*, **53**, 16338–16346.

Oluwasesan, A., Antunes, E., and Nyokong, T. (2013). Nanoconjugates of CdTe@ZnS quantum dots with cobalt tetraamino-phthalocyanine: characterization and implications for the fluorescence recognition of superoxide anion, *J. Photochem. Photobiol. A Chem.*, **257**, 11–19.

Peng, X., Schlamp, M. C., Kadavanich, A. V., and Alivisatos, A. P. (1997). Epitaxial growth of highly luminescent CdSe/CdS core/shell nanocrystals with photostability and electronic accessibility, *J. Am. Chem. Soc.*, **119**, 7019–7029.

Pennycook, T. J., McBride, J. R., Rosenthal, S. J., Pennycook, S. J., and Pantelides, S. T. (2012). Dynamic fluctuations in ultrasmall nanocrystals induce white light emission, *Nano Lett.*, **12**, 3038–3042.

Petrov, E. P., Cichos, F., and Borczyskowski, C. von (2006). Intrinsic photophysics of semiconductor nanocrystals in dielectric media: formation of surface states, *J. Lumin.*, **119–120**, 412–417.

Quagliano, L. G. (2004). Observation of molecules adsorbed on III-V semiconductor quantum dots by surface-enhanced raman scattering, *J. Am. Chem. Soc.*, **126**, 7393–7398.

Ren, T., Mandal, P., Erker, Liu, W. Z., Avlasevich, Y., Puhl, L., Muellen, K., and Basche, T. A. (2008). Simple and versatile route to stable quantum dot-dye hybrids in nonaqueous and aqueous solutions, *J. Am. Chem. Soc.*, **130**, 17242.

Riley, E. A., Hess, C. M., and Reid, P. J. (2012). Photoluminescence intermittency from single quantum dots to organic molecules: emerging themes, *Int. J. Mol. Sci.*, **13**, 12487–12518.

Rogach, A. L. (2008). *Semiconductor Nanocrystal Quantum Dots: Synthesis, Assembly, Spectroscopy and Applications*, Springer, Wien.

Sagar, D. M., Cooney, R. R., Sewall, S. L., Dias, E. A., Barsan, M. M., Butler, I. S., and Kambhampati, P. (2008). Size dependent, state-resolved studies of exciton-phonon couplings in strongly confined semiconductor quantum dots, *Phys. Rev. B*, **77**, 235321.

Salman, A. Al, Tortschanoff, A., Mohamed, M. B., Tonti, van Mourik, D. F., and Chergui, M. (2007). Temperature effects on the spectral properties of colloidal CdSe nanodots, nanorods, and tetrapods, *Appl. Phys. Lett.*, **90**, 093104-093104-3.

Salman, A. Al, Tortschanoff, A., van der Zwan, G., van Mourik, F., and Chergui, M. (2009). A model for the multi-exponential excited-state decay of CdSe nanocrystals, *Chem. Phys.*, **357**, 96–101.

Salvador, M. R., Graham, M. W., and Scholes, G. D. (2006). Exciton-phonon coupling and disorder in the excited states of CdSe colloidal quantum dots, *J. Chem. Phys.*, **125**, 184709.

Schlegel, G., Bohnenberger, J., Potapova, I., and Mews, A. (2002). Fluorescence decay time of single semiconductor nanocrystals, *Phys. Rev. Lett.*, **88**, 1374011.

Schmelz, O., Mews, A., Basché, T., Herrmann, A., and Müllen, K. (2001). Supramolecular complexes from CdSe nanocrystals and organic fluorophors, *Langmuir*, **17**, 2861.

Schmidt, R., Krasselt, C., and Borczyskowski, C. von (2012). Change point analysis of matrix dependent photoluminescence intermittency of single CdSe/ZnS quantum dots with intermediate intensity levels, *Chem. Phys*, **406**, 9–14.

Schmidt, R., Krasselt, C., Göhler, C., and Borczyskowski, C. von (2014). The fluorescence intermittency for quantum dot is not power-law distributed: a luminescence intensity resolved approach, *ACS Nano*, **8**, 3506–3521.

Schreuder, M. A., McBride, J. R., Dukes III, A. D., Sammons, J. A., and Rosenthal, S. J. (2009). Control of surface state emission via phosphonic acid modulation in ultrasmall cdse nanocrystals: the role of ligand electronegativity, *J. Phys. Chem. C*, **113**, 8169–8176.

Schuster, J., Brabandt, J., and Borczyskowski, C. von (2007). Discrimination of photoblinking and photobleaching on the single molecule level, *J. Lumin.*, **127**, 224.

Sharon, E., Freeman, R., and Willner, I. (2010). CdSe/ZnS quantum dots-G-quadruplex/hemin hybrids as optical DNA sensors and aptasensors, *Anal. Chem.*, **82**, 7073–7077.

Smith, A. M., and Nie, S. (2010). Semiconductor nanocrystals: structure, properties, and band gap engineering, *Acc. Chem. Res.*, **43**, 190–200.

Tagliazucchi, M., Tice, D. B., Sweeney, C. M., Morris-Cohen, A. J., and Weiss, E. A. (2011). Ligand-controlled rates of photoinduced electron transfer in hybrid CdSe nanocrystal/poly(viologen) films, *ACS Nano*, **5**, 9907–9917.

Takagahara, T. (1993). Electron-phonon interactions and excitonic dephasing in semiconductor nanocrystals, *Phys. Rev. Lett.*, **71**, 3577–3580.

Talapin, D.V., Lee, J.-S., Kovalenko, M. V., and Shevchenko, E. V. (2010). Prospects of colloidal nanocrystals for electronic and optoelectronic applications, *Chem. Rev.*, **110**, 389–458.

Tanne, J., Scheafer, D., Khalid, W., Parak, W. J., and Lisdat, F. (2011). Light-controlled bioelectrochemical sensor based on CdSe/ZnS quantum dots, *Anal. Chem.*, **83**, 7778–7785.

Tripathi, S. K., Kaur, R., and Sharma, M. (2015). Synthesis and characterization of TGA-capped CdTe nanoparticles embedded in PVA matrix, *Appl. Phys. A*, **118**, 1287–1295.

Valerini, D., Creti, A., Lomascolo, M., Manna, L., Cingolani, R., and Anni, M. (2005). Temperature dependence of the photoluminescence properties of colloidal CdSe/ZnS core/shell quantum dots embedded in a polystyrene matrix, *Phys. Rev. B*, **71**, 235409.

Voznyy, O. (2011). Mobile surface traps in CdSe nanocrystals with carboxylic acid ligands, *J. Phys. Chem. C*, **115**, 15927–15932.

Voznyy, O., and Sargent, E. H. (2014). Atomistic model of fluorescence intermittency of colloidal quantum dots, *Phys. Rev. Lett.*, **112**, 157401.

Woggon, U. (2006). *Optical Properties of Semiconductor Quantum Dots*, Springer, Berlin.

Wuister, S. F., Mello Donega, C. de, Bode, C., and Meijerink, A. (2004a). Luminescence temperature antiquenching of water-soluble CdTe quantum dots: role of solvents, *J. Am. Chem. Soc.*, **126**, 10397–19402.

Wuister, S. F., van Houselt, A., Mello Donega, C. de, Vanmaekelbergh, D., and Meijerink, A. (2004b). Temperature antiquenching of the luminescence from capped CdSe quantum dots, *Angew. Chem. Int. Ed.*, **43**, 3029–3033.

Zenkevich, E., Shulga, A., Cichos, F., Petrov, E. P., Blaudeck, T., and Borczyskowski, C. von (2005). Nanoassemblies designed from quantum dots and molecular arrays, *J. Phys. Chem. B*, **109**, 8679–8692.

Zenkevich, E. I., Sagun, E. I., Knyukshto, V. N., Stasheuski, A. S., Galievsky, V. A., Stupak, A. P., Blaudeck, T., and Borczyskowski, C. von (2011). Quantitative analysis of singlet oxygen ($^1O_2$) generation via energy transfer in bioconjugates based on semiconductor quantum dots and porphyrin ligands, *J. Phys. Chem. C*, **115**, 21535–21545.

Zenkevich, E. I., Stupak, A. P., Kowerko, D., and Borczyskowski, C. von (2012). Influence of single dye molecules on temperature and time dependent optical properties of CdSe/ZnS quantum dots: ensemble and single nanoassembly detection, *Chem. Phys.*, **406**, 21–29.

Zenkevich, E. I., Stupak, A. P., Göhler, C., Krasselt, C., and Borczyskowski, C. von (2015). Tuning electronic states of a CdSe/ZnS quantum dot by only one functional dye molecule, *ACS Nano*, **9**, 2886–2903.

Zhang, K., Chang, H., Fu, A., Alivisatos, A. P., and Yang, H. (2006). Continuous distribution of emission states from single CdSe/ZnS quantum dots, *Nano Lett.*, **6**, 843–847.

Chapter 3

Temperature-Dependent Exciton–Phonon Coupling in Quantum Dots

Eduard Zenkevich,[a] Alexandre Stupak,[b] and Christian von Borczyskowski[c]
[a]*Department of Information Technologies and Robototechnique, National Technical University of Belarus, Prospect Nezavisimosti 65, 220013 Minsk, Belarus*
[b]*B.I. Stepanov Institute of Physics, National Academy of Sciences of Belarus, Prospect Nezavisimosti 70, 220072 Minsk, Belarus*
[c]*Institute of Physics, Center for Nanostructured Materials and Analytics, Technische Universität Chemnitz, Reichenhainer Str. 70, D-09107 Chemnitz, Germany*
zenkev@tut.by, stupak@imaph.bas-net.by, borczyskowski@physik.tu-chemnitz.de

3.1 Introduction

It is well documented that in semiconductor quantum dots (QDs), not only the electronic energy levels but also the lattice vibrational modes (phonons) become discrete due to the 3D confinement [Takagahara, 1996; Sagar et al., 2008]. One long-standing question in QD physics is the effect of quantum confinement on the strength of exciton–phonon coupling and phonon frequency, a topic that

Tuning Semiconducting and Metallic Quantum Dots: Spectroscopy and Dynamics
Edited by Christian von Borczyskowski and Eduard Zenkevich

ISBN 978-981-4745-24-6 (Hardcover), 978-981-4745-25-3 (eBook)
www.panstanford.com

has received considerable experimental [Klimov, 2003; Anni et al., 2004; Son et al., 2006; Liptay et al., 2007; Nonoguchi et al., 2008] and theoretical [Takagahara, 2002; Ehlert et al., 2006; Kelley, 2010; Kelley, 2011] attention. The strength of exciton–phonon coupling bears relevance to several fundamental issues of nanoscale physics: exciton relaxation dynamics, homogeneous line widths, and electronic structure. Typically, the study of temperature dependence of exciton–phonon coupling has been carried out for QDs in rigid matrices [Klimov, 2003; Son et al., 2006; Liptay et al., 2007; Nonoguchi et al., 2008], or on Si-SiO_2 substrates [Anni et al., 2004; Morello et al., 2007]. Noteworthy, the interaction of electrons and holes with phonons is responsible for Raman scattering [Miranda et al., 2006; Yaremko et al., 2007; Dzhagan et al., 2007; Dzhagan et al., 2012]. Direct optical access to exciton–phonon coupling is in principle based on two types of experimental observations, namely (i) the intensity ratios of the spectrally resolved pure (phonon-less) electronic transition intensities and phonon side bands and (ii) overall optical transition intensities, contributions to the homogeneous spectral line widths, photoluminescense (PL) decay times, and energies of absorption and/or PL emission, together with experimentally determined Stokes shifts [Kelley, 2010; Lo et al., 2009] (between absorption and PL bands).

Since phonon energies of CdSe QDs are small both for acoustic (<6 meV) and optical (<28 meV) phonons, direct experimental access to spectroscopic data is difficult to realize for a broad temperature range and rules out case (i), at least at elevated temperatures. Additionally, in ensembles of QDs spectral resolution is limited by inhomogeneities mainly caused by the remaining size distribution, crystal deformations, and surface defects. In the case of single-QD detection, spectral diffusion [Salvador et al., 2006] hampers the evaluation of the homogeneous line width and the phonon-related dephasing processes.

However, a few techniques such as three PEPS [Salvador et al., 2006] are able to circumvent inevitable inhomogeneous broadening. Acoustic phonons will predominantly contribute to the homogeneous line width via deformation potentials and piezoelectric coupling [Takagahara, 1993]. Coupling to LO optical phonons is assumed to be due to polar Fröhlich coupling, in which the excitonic charge density couples via Coulomb interaction to the induced charge distribution generated by lattice vibrations. These LO modes (~27 meV for CdSe phonons) couple weakly to the exciton characterized by a Huang Rhys factor *S* of about 0.12 [Salvador et al., 2006]. A convenient approach to get access to exciton–phonon coupling, despite the mentioned complications, is to vary the temperature [Nomura and Kobayashi, 1992; Takagahara, 1993; Zhang et al., 2002; Valerini et al., 2005; Joshi et al., 2006; Ehlert et al., 2006; Salvador et al., 2006; Al Salman et al., 2007; Morello et al., 2007; Lo et al., 2009; Narayanaswamy et al., 2009; Chon et al., 2011]. However, below about 75 K at least PL properties may depend on thermalization of the QD fine structure and acoustic phonon modes [Salvador et al., 2006].

It has been shown [Melnikov and Fowler, 2001; Yaremko et al., 2007] that the quantum size effect results in some enhancement of the electron–phonon coupling in the case of QDs embedded in a borosilicate glass matrix. However, for individual QDs in various solvents or in the case of QD-dye nanoassemblies, the specificity of the solvent parameters and/or details of QD–dye interactions, together with the influence of trap states (in the band gap [Frenzel et al., 2007]), may result in different relaxation pathways and rates, the differences among which can be reconciled by a multichannel picture of exciton relaxation dynamics (as was shown recently [Cooney et al., 2007]), while the strength of the exciton–phonon coupling is relevant for one of the manifold of pathways by which an exciton in a QD may relax. Correspondingly, a detailed understanding of the photochemical and photophysical properties of individual QDs and QD-dye nanoassemblies upon temperature lowering is essential for understanding the physics of quantum-confined nanoassemblies as well as for their future applications.

Upon studies of exciton–phonon interactions in semiconductor QDs, the majority of previous temperature experiments have been carried out for QDs dissolved in rigid matrixes (polymers, resins, etc.), while only few data have been obtained for QDs in typical solvents (like toluene) for CdSe QDs without a ZnS shell [Mello Donega et al., 2006]. In the latter case, a possible phase transition of the capping ligand layer should be taken into account. This phase transition is accompanied by surface relaxation or reconstruction, which might change both the spatial position of surface traps and the energies of trap states. As a result, the temperature dependence of QD PL (intensities, line widths [full-width at half-maximum (FWHM)], and energies) have to be explained not only by the temperature influence via exciton–phonon coupling but also by the spatial and energetic reordering of trap states caused by the phase transition of capping ligand layers such as trioctylphosphine oxide (TOPO) or long-chain amines (AMs) (for CdSe QDs with and without ZnS in toluene [Mello Donega et al., 2006; Zenkevich et al., 2012; Zenkevich et al., 2015] and for water-soluble aminoethanethiol [AEP]-capped CdTe QDs [Wuister et al., 2004]). For InP/ZnS core-shell QDs in toluene, PL changes caused by the phase transition were not observed in the temperature range of 2–310 K [Narayanaswamy et al., 2009].

Phonons of the CdSe core constitute the dominant contribution to exciton–phonon coupling in CdSe/ZnS QDs. However, already the influence of a phase transition of the ligand layer upon PL emission and absorption [Mello Donega et al., 2006; Zenkevich et al., 2012; Zenkevich et al., 2015; Wuister et al., 2004] indicates that properties of the interfaces have to be taken into account. Since ZnS phonons are different from those of CdSe, surface and interface phonons have also to be considered [Baranov et al., 2003; Dzhagan et al., 2007; Dzhagan et al., 2013]. The presence of such phonons has been repeatedly reported, for example, by (resonant) Raman experiments. Whether they also show up for CdSe/ZnS QDs in exciton–phonon coupling has not yet been discussed in the literature so far. However, recent experiments on the PL of Si QDs with a capping SiO_2 shell clearly show coupling to Si-O phonons [Martin et al. 2004; Martin et al., 2008; El-Kork et al., 2011; Martin et al., 2012; Schmidt et al., 2012]. Also the luminescence of very small Ag clusters shows evidence of exciton–phonon coupling to the embedding environment [Krause et al., 2013; Krause et al., 2014]. Exciton–phonon coupling to phonons beyond the CdSe core (e.g., to the ZnS shell) of a QD relies on the extension of the exciton wavefunction beyond the core of the QD. Evidence has been reported for such tunneling of the wavefunction [Blaudeck et al., 2008]. General concepts of exciton–phonon coupling are discussed in Chapter 1. In the following paragraphs, we will discuss temperature-dependent exciton–phonon coupling in CdSe/ZnS QDs capped by TOPO ligands and embedded into a glass-forming liquid. We will use absorption and PL data, a combination which is to the best of our knowledge not yet reported in the literature. We will investigate the interplay of phonons in the CdSe core and ZnS shell and the relation to specific excitonic states of the QD. Additionally, we will analyze the exciton–phonon coupling for CdSe/ZnS QDs capped by long-chain AMs, which have also been investigated by us.

3.2 Results

Experiments have been performed on CdSe/ZnS QDs (diameter d = 3 nm) capped by TOPO ligands. Optical spectra have been obtained in a 6:1 solution of methylhexane:toluene, which guarantees an optical transparent matrix across the temperature range from 90 K to 300 K. Details of the experimental conditions have been described elsewhere [Zenkevich et al., 2015]. Some experiments have been performed under otherwise similar conditions for CdSe/ZnS QDs (diameter d = 3.3 nm) capped by long-chain AMs [Kowerko et al., 2010].

3.2.1 Temperature-Dependent Optical Absorption

Figure 3.1A shows the temperature-dependent optical absorption spectra of CdSe/ZnS QDs with TOPO ligands dissolved in an optical transparent solvent mixture (methylcyclohexane:toluene = 6:1), which becomes a glass upon lowering of temperature. The optical density (OD) at an excitation of 410 nm is nearly temperature independent, while we observe changes in the OD line width and spectral position in the long-wavelength range (see also Chapter 2).

Spectral deconvolution of the lowest-energy absorption band reveals for two different samples a nearly continuous shift of the energetic position to lower energies with increasing temperature (Fig. 3.2A). A sudden phase-transition-type change of the absorption energy occurs at about T_{crit} = 215 K (curve 2, Fig. 3.1B), which has recently been assigned to a phase transition of the TOPO ligand shell [Mello Donega et al., 2006; Zenkevich et al., 2012; Zenkevich et al., 2015; Wuister et al., 2004]. A similar phase-transition-type behavior is observed (see Fig. 3.4) for PL at 225 (3) K, 225 (1) K, and 227 (5) K

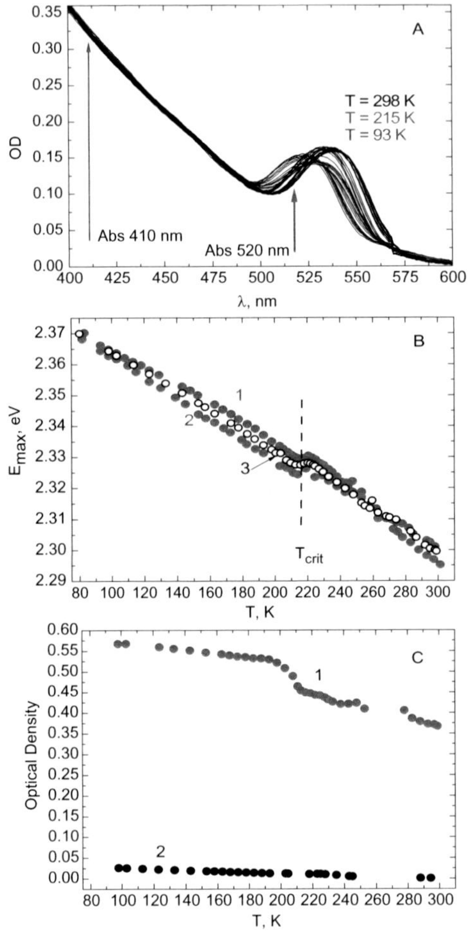

Figure 3.1 Temperature dependence of the absorption spectra (A), energy E_{max} of the energetically lowest absorption band maximum of TOPO-capped CdSe/ZnS QDs in the solvent mixture (methylcyclohexane:toluene = 6:1) (B), and optical densities of the solvent at λ_{det} = 410 nm (C). (A) Spectra at high, low, and phase transition temperatures are picked out by black, blue, and red colors, correspondingly. (B) Curves (1) and (2) correspond to the temperature dependence following a temperature decrease (1) or increase (2), respectively; curve (3) shows the temperature dependence for a second sample (temperature increase). (C) Change of the optical density at 410 nm as a function of temperature for the QD in the solvent (1, red points) in comparison to the effect of temperature upon the optical density of the solvent (2, black points), respectively.

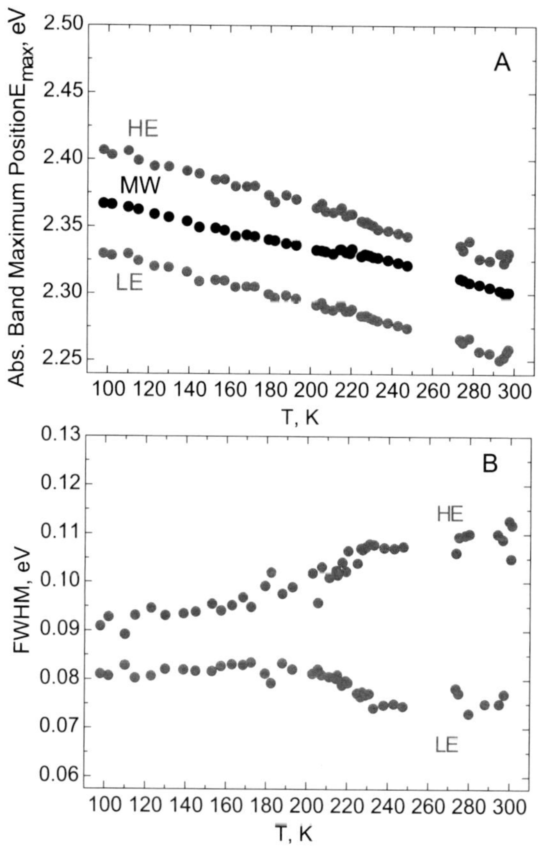

Figure 3.2 Spectrally deconvoluted absorption band of TOPO-capped CdSe/ZnS QDs into two Gaussian subbands with the respective energy maxima for low-energy (LE) component and high-energy (HE) component (A) and related band half-widths (FWHM) (B) on temperature change in the solvent mixture (methylcyclohexane:toluene = 6:1). MW corresponds to the temperature dependence of the lowest absorption energy resulting from the two mean weighted Gaussian components: $E = E_{HE} \times [\Sigma_{HE}/(\Sigma_{HE} + \Sigma_{LE})] + E_{LE} \times [(\Sigma_{LE}/(\Sigma_{HE} + \Sigma_{LE})]$.

for excitation at 410 nm, 450 nm, and 520 nm, respectively. T_{crit} related to absorption data is always lower by about 10 K compared to the one identified via PL. It follows from Fig. 3.1B that above T_{crit}, the shift of the E_{max} position for the lowest-energy absorption band depends, within the experimental and fitting error, neither on the sample nor on the temperature cycle. At the lowest-investigated temperature all absorption energies amount to 2.370 eV, while the difference in energy between 130 K and T_{crit} is within about 5 meV. Figure 3.1C shows a change in the OD at 410 nm as a function of temperature in comparison to the effect of temperature upon the OD of the solvent. Taking the change in the baseline into account the OD is constant within ≤10%. Despite that we observe at the phase transition a jump in OD by ~20% upon passing from below to above the phase transition temperature.

Figure 3.2 shows a spectral deconvolution of the lowest absorption band into two subbands, low-energy (LE) component (LE) and high-energy (HE) component of different band half-widths (FWHM), and the dependence of their properties on temperature. Surprisingly, the apparent phase transition is nearly absent with respect to absorption energies (Fig. 3.2A) but shows up in the FWHM (Fig. 3.2B). For details see Chapter 2. The temperature dependence of the absorption energies is identical for both components, LE and HE, and is nearly linear across the total temperature range.

3.2.2 Temperature-Dependent Photoluminescence

Figure 3.3 shows the temperature dependence of the PL spectra of CdSe/ZnS (TOPO) at three different excitation wavelengths. Again, a phase transition becomes apparent at about T_{crit} = 220 K, as can be seen from the deconvoluted parameters for PL energy and the FWHM, as shown in Fig. 3.4. An Arrhenius-type presentation of integrated PL intensity versus $1/T$ shows in Fig. 3.5 that the related temperature dependence is not a simple temperature-activated process. Moreover, the temperature dependence of the PL intensity depends noticeably on the excitation wavelength.

Remarkably, the signature of the phase transition is most clearly observed at an excitation wavelength of 450 nm and is reduced at excitation wavelengths below and above 450 nm. This might be explained by resonant excitation of coupled ligand-exciton states, which transfers excitation from the QD to the ligands [Kilina et al., 2009; Fischer et al., 2012]. Finally, Fig. 3.4 shows that the signature of the phase transition is absent practically for CdSe/ZnS QDs spin-coated on a quartz substrate, contrary to TOPO- or AM-capped CdSe/ZnS QDs in glass-forming solutions [Zenkevich et al., 2015].

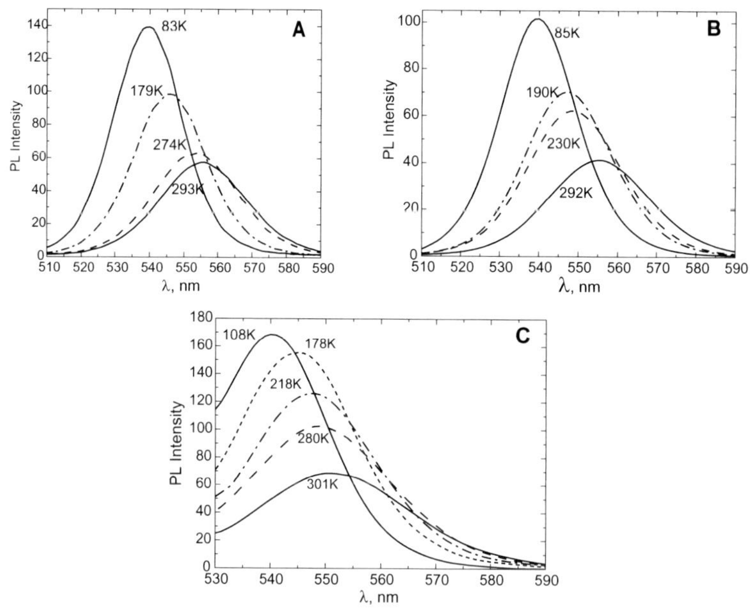

Figure 3.3 Typical temperature-dependent PL spectra of TOPO-capped CdSe/ZnS QDs in the solvent mixture (methylcyclohexane:toluene = 6:1) upon excitation at 410 nm (A), 450 nm (B), and 520 nm (C).

The PL spectra can also be deconvoluted into at least two PL bands, which respond differently on temperature variation. Details of this behavior have been published recently [Zenkevich et al., 2015]. With respect to exciton–phonon coupling we will refer here only to nondeconvoluted PL spectra since energy separations are much smaller than those observed in absorption.

It is obvious that the magnitude of the PL intensity change at the phase transition depends both on the excitation wavelength and the matrix. At λ_{exc} = 520 nm, excitation of QDs with large crystal diameters within the remaining size distribution is predominantly selected, whereas at λ_{exc} = 410 nm size selection is probably negligible. In this regard it should be added that the shift of the phase transition temperature in absorption compared to PL of about ~10 K for QDs in glass-forming solutions might also be due to size-related effects.

The dependence of PL energy and the FWHM shown in Fig. 3.4 also reveals evidence of phase transition.

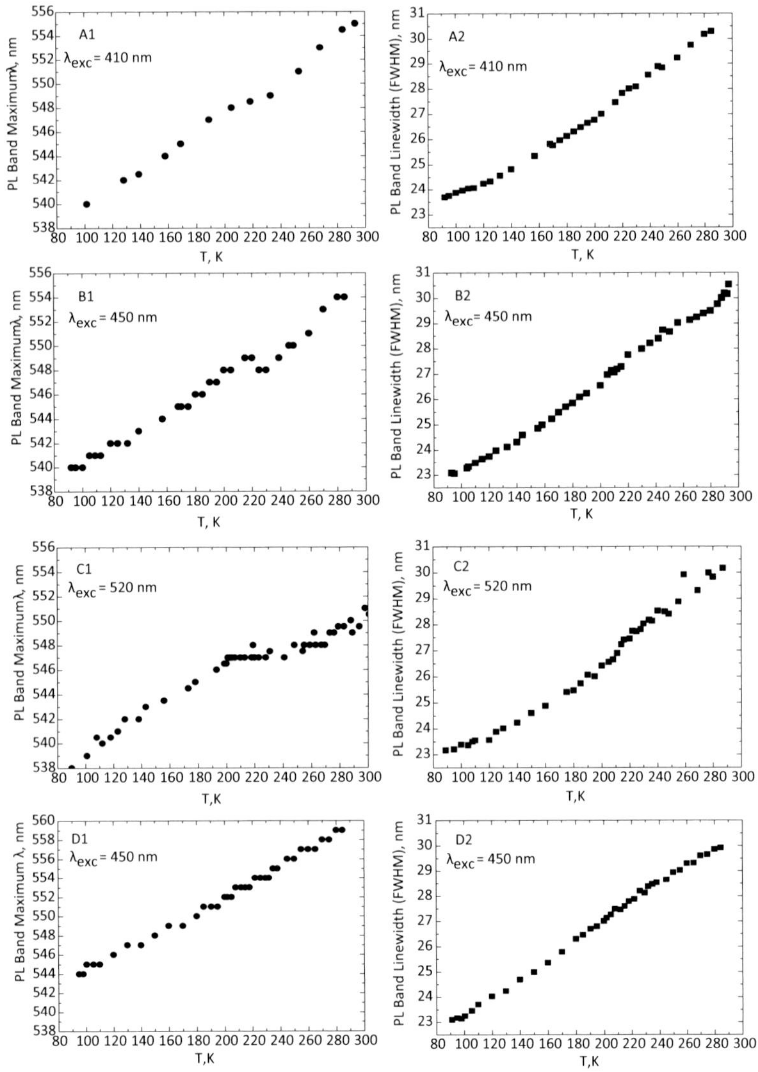

Figure 3.4 Temperature dependence of PL band maximum position (1) and PL band line width (FWHM) (2) for TOPO-capped CdSe/ZnS QDs in the solvent mixture (methylcyclohexane:toluene = 6:1) upon excitation at 410 nm (A), 450 nm (B), and 520 nm (C), as well as for QDs on a quartz substrate at λ_{exc} = 450 nm (D).

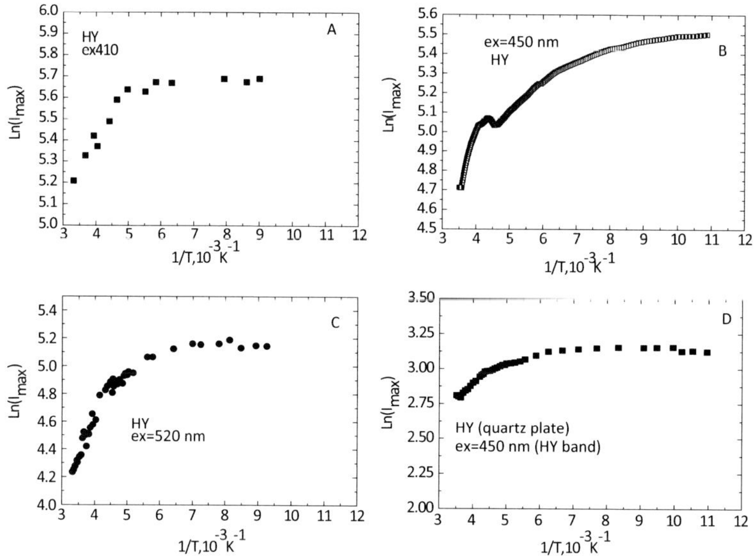

Figure 3.5 Arrhenius plots of integrated intensities of PL bands for TOPO-capped CdSe/ZnS QDs in the solvent mixture (methylcyclohexane:toluene = 6:1) for three different excitation wavelengths at 410 (A), 450 (B), and 520 nm (C), as well as on a quartz substrate at λ_{exc} = 450 nm (D).

The PL excitation data for CdSe/ZnS (AM) in Fig. 3.6 show that a considerable structure remains in the excitation spectra at detection wavelengths above the maximum of the PL maximum energy. We suggest that these effects might be related to the fact that either small QDs are more sensitive to the phase transition or alternatively excitation close to 450 nm results in excitation energy transfer from QDs to ligands [Kilina et al., 2009; Fischer et al., 2011]. With excitation at a given wavelength we select preferentially a certain QD size (at 450 nm small QDs, and at 520 nm large QDs).

In absorption we scan through the size distribution, while the OD decreases for small QDs with respect to larger ones (see Chapter 1). We propose that the phase transition temperature is lower for larger QDs than for smaller ones (450 nm excitation selects small QDs, while absorption is more sensitive to large QDs). The size dependence of the phase transition temperature is also reflected by the fact that the transition is not really sharp but is extended across about 10 K. It is narrower in absorption (with a width of 5 K). Additionally, the PL intensity of, compared to TOPO-capped QDs, slightly larger AM-capped CdSe/ZnS QDs reveals, as shown in Fig. 3.7, a comparable temperature dependence as for TOPO-capped ones [Zenkevich et al., 2012; Zenkevich et al., 2015]. However, the phase transition temperature is increased from 220 K to 240 K.

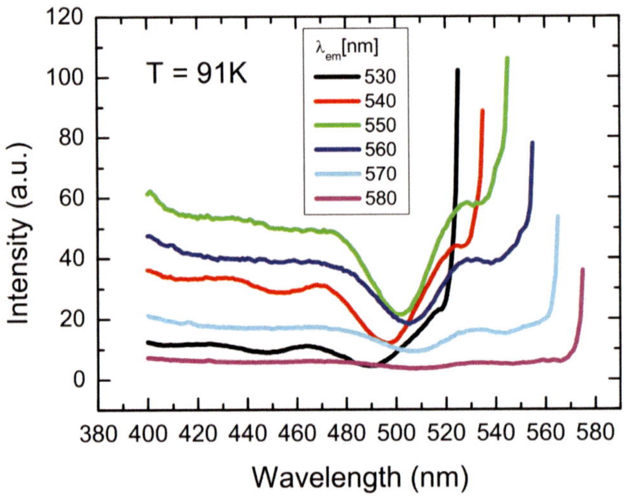

Figure 3.6 PL excitation data for CdSe/ZnS (AM) in the solvent mixture (methylcyclohexane:toluene = 6:1) at T = 91 K. Interesting is that high-energy detection results in a noticeable structure in excitation at high-energy QD states which shift energetically due to size selection.

Figure 3.7 also includes the temperature dependence of a nanoassembly of CdSe/ZnS (AM) with a functionalized (pyridyl)$_2$-perylene diimide (DPP) molecule (data 2, solid circles), which induces a kind of amplification of the PL intensity jump at T_{crit}, which is accompanied with a shift of T_{crit} by about 10 K toward lower temperatures. Even stronger amplifications have been reported recently for CdSe/ZnS (TOPO) QD-porphyrine nanoassemblies [Zenkevich et al., 2012; Zenkevich et al., 2015]. However, in the latter case we did not observe an influence of assembly formation on T_{crit}. We tentatively explain this difference by the fact that DPP removes upon surface attachment more ligands than H_2P does [Kowerko et al. 2010; Blaudeck et al. 2012], thus having a larger impact on the phase transition temperature.

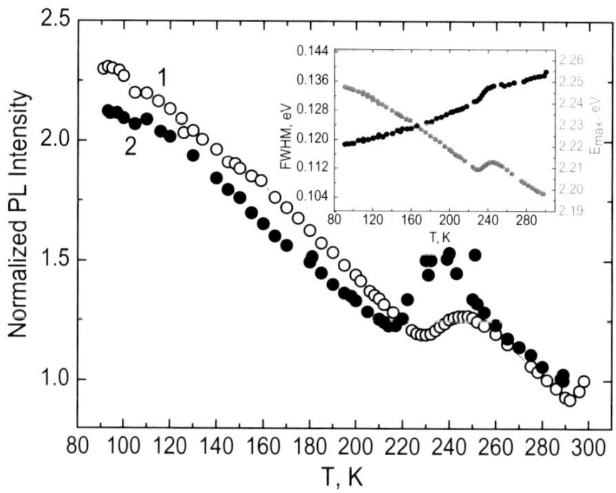

Figure 3.7 Temperature dependence at the maximum of the PL band (λ_{exc} = 450 nm) normalized PL intensity for CdSe/ZnS-AM QDs (d_{CdSe} = 3.3 nm, two ZnS monolayers) capped by long-chain amines (1, open circles) and for nanoassemblies based on CdSe/ZnS-AM QDs and (pyridyl)$_2$-perylene diimide (DPP) dye molecules at molar ratio $x = [C_{DPP}]/[C_{QD}]=1$ (2, solid circles) in a methylcyclohexane/toluene (6:1) mixture. The inset shows the temperature dependence of the PL band maximum E_{max} (gray) and the related PL band line width (FWHM) (black) for AM-capped CdSe/ZnS QDs without DPP.

To examine whether the phase transition is an inherent property of CdSe/ZnS, we have deposited TOPO capped QDs (solvated in toluene) onto a quartz substrate. Measurements have been performed after evaporation of the initial toluene solvent. While temperature changes of the PL parameters are similar to those of CdSe/ZnS in a glass, the clear fingerprint of the phase transition is missing (see Fig. 3.4D). We tentatively assign this effect to depletion of the TOPO ligand shell following sample deposition. This is supported by recent findings of Graaf et al. (2007) following nanolithography experiments (see also Graaf and Baumgaertel, 2016).

3.3 Discussion

We will concentrate our discussion in Section 3.3.1 with respect to exciton–phonon coupling in CdSe/ZnS QDs to a large extent on the temperature-dependent absorption properties. Though PL emission might in principle also provide similar information the value of it is limited since PL is to some extent (especially below the ligand phase transition temperature) related to surface traps [Zenkevich et al., 2015] and thermalization of exciton fine structure energies. Nevertheless we will present PL-related data since such are commonly reported in the literature [Nomura and Kobayashi, 1992; Chon et al., 2011; Takagahara, 1993; Valerini et al., 2005; Joshi et al., 2006; Ehlert et al., 2006; Salvador et al., 2006; Al Salman et al., 2007; Morello et al., 2007; Lo et al., 2009]. Moreover, since the (temperature-dependent) Stokes shift might also reflect exciton–phonon coupling properties [Kelley, 2010; Kelley, 2011] we will compare in Section 3.3.2 temperature-dependent absorption and PL and finally discuss the Stokes shift in Section 3.4.

3.3.1 Temperature Dependence of Absorption Properties

We would like to describe the temperature dependence of absorption over the total temperature range investigated in the present experiments. In semiconductor nanocrystals, the temperature dependence of the energy gap is usually similar to the bulk semiconductor one, except for a temperature-independent energy offset due to the quantum confinement [Gaponenko, 1998; Baranov et al., 2003]. For wide-band materials, such as CdSe/ZnS QDs, where the exciton binding energy is much smaller than the band-gap energy (E_g), the corresponding E_g values may be approximated as the energy of the first exciton peak (the band-edge exciton) measured in optical absorption spectra at various temperatures. Using this approach, the band gap has recently been investigated as a function of temperature (10÷300 K) for TOPO-capped CdSe/ZnS core/shell QDs with different core diameters deposited into a resin matrix [Joshi et al., 2006]. Temperature dependence of E_g has been modeled by the empirical Debye or Einstein model for electron–phonon coupling and is described by the Debye model [Varshni, 1967] as

$$E_g(T) = E_g(0) - \alpha T^2/(T + \theta_D) \tag{3.1}$$

where $E_g(0)$ is the band gap at 0 K, α is a constant, and θ_D is the Debye temperature. An alternative approach follows the Einstein model [Cody, 1984] resulting in

$$E_g(T) = E_g(0) - k/[\exp(\theta_E/T) - 1] \quad (3.2)$$

where k is a constant and θ_E is the Einstein temperature.

Alternative descriptions have been also reported by Paessler [Paessler, 1999; Paessler, 2001] and O′Donnell and Chen [O′Donnell and Chen, 1991]. It has been shown, however, that a clear discrimination between these different models needs a temperature range down to a few Kelvin [Lo et al., 2009]. We therefore stay with the most common approximations given in Eqs. 3.1 and 3.2.

Joshi et al. [2006] report θ_D values which are in good agreement with the data presented for capped CdSe/ZnS core/shell QDs in a rigid polystyrene matrix [Valerini et al., 2005]. Both models gave very similar results, where θ_E and θ_D correspond to the average phonon temperature. For comparison, a LO (CdSe) phonon energy of 25 meV corresponds to a phonon temperature of 290 K. The above cited results have been obtained in rigid matrixes for capped CdSe/ZnS [Joshi et al., 2006; Valerini et al., 2005]. In our case, one should take the phase transition of the capping ligand layer into account when applying the above empirical expressions. In addition, since the Einstein approximation (Eq. 3.2) appears to provide more accurate results (due to smaller errors obtained by the fitting procedure for θ_E), we prefer to use this approximation [Joshi et al., 2006].

The results of the fitting procedures for the optical absorption band energies at various temperatures are shown in Fig. 3.8. Both approaches fit the data reasonably well besides the deviation at T_{crit}. Fitted data are given in Tables 3.1 and 3.2.

Though it can be seen from Fig. 3.8 that the fit curves reproduce rather well the experimental data below and above the phase transition region, we carried out fits for TOPO-capped CdSe/ZnS QDs separately below and above the phase transition temperature of the solvent mixture (Fig. 3.9). It is seen that in the latter case the simulation gives slightly different θ_E values 165 K and 158 K above and below T_{crit}, compared to θ_E = 207 K fitting across the total temperature range (see Tables 3.1 and 3.2). We would like to mention that the fitting results obtained on the basis of Eq. 3.1 gave θ_D values of the same order of magnitude with the same tendencies with respect to the phase transition. A severe drawback of our data is, however, that the temperature dependence is only very smooth across the investigated temperature range. Future experiments should be performed toward much lower temperatures.

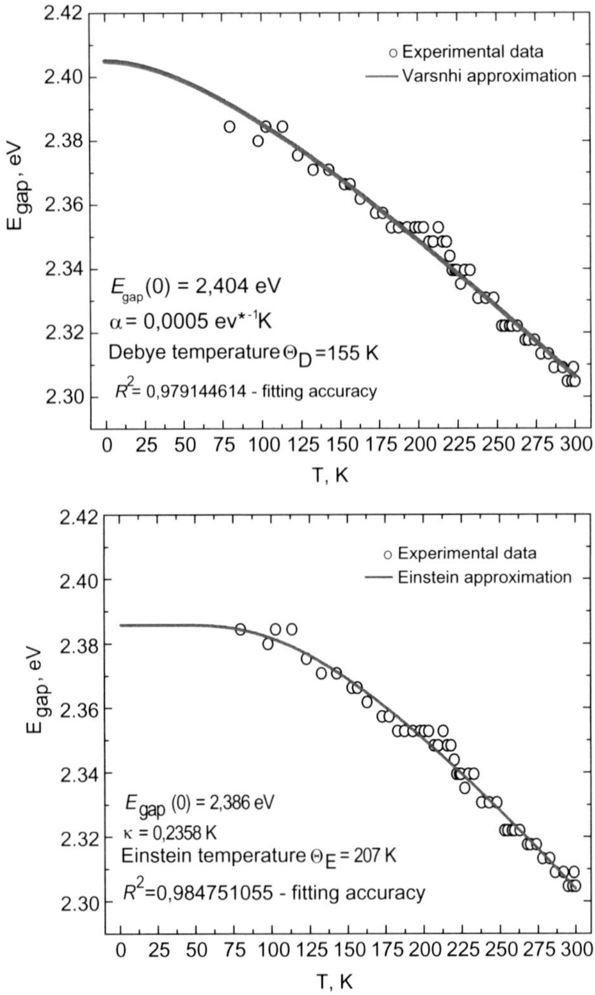

Figure 3.8 Temperature dependence of energy gap values $E_{gap} = E_g(T)$ based on absorption measurements and fitting according to Varshni Eq. 3.1 (top) and Einstein Eq. 3.2 (bottom) for TOPO-capped CdSe/ZnS QDs in the solvent mixture (methylcyclohexane:toluene = 6:1).

All data obtained at λ_{exc} = 450 nm for CdSe/ZnS (TOPO) are included in Table 3.1, making use of Eq. 3.1 [Varshni, 1967] or in Table 3.2 for the Einstein approximation according to Eq. 3.2 [Cody, 1984]. Taking the large fitting errors into account, the values are in satisfying agreement with each other, observed for either Einstein or Debye fits. However, data reported by Al Salman et al. [Al Salman et al., 2007] are apparently resulting in a too low Debye temperature. In all fitting approximations with respect to the temperature, ranges θ_D and θ_E are always higher in the case of PL detection (discussed later) compared to absorption, which indicates a higher phonon energy relevant in the case of PL detection.

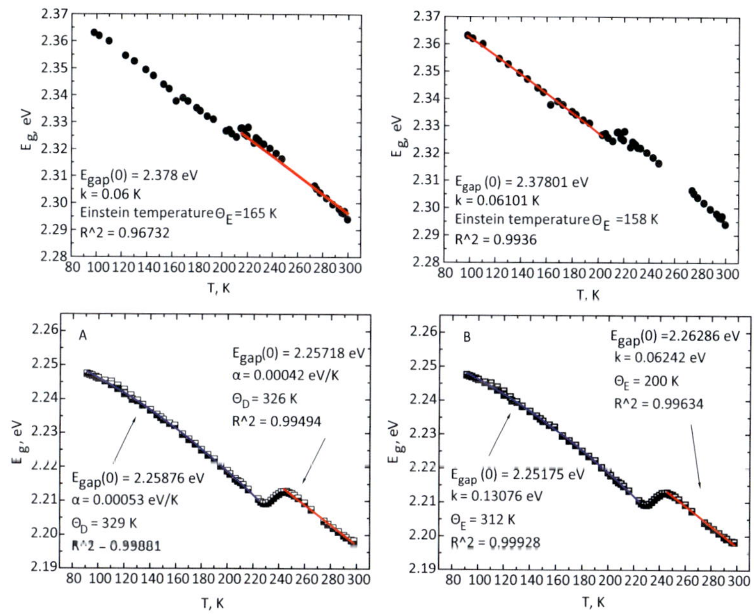

Figure 3.9 Temperature dependence of energy gap values $E_g(T) = E_{gap}$ for CdSe/ZnS QDs in the solvent mixture (methylcyclohexane:toluene = 6:1). Top: Fits are based on PL band measurements and fitting according to Einstein model (Eq. 3.2) for TOPO-capped QDs. Experimental data present energy E_{max} of the energetically lowest absorption band maximum derived from Gaussian fits. Red lines show simulations based on Eq. 3.2 above (left) and below (right) the phase transition temperature of the solvent mixture. Bottom: Data are based on PL band measurements and fitting according to Varshni Eq. 3.1 (A) and Einstein Eq. 3.2 (B) for AM-capped CdSe/ZnS QDs. Experimental points present energy $E_g = E_{max}$ of the PL band maximum. (A, B) Red and blue lines show simulations based on Eq. 3.1 and 3.2 above and below the phase transition temperature (~240 K), correspondingly.

Table 3.1 Comparison of temperature-dependent absorption and PL data for CdSe/ZnS (TOPO) according to Varshni Eq. (3.1)[a]

	Valerini[b]	**Joshi[c]**	**Salman[d]**	**Results of this study**						**Bulk[e]**
QD Diameter	CdSe/ZnS 5 nm	CdSe/ZnS 1.9/3.6 nm	CdSe 3 nm	CdSe/ZnS 3 nm						CdSe
	PL total	PL total		PL total	Abs LT	PL LT	PL HT	Abs LT	Abs HT	
E_0/eV	2.14	2.683/2.198	2.17	2.404	2.379 (1)	2.326 (70)	2.308 (1)	2.350 (7)	2.346 (5)	
$\alpha/(10^{-4})$ eV/K	3.2 (0.2)	4.4/4.5 (0.5)	3.2 (0.5)	5 (1)		6 (10)	4.8 (3)	4.8 (1.7)		2.8 – 4.1
θ_D/K	220 (30)	131/195 (43)/(70)	78 (35)	155	173 (10)	251 (139)	213 (35)	135 (158)	167 (29)	181-315

[a]LT and HT correspond to fits across the low- and high-temperature range, respectively, while "total" refers to the total temperature range, neglecting effects at the phase transition temperature. Errors of the last digits are given in parentheses. Excitation at 450 nm.
[b]Valerini et al. (2005).
[c]Joshi et al. (2006).
[d]Al Salman et al. (2007).
[e]Hellwege (1982).

Table 3.2 Comparison of temperature-dependent absorption and PL for CdSe/ZnS (TOPO) according to Einstein[a]

	Joshi[b]		**Results of this study**		
QD	CdSe/ZnS	CdSe/ZnS	CdSe/ZnS	CdSe/ZnS	CdSe/ZnS
Diameter	1.9/5.2 nm	3 nm	3 nm	3 nm	3 nm
Parameter	PL total	PL total	PL LT	Abs LT	Abs HT
E_0/eV		2.386	2.305 (2)	2.378 (7)	2.378 (4)
κ/meV	60 /67 (13)	236	63 (10)	61 (39)	
θ_E/K	155/184 (25)/(29)	207	180 (23)	158 (86)	165 (7)

[a]LT and HT correspond to fits across the low- and high-temperature range, respectively, while "total" refers to the total temperature range, neglecting effects at the phase transition temperature. Errors of the last digit are given in parentheses. Excitation at 450 nm.
[b]Joshi et al. (2006).

Concluding, the present results are consistent with the corresponding data obtained for bulk CdSe [Hellwege, 1982], where $\theta_D \approx 181 \div 315$ K and $\alpha \approx (2.8 \div 4.1) \times 10^{-4}$ eV/K. This means that the energy shift upon temperature lowering is due to the temperature-dependent phonon-induced band-gap shrinkage of the CdSe core material. According to Valerini et al. (2005), such a result can be understood in a simple 3D potential well model with finite barriers. As the sample temperature increases, the barrier height changes, due to the different temperature dependence of the CdSe and the ZnS energy gap, and the potential well becomes larger due to thermal expansion of the QD.

We did not investigate the influence of exciton–phonon coupling on the absorption line width since the inhomogeneous width is extremely broad due to the superposition of two bands, namely the one at low energies (LE) containing CdSe exciton fine structure levels and a hole-related absorption band at higher energies (HE), as has been discussed in Chapter 2. In addition, even at λ_{exc} = 450 nm the experimental shift of the QD PL band may be partly disturbed by the influence of the size selection effect. However, in the case of PL we will also study in Section 3.3.2 the influence of temperature on the line width.

3.3.2 Temperature Dependence of PL Energy and FWHM

Recently we have shown [Zenkevich et al., 2015] and outlined in more detail in Chapter 2 that the PL of TOPO-capped CdSe/ZnS QDs (and also AM-capped ones) is considerably influenced by near-band-edge trap emission, which complicates the analysis with respect to exciton–phonon coupling. Additionally, it has been shown that a kind of ligand phase transition abruptly changes at T_{crit} the nature of the emitting states [Zenkevich et al., 2015]. Under these circumstances relating the exciton–phonon coupling to specific PL emission is not a straightforward task. However, the general trend of PL temperature dependence is quite similar to the one observed for absorption (see Tables 3.1 and 3.2). Moreover, since the inhomogeneous FWHM is at low temperatures much smaller than the one for absorption, even the temperature-dependent line width (FWHM) of the PL will provide information on exciton–phonon coupling in the specific photoluminescent states.

Figure 3.9 shows the temperature dependence of QD PL energies. Data have been fitted by Eqs. 3.1 and 3.2. The fitting results are shown in Tables 3.1 and 3.2 and are in general agreement with the data obtained from absorption, though, as already mentioned, θ_E and θ_D are generally higher from PL data than those obtained from absorption data. Taking average overall fits we obtain from PL $\theta_{D,PL}$ = 206 K and $\theta_{E,PL}$ = 194 K, while absorption reveals $\theta_{D,abs}$ = 158 K and $\theta_{E,abs}$ = 162 K. Though fitting errors are generally large, the observed trend when comparing absorption and PL data seems to be a systematic one. Thus fitted temperatures are by Eqs. 1.2–1.3 higher from PL compared to those from absorption.

As already mentioned exciton–phonon coupling adds to dephasing processes, resulting in homogeneous line width broadening. In case the intrinsic (temperature-dependent) inhomogeneous broadening of the FWHM is comparable to these homogeneous contributions, exciton–phonon coupling can be deduced at least from PL line width data. Figure 3.10 shows the temperature-dependent broadening of the FWHM.

Equation 3.3 corresponds to an approximation of the exciton–phonon-induced broadening of the line width $\Gamma(T)$ given by Lee et al. (1986) according to

$$\mathrm{FWHM}(T) = \Gamma(T) = \Gamma_{inh} + \sigma T + \Gamma_{LO}/[\exp(E_{LO}/k_B T) - 1]. \qquad (3.3)$$

Γ_{inh} is the inhomogeneous broadening, which is assumed to be temperature independent and is due to fluctuations in size, shape, and composition. Γ_{LO} represents the exciton-LO-phonon coupling strength, and E_{LO} is the LO-phonon energy. Additionally, a linear dependence on temperature follows a coupling constant σ. The temperature-dependent terms represent a T linear term (direct dephasing processes via acoustic phonons), while the third term is an Orbach-type dephasing process caused by LO phonons.

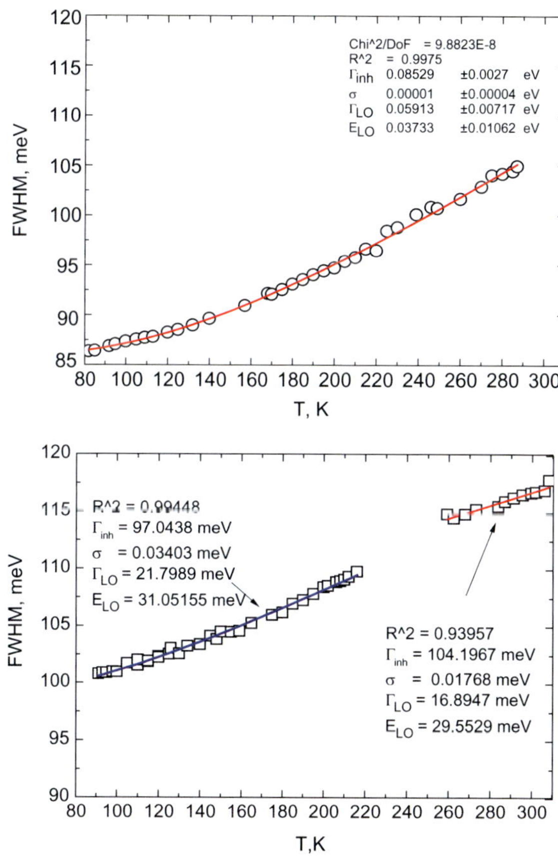

Figure 3.10 Fit of the temperature-dependent PL line widths (FWHM) for CdSe/ZnS QDs in the solvent mixture (methylcyclohexane:toluene = 6:1) according to Eq. 3.3. Fit parameters are included. Top: TOPO-capped QDs (λ_{exc} = 450 nm). Bottom: AM-capped QDs (λ_{exc} = 450 nm). Lines show fits above (red) and below (blue) the phase transition temperature (~240 K).

Table 3.3 summarizes the fitting results. The most interesting fitting parameter is the energy E_{LO} of the respective phonon responsible for the coupling. Without taking the weakly manifested phase transition into account, we obtain from the total PL FWHM across the complete temperature range a value for the inhomogeneous width of $\Gamma_{in} \approx 85$ meV and a LO phonon energy of $E_{LO} \approx 37$ meV. This latter value is very close to the LO (ZnS) phonon of 39 meV [Dzaghan et al., 2012]. Comparison with results from the literature reveals that E_{LO} (CdSe) is expected to be close to 25 meV [Dzaghan et al., 2012]. The acoustic phonon coupling constant σ shows a large fitting error, especially in the case where the fitted temperature range is limited. However, its value hardly influences the overall fitting accuracy. This is probably related to the limited temperature range. A linear coupling should be especially important at very low temperatures.

We have extensively shown that PL has to be deconvoluted energetically into at least two PL bands differing in energy by about 5–20 meV [Zenkevich et al., 2015]. Following this assignment we performed fits of the FWHM for the respective LE and HE components below (LT) and above (HT) T_{crit}. The resulting fit parameters are also included in Table 3.3. Inspection of the results shows that the energetically low component LE reveals across the total temperature range a phonon energy E_{LO} (LE) ≈ (39–53 meV), which is more close to the LO (ZnS) phonon compared to the phonon energy for fitting the energetically high component HE, which is with E_{LO} (HE) ≈ 27 meV very close to the LO (CdSe) phonon energy. The less specific fitting parameter Γ_{LO} is close to the reported experimental findings (see Table 3.3).

A considerable variance is observed with respect to the direct (acoustic) phonon-coupling constant σ. Our fitting results are characterized compared to literature reports by a considerable fitting error. This is probably due to the fact that our data start only above $T = 85$ K, while direct (linear) processes are more clearly seen in the very low temperature range. Though our findings will not be discussed in detail, it is, however, quite remarkable that the σ(HE) fit is close to reported theoretical and experimental values, while σ(LE) values are completely different from reported ones. We tentatively attribute this finding to the fact that acoustic phonons contribute much less or ill-defined to LE spectral bands compared to HE bands,

Table 3.3 Temperature-dependent PL line width (FWHM)[a] parameters obtained for TOPO-capped CdSe/ZnS QDs according to Eq. 3.3

	Valerini[b]	**Theory**	**Salman[c]**	**Results of this study**							**Bulk**
QD	CdSe/ZnS		CdSe				CdSe/ZnS				CdSe
Diameter (nm)	5		3				3				
T/K Energy				PL total total	PL LT LE	PL HT LE	PL LT HE	PL LT total	PL[h] LT total	PL[h,i] HT total	
Γ_{inh} /meV	85.5 (7)		79 (0.4)	85.3	146	49	78	84.1	97	104 **98**	
σ /μeV/K	71 (9)	35[d]	20 (3)	11 (36)	3.7	528	27	26	34	18 **13**	8[e]
Γ_{LO} /meV	21 (7)		30 (5)	59 (7)	58	48	33	55	22	17 **34**	100[f]
E_{LO} /meV	24.5 (2)	23.8–26[g)]	25	37.4 (11)	39	53	27	41 (10)	31	30 **30**	26.1[g]

[a]LT and HT correspond to fits across the low- and high-temperature range, respectively, while "total" refers to the total temperature range, neglecting effects at the phase transition temperature. Errors of the last digits are given in brackets. Excitation at 450 nm.
[b]Valerini et al. (2005); [c]Al Salman et al. (2007).
[d](Takagahara (1993), Takagahara (1996); [e]Gindele et al. (2000); [f]Cerullo et al. (1999).
[g]Hellwege (1982); [h]For AM-capped CdSe/ZnS QDs with four variable parameters.
[i]Bold numbers from two independent fits within the total temperature range.

the latter of which agree obviously more closely with the acoustic phonon energy of the CdSe-type character. An additional reason for this difference may be explained by the fact that upon temperature increase or decrease, excited QD states of different nature [Zenkevich et al., 2015] interact with acoustic phonons, thus giving rise to a modification of the acoustic coupling constant σ across the whole temperature range.

3.3.3 Temperature-Dependent PL Intensities

From Fig. 3.11A,B it is evident that also the PL intensity depends drastically on temperature. Valerini et al. (2005) have related the temperature-dependent PL decrease with increasing T of CdSe QDs to a superposition of a temperature-activated increase of nonradiative channels (e.g., among a manifold of traps) and exciton–phonon-related activation of nonradiative channels. According to this approach [Morello et al., 2007], assuming a temperature-independent radiative lifetime due to the strong confinement regime, the temperature dependence of the PL intensity follows Eq. 3.4 according to

$$I_{\mathrm{PL}}(T) = \frac{I_0}{1 + a(e^{-E_{\mathrm{a}}/k_{\mathrm{B}}T}) + b(e^{E_{\mathrm{LO}}/k_{\mathrm{B}}T} - 1)^{-m}} \qquad (3.4)$$

where I_0 is the PL intensity at $T = 0$, E_a is the activation energy for excitation of nonradiative traps, E_{LO} the energy of either CdSe or ZnS phonons, and m is the number of phonons needed to populate the next higher excitonic QD state which promotes exciton–phonon-induced radiationless transitions. a and b correspond to the relative contribution of these two processes.

We did not fit phonon energies and kept them constant across the total temperature range. Unfortunately, the lowest temperature of our experiment is limited to 100 K. For this reason, I_0 has also to be fitted (according to known experimental PL dependences down to 4.2 K for CdSe and CdSe/ZnS QDs [Mello Donega et al., 2006; Crooker et al., 2003; Morello et al., 2007]), which reduces the fit accuracy. Figure 3.11 contains fits of the temperature-dependent PL intensities at phonon energies of 27 meV and 39 meV for CdSe and ZnS, respectively. Fit accuracy is slightly higher for the latter case.

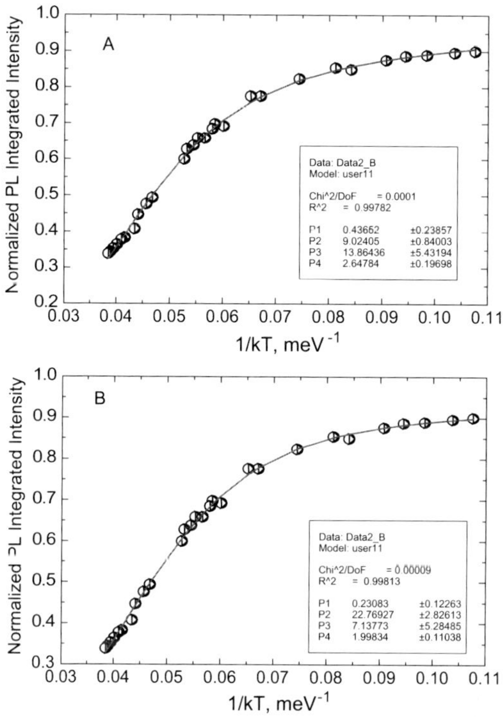

Figure 3.11 Fit of the temperature-dependent normalized PL integrated intensities according to Eq. 3.4 [Morello et al., 2007] for TOPO-capped CdSe/ZnS QDs in the solvent mixture (methylcyclohexane:toluene – 6:1) at λ_{exc} = 520 nm (experimental data at every temperature have been extracted from the results shown in Fig. 3.5C). Parameters a, b, E_a, and m have been variable, and phonon energies E_{LO} = 27 meV (A) and 39 meV (B) for CdSe and ZnS phonons, respectively, have been fixed. The fitting procedure has been performed at ratio I/I_0 = 0.9 for normalized PL integrated intensities.

For the ratio $I_{PL}(T)/I_0 = I_{PL}(T)/I_{max}$ = 0.93 to 0.85, all fits give realistic parameters. The best results are obtained for I/I_0 = 0.9 and are collected in Table 3.4 in comparison to data obtained by Valerini et al. (2005) and Morello et al. (2007). Considering all reasonable fitting approaches we find activation energies $5 < E_a < 20$ meV and number of photons $2.0 < m < 2.7$. Activation energies have fitting errors up to 50%, while errors for m are <10%. In comparison, activation energies are in a similar range, as those determined by Valerini et al. Activation energies of the reported magnitude are in the same range, as recently determined energies among different trap-related states [Zenkevich et al., 2015]. The number of phonons is according to Table 3.4 by a factor of 2 lower than reported for CdTe, though the product $m(E_{LO})$ is quite similar. From spectroscopy it is known that energy separation of the two lowest excitonic states is about 200 meV for CdSe/ZnS QDs investigated in the present experiment (see Chapter 2). This is more than the factor of 2 larger than we obtain from fitting (73 and 78 meV, respectively), which is quite stable with respect to *m*. In Chapter 2 we have shown via absorption experiments that a probably hole-related state is found about 80 meV above the lowest excitonic absorption. Therefore we tentatively assign the state which induces temperature-activated electron–phonon relaxation to the recently reported hole state. Finally, it should be mentioned that the assumption of a single thermally activated process is the simplest to describe the observed exponential decrease of the PL intensity in the range of 40–170 K [Morello et al. 2007]. It is not excluded that a similar PL temperature dependence could result also from more complicated processes, like the thermal activation of trapping in a distribution of defects at slightly different energies.

Table 3.4 Fit values obtained from temperature-dependent integrated PL intensities (λ_{exc} = 520 nm) for TOPO-capped CdSe/ZnS QDs according to Eq. 3.4[a]

	CdSe/ZnS		CdSe/ZnS[b]	CdTe[a,c]
Ligand/ Diameter	TOPO 3 nm	TOPO 3 nm	(TOPO) 5 nm	(TOPO) 4.2–5.9 nm
E_{LO}/meV	27[e]	39[e]	24.5	20
E_a/meV	13.5 (±5.0) 15[d]	7.0 (±5.0) 15[d]	15 (±4)	17.6
M	2.7 (±0.2) 2.8[d]	2.0 (±0.1) 2.3[d]	4.1 (±0.5)	4.8
b/*a*	20.6 (±20) 21.6[d]	98.6 (±50) 67.6[d]	296	
$m{\cdot}E_{LO}$/meV	73 76[d]	78 89.7[d]	100	99

[a]Morello et al. (2007); [b]Valerini et al. (2005).

[c]Data have been obtained by fitting results from three different samples on QDs with different diameters.

[d]Fit results for λ_{exc}= 410 nm.

[e]Values have been fixed.

3.4 Temperature-Dependent Stokes Shifts

In case that absorption and PL emission are of the same nature (i.e., neglecting internal relaxation processes, for example, from near-band-edge to trap states), the experimentally observed Stokes shift Δ_{ST} (which corresponds to the reorganization energy [Kelley, 2011] amounts in case that $n = 1$ phonons are involved to [Kelley, 2011]

$$\Delta_{ST} = SE_{LO}, \tag{3.5}$$

where S corresponds to the Huang–Rhys factor and E_{LO} to the corresponding phonon energy. Note that according to the literature several authors use "2S instead of "S" [Alivisatos et al., 1989; Joshi et al., 2006]. Since exciton–phonon coupling is weak in colloidal QDs [Sagar et al., 2008; Kelley, 2010] only one phonon has to be taken into account. Due to the exciton fine structure it is often not straightforward to determine the correct Stokes shift, as is outlined in more detail in Chapter 2. For this reason exciton–phonon coupling determined by Stokes shifts might result in too large Huang–Rhys factors.

Figure 3.12 (top) shows direct comparison of the temperature dependence of absorption (1) and PL emission (2), while the bottom part shows the Stokes shift directly. In the low-temperature range (below T_{crit}) the value of Δ_{ST} is slightly temperature dependent, which we attribute to the thermalization of the excitonic fine structure, as outlined in detail in Chapter 2. At 180 K (the temperature-independent) shift Δ_{ST} is ~32 meV. According to Eq. 3.5 such a Stokes shift gives a Huang–Rhys factor of 1.19 in the case of CdSe phonons and 0.8 in the case of ZnS LO phonons. These values decrease by roughly a factor of 1.5 to 21 meV (with $S = 0.79$ and 0.54, respectively) according to the (temperature-independent) Stokes shift at high temperatures. PL states at high temperatures are more likely related to CdSe near-band-edge states than those at low temperatures [Zenkevich et al., 2015].

Recently Kelley has reported a force field calculation of the size dependence of exciton–phonon coupling in CdSe QDs [Kelley, 2011]. Besides these calculations this paper is an excellent summary of recent experimental and theoretical results in the field of exciton–phonon coupling. Kelley calculates for CdSe ($d = 3$ nm) a Stokes shift Δ_{ST} of 10 meV in the range of LO phonons and 22 meV in the range of acoustic phonons. The corresponding Huang–Rhys factors are 0.95 and 5.5 [Kelley, 2011]. According to calculations, S and Δ_{ST} decrease with QD size and with excitation into higher exciton states in the range of quantum confinement. From these calculations it becomes obvious that both types of phonons contribute nearly equally to the Stokes shift. Moreover, calculated shifts and thus Huang–Rhys factors are very close to those reported in this contribution.

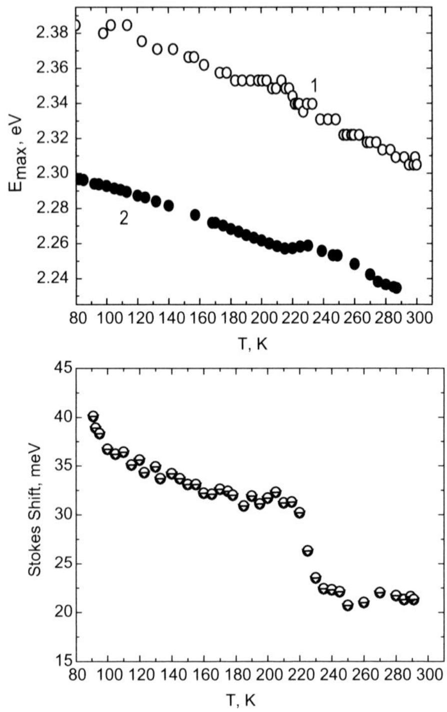

Figure 3.12 Stokes shift analysis for TOPO-capped CdSe/ZnS QDs in the solvent mixture (methylcyclohexane:toluene = 6:1). Top: Comparison of temperature dependence of (1) the first absorption peak (mean weighted data for two Gaussians lines) and (2) PL maximum at λ_{exc} = 450 nm (data taken from Zenkevich et al. [2012]). Bottom: Temperature-dependent Stokes shift obtained between the PL band fitted by one Gaussian component and the lowest absorption band (data taken from Zenkevich et al. [2015]).

Note that in the Kelley paper, S and Δ_{ST} have been integrated over all modes either in the acoustic or in the optical range. All values become much smaller when fixing the data evaluation to exactly one phonon frequency in each range ($S_{LO} \approx 0.03$; $S_{ac} = 0.25$). According to Eq. 3.5 this will result in negligible Stokes shifts. Moreover, calculations have been performed in vacuum on spherical QDs, omitting capping semiconductor layers and ligands. Recent results show that the kind of QD phase (and thus implicitly also exciton–phonon coupling) might depend on the conditions of preparation [Karakoti et al., 2011].

Though there is reasonable agreement between our findings and Kelley's calculations, in most cases reported experiments result in much smaller Huang–Rhys factors in quantum-confined CdSe QDs compared to bulk experiments [Kelley, 2011]. Qualitatively the latter is understandable since the separation of the electron and hole is much smaller than in bulk material, thus creating a much smaller dipole moment. Sagar et al. (2008) found from femtosecond pump-probe experiments that Stokes shifts are in the range of 0.2–0.4 meV accompanied by $S_{ac} = 0.15$ and $S_{LO} = 0.016$. These values are much smaller than deduced from the present (wavelength-resolved) continuous-wave (cw) experiments. Sagar et al. argue that cw experiments depend on time-dependent fluctuations of exciton–phonon coupling over a wide range. This is supported by observations of fluctuations of phonon side bands of single QDs [Empedocles et al., 1996]. Such fluctuations may be closely related to PL intermittency caused by charging events [Cichos et. al., 2007]. Additionally, it has been shown via low-temperature experiments on single QDs that the contribution of acoustic phonons varies considerably both in energy and coupling strength [Fernee et al., 2008].

Since calculations of Kelley already take (static) fluctuations into account, our experimental findings in agreement with the reported calculations make us believe that we can determine an effective (time-averaged) exciton–phonon coupling from experimental Stokes shifts. From this we arrive at the conclusion that we observe predominantly a coupling to ZnS phonons (with $S = 0.8$) below T_{crit} and to CdSe ones (with $S = 0.79$) above the phase transition temperature. The phase transition leaves the coupling almost constant but changes between the influence of CdSe core phonons and ZnS shell phonons, thus giving rise to a jump of the Stokes shift of ~9 meV at the phase transition. This is supported by a comparison of the temperature dependence via absorption and PL experiments (see Sections 3.2 and 3.3). Moreover, surface, interface, and transversal phonons will also add to a spectral broadening of phonon-coupled transitions [Dzhagan et al., 2012].

Additionally we would like to remark that the apparent Stokes shift might by enlarged by an energy cascade from near-band-edge states to traps close to the band edge. As a consequence determination of exciton–phonon coupling from Stokes shifts will generally result in too large couplings.

3.5 Exciton–Phonon Coupling in AM-Type QDs

Exciton–phonon coupling has also been investigated for CdSe/ZnS QDs capped by long-chain AMs. The diameter of QDs is slightly larger than the TOPO-capped one. However, this should not have a noticeable impact [Takagahara, 1993]. According to the spectral deconvolution of PL bands [Zenkevich et al., 2012; Zenkevich et al., 2015] of the lowest PL energies, AM-capped QDs are different from TOPO-capped ones. Since we have shown that spectral resolution is related to distributions of trap states at the exciton band edge, exciton–phonon coupling might also be changed (see the discussion above). Figure 3.13 shows the PL energy and FWHM for AM-capped QDs. It includes fits to Eq. 3.1 and 3.2 in the low- and high-temperature range with regard to phase transition.

The data of the fits are collected in Table 3.5. A comparison with corresponding fits of TOPO-capped CdSe/ZnS in Tables 3.1 and 3.2 shows that the respective coupling parameters α and κ are comparable to each other. However, we observe differences in the respective temperatures. The average value of θ_D is 182 K and 310 K, while the average θ_E value is 178 K and 320 K for TOPO-capped and AM-capped QDs, respectively. Though fitting depends strongly on the fitted temperature range, these differences in temperatures are systematic. Debye and Einstein temperatures are related to the heat capacity of solids. Roughly speaking, higher temperatures reflect a higher rigidity of the crystal lattice. With regard to the present results, this would imply that AM-capped QDs of very similar (but not exactly the same) size are somewhat more rigid than TOPO-capped ones. It is known that TOPO can saturate only about 30% of Zn^{2+} lone pairs due to its pyramidal structure [Kowerko et al., 2010; Blaudeck et al., 2012]. Contrary, the nearly linear AM can cap much more surface sites. Though the binding energy of AM is lower than that of TOPO [Blaudeck et al., 2012], this can be neglected since ligand exchange is probably slowed down considerably in the viscous glass matrix. Figure 3.7 shows that the phase transition temperature of AM-capped QDs is higher by about 10 K than the one of TOPO-capped ones. Moreover, Mello Donega et al. (2006) missed a phase transition of TOPO-capped QDs compared to AM-capped ones. These arguments make us believe that ligands increase the rigidity of a QD crystal lattice by forming a cage exerting strain to the crystal lattice. Since we observe in PL strong contributions from PL at the ZnS/CdSe interface (see Chapter 2), we suggest that ligand-dependent exciton–phonon coupling might increase the Debye and Einstein temperatures.

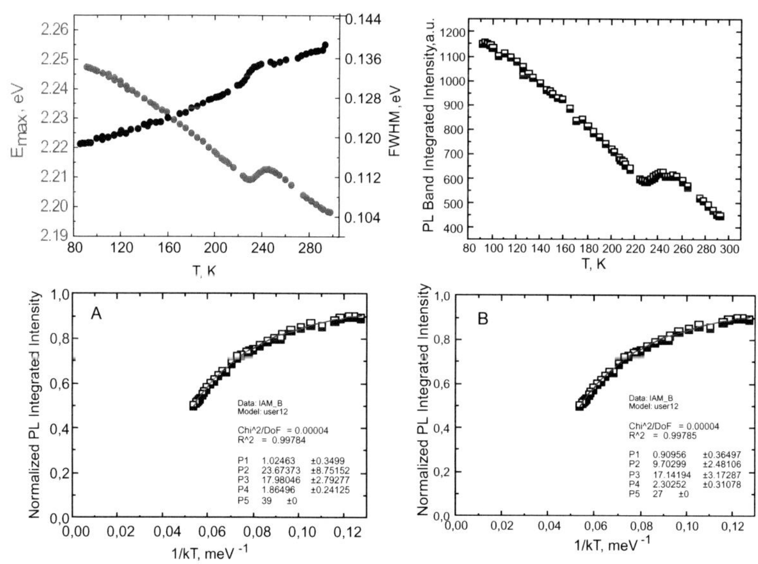

Figure 3.13 Top: Temperature-dependent PL energies, spectral widths FWHM (left), and PL intensities (right) of CdSe/ZnS capped by amine (in 6:1 methylcyclohexane:toluene; $\lambda_{exc} = 450$ nm). Bottom: Fit of the temperature-dependent normalized PL integrated intensities according I to Eq. 3.4 [Morello et al., 2007] below the phase transition temperature (~240 K) with fixed phonon energies $E_{LO} = 39$ meV (A) and 27 meV (B) for ZnS and CdSe/ZnS phonons, respectively. Fits are restricted to the temperature range below the phase transition. The fitting procedure has been performed at the ratio $I/I_0 = 0.9$. Note the $1/kT$ presentation.

Table 3.5 Comparison of temperature-dependent PL for CdSe/ZnS (AM) according to Einstein[a] or Varshni[b]

	Einstein		Varshni		
QD/AM	**CdSe/ZnS**	**CdSe/ZnS**	**CdSe/ZnS**	**CdSe/ZnS**	**CdSe/ZnS**
Parameter	LT	HT	LT	LT[d]	HT
E_0/eV	2.252 (1)	2.257[c]/2.250 (1)	2.259 (1)	2.258 (1)	2.257[c] (4)
κ/meV	131 (7)	86[c]/130[c]			
$\alpha/(10^{-4})$ eV/K			5.3[c]/4.8[c]	4.8[c]	4.22 (3)
θ_E/K	320 (11)	266/374 (11/5)			
θ_D/K			273/330 (4)	312 (16)	326 (40)

[a]LT and HT correspond to fits across the low- and high-temperature range, respectively, while "total" refers to the total temperature range, neglecting effects at the phase transition temperature. Errors of the last digit are given in parentheses. Excitation at 450 nm.

[b]Varshni (1967).

[c]Fixed during fitting.

[d]Temperature range 100–150 K.

However, according to fits of the temperature-dependent FWHM (see Table 3.3), we do not observe the same tendency. On the contrary, the phonon-coupling constant Γ_{LO} seems to be on average somewhat smaller compared to TOPO-capped QDs. This might be due to the fact that it is only possible to fit FWHM data below and above T_{crit} separately due to the pronounced jump at this temperature. For this reason fitting errors are much larger compared to TOPO-capped QDs. Nevertheless, also in this case E_{LO} is larger than expected for CdSe phonons. Evaluation of exciton–phonon coupling from the FWHM according to Eq. 3.3 results in Γ_{in} = 97 meV, σ = 30 μeV, Γ_{LO} = 22 meV, and E_{LO} = 31 meV.

Finally we compare fits of the temperature-dependent PL intensity for TOPO-capped and AM-capped CdSe/ZnS QDs in Table 3.6. All fitted values are close to those for TOPO-capped CdSe/ZnS.

Table 3.6 Comparison of exciton–phonon coupling[a] obtained from temperature dependent integrated PL intensities (λ_{exc} = 520 nm) for TOPO- and AM-capped CdSe/ZnS QDs

	CdSe/ZnS		CdSe/ZnS	CdSe/ZnS
Ligand	**TOPO**	**TOPO**	**(AM)**	**(AM)**
Diameter	**3 nm**	**3 nm**	**3 nm**	**3 nm**
E_{LO}/meV[b]	27	39	27	39
E_a/meV	13.5 (±5.0)	7.0 (±5.0)	17 (±3)	18 (±3)
m	2.7 (±0.2)	2.0 (±0.1)	2.3 (±0.3)	1.9 (±0.3)
b/a	20.6 (±20)	98.6 (±50)	11	23
$m \cdot E_{LO}$/meV	73	78	62	74

[a]According to Eq. 3.4 following Morello et al. (2007).
Data have been obtained by fitting results from three different samples on QDs with different diameters.
[b]Values have been fixed.

3.6 Conclusion

Our temperature-dependent optical experiments on CdSe/ZnS QDs are in almost quantitative agreement with experimental reports on the PL-detected influence of electron–phonon coupling on optical properties. While in most of those reported experiments, PL data (PL energy, PL band width, or Stokes shifts) have been used to determine electron–phonon coupling parameters (for a comparison, see also Tables 3.1 and 3.2 presented above), we used in parallel temperature-dependent absorption data, which provided in combination with PL emission the identification of the temperature dependence of the Stokes shift, which in return reveals temperature-dependent Huang–Rhys factors. Temperature-dependent data from absorption are slightly but definitely different from those obtained by PL. We explained this finding by the analysis that absorption is more sensitive to core CdSe LO optical phonons, while PL also contains coupling to ZnS phonons of the shell. This supports the earlier posed assumption [Zenkevich et al., 2015] that PL emission is to a large extent due to QD surface states.

An interesting finding is related to the presence of a phase-transition-type change in absorption and PL [Zenkevich et al.,

2015]. From the temperature-dependent Stokes shift we learn that exciton–phonon coupling below the phase transition temperature T_{crit} is governed more efficiently by coupling to ZnS phonons, while above coupling to CdSe increases. This finding is supported by the observation that the (size-dependent) phase transition appears at lower temperatures (−10 K) in absorption (predominant excitation of small QDs) compared to the one detected in PL from dominantly surface states.

With respect to methodological aspects we state that determination of exciton–phonon coupling via Stokes shifts needs a temperature-dependent investigation to reduce (i) contributions from the thermalization of the excitonic fine structure levels, (ii) influence of differently contributing near-band-edge states and shallow traps, and (iii) identification of the "true" first absorption band. Neglecting these influences will result in too large Stokes shifts and exciton–phonon coupling constants.

Due to the limited temperature range (>80 K) our information on the contribution of coupling to acoustic phonons is very limited. Further experiments are needed to cover this temperature range in order to achieve a separation of optical and acoustic phonon contributions (e.g., to the Stokes shift).

Investigation of temperature-dependent PL intensities reveals that exciton–phonon coupling is not the only mechanism to cause changes in optical properties. As discussed with respect to Stokes shifts these additional relaxation processes might also influence PL energies and line widths. The activation energies are in the range of trap-state distributions, as determined very recently [Zenkevich et al., 2015]. Most interesting, however, is the finding that contrary to results reported in the literature, temperature-activated coupling to optical phonons proceeds not via excitation to the second excitonic state but to a recently identified hole state (see Chapter 2), which has often been overlooked in several reported investigations.

Interestingly, the kind of ligand has a noticeable impact on exciton–phonon coupling. On average, it is found to be larger for AMs compared to TOPO ligands, which might have two reasons, namely, firstly, the large contribution of PL from surface or interface states and, secondly, the higher coverage of the surface with AM ligands compared to TOPO ones. Additionally, AM ligands have a larger impact on the ligand-related phase transition.

Comparing our (and comparable) results from wavelength-detected exciton–phonon coupling parameters with theoretical and time-resolved experiments reveals that the magnitude of the coupling strength to phonons is overestimated. The origin of this mismatch is probably related to the fact that QDs (and other quantum systems) undergo on slow timescales (on milliseconds and longer) charging processes and modifications of the coupling to the environment. These fluctuations are sampled by wavelength-based experiments. It would be interesting to get further information on such processes via analysis of blinking and spectral diffusion experiments with respect to exciton–phonon coupling [Schmidt et al., 2014; Zenkevich et al., 2015].

Finally, we found indications that selective optical excitation might influence the ligand-induced phase transition. This might be explained by electronic coupling of ligand and intrinsic QD states, as has been predicted by recent theoretical calculations [Fischer et al., 2012].

Recently, Lin et al. (2015) published a detailed resonance Raman study on electron–phonon coupling in CdSe/CdS core/shell QDs (Lin et al., 2015). They found that optical phonon energies of both the core and the shell depend noticeably on the excitation energy which provides information about the kind of localization of the excited excitons.

References

Alivisatos, A. P., Harris, T. D., Carrol, P. J., Steigerwald, M. L., and Brus L. E. (1989). Electron-vibration coupling in semiconductor clusters studied by resonance Raman spectroscopy, *J. Chem. Phys.*, **90**, 3463–3468.

Al Salman, A., Tortschanoff, A., Mohamed, M. B., Tonti, D., van Mourik, F., and Chergui, M. (2007). Temperature effects on the spectral properties of colloidal CdSe nanodots, nanorods, and tetrapods, *Appl. Phys. Lett.*, **90**, 093104.

Anni, M., Manna, L. Cigolani, R., Valerini, D., Creti, A., and Lomascolo, M. (2004). Förster energy transfer from blue-emitting polymers to colloidal CdSe/ZnS core shell quantum dots, *Appl. Phys. Lett.*, **85**, 4169.

Baranov, A. V., Rakovich, Yu. P., Donegan, J. F., Perova, T. S., Moore, R. A., Talapin, D. V., Rogach, A. L., Masumoto, Y., and Nabiev, I. (2003). Effect of ZnS shell thickness on the phonon spectra in CdSe quantum dots, *Phys. Rev. B*, **68**, 165306.

Blaudeck, T., Zenkevich, E. I., Cichos, F., and Borczyskowski, C. von (2008). Probing wave functions at semiconductor quantum-dot surfaces by non-FRET photoluminescence quenching, *J. Phys. Chem. C*, **112**, 20251–20257.

Blaudeck, T., Zenkevich, E. I., Abdel-Mottaleb, M., Szwaykowska, M. K., Kowerko, D., Cichos, F., and Borczyskowski, C. von (2012). Formation principles and ligand dynamics of nanoassemblies of CdSe quantum dots and functionalised dye molecules, *ChemPhysChem*, **13**, 959–972.

Cerullo, G., De Silvestri, S., and Banin, U. (1999). Size-dependent dynamics of coherent acoustic phonons in nanocrystal quantum dots, *Phys. Rev. B*, **60**, 1928.

Chon, B., Bang, J., Park, J., Jeong, C., Choi, J. H., Lee, J-B., Joo, T., and Kim, S. (2011). Unique temperature dependence and blinking behavior of CdTe/CdSe (core/shell) type-II quantum dots, *J. Phys. Chem. C*, **115**, 436–442.

Cichos, F., Borczyskowski, C. von, and Orrit, M. (2007). Power-law intermittency of single emitters, *Curr. Opin. Colloid Interface Sci.*, **12**, 272–284.

Cody, G. D. (1984). In *Hydrogenated Amorphous Silicon, Semiconductors and Semimetals* (Pankove, J. I., ed), Academic, New York, Vol. 21, pp. 11–79.

Cooney, R. R., Sewall, S. L., Dias, E. A., Sagar, D. M., Anderson, K. E. H., Butler, J. S., and Kambhampati, P. (2007). Unified picture of electron and hole relaxation pathways in semiconductor quantum dots, *Phys. Rev. B*, **75**, 2453111.

Dzhagan, V. M., Valakh, M. Ya., Raevskaya, A. E., Stroyuk, A. L., Kuchimiy, S. Y., and Zahn, D. R. T. (2007). Resonant Raman scattering study of CdSe nanocrystals passivated with CdS and ZnS, *Nanotechnology*, **18**, 285701.

Dzhagan, V., Valakh, M., Melnik, N., Rayevska, O., Lokteva, I., Kolny-Olesiak, J., and Zahn, D. R. T. (2012). Phonon spectra of small colloidal II-VI semiconductor nanocrystals, *Int. J. Spectr.*, 532385.

Dzhagan, V. M., Valakh, M. Y., Milekhin, A. G., Yeryukov, N. A., Zahn, D. R. T., Cassette, E., Pons, T., and Dubertret, B. (2013). Raman- and IR-active phonons in CdSe/CdS core/shell nanocrystals in the presence of interface alloying and strain, *J. Phys. Chem. C*, **117**, 18225–18233.

Ehlert, O., Tiwari, A., and Nann, T. (2006). Quantum confinement of the thermodynamic functions for the formation of electrons and holes in CdSe nanocrystals, *J. Appl. Phys.*, **100**, 074314.

El-Kork, N., Huisken, F., and Borczyskowski, C. von (2011). Dielectric effects on the optical properties of single silicon nanocrystals, *J. Appl. Phys.*, **110**, 074312.

Frenzel, J., Joswig, J.-O., and Seifert, G. (2007). Optical excitations in cadmium sulfide nanoparticles, *J. Phys. Chem. C*, **111**, 10761–10770.

Empedocles, S. A., Norris, D. J., and Bawendi, M. G. (1996). Photoluminescence spectroscopy of single CdSe nanocrystallite quantum dots, *Phys. Rev. Lett.*, **83**, 3873.

Fernee, M. J., Littleton, B. N., Cooper, S., Rubinsztein-Dunlop, H., Gomez, D. E., and Mulvaney, P. (2008). Acoustic phonon contributions to the emission spectrum of single CdSe nanocrystals, *J. Phys. Chem. C*, **112**, 1878–1884.

Fischer, S. A., Crotty, A. M., Kilina, S. V., Ivanov, S. A., and Tretiak, S. (2012). Passivating ligand and solvent contributions to the electronic properties of semiconductor nanocrystals, *Nanoscale*, **4**, 904–914.

Gaponenko, S. V. (1998). *Optical Properties of Semiconductor Nanocrystals*, Cambridge University Press, Cambridge, New York.

Gindele, F., Hild, K., Langbein, W., and Woggon, U. (2000). Temperature-dependent line widths of single excitons and biexcitons, *J. Lumin.*, **87–89**, 381.

Graaf, H., Vieluf, M., and Borczyskowski, C. von (2007). Selective binding of dye molecules and CdSe nanocrystals on nanostructures generated by AFM lithography of silicon surfaces, *Nanotechnology*, **18**, 265306.

Graaf, H., and Baumgaertel, T. (2016). Chapter 6, in Self-Assembled Organic-Inorganic nanostructures (Zenkevich, E., and Borczyskowski, C. von, eds.), Pan Stanford, Singapore.

Hellwege, H. (ed.) (1982). *Numerical Data and Functional Relationship in Science and Technology*, Landolt-Bornstein, New Series, Group III, Vol. 17, Pt. B Springer-Verlag, Berlin.

Joshi, A., Narsingi; K.Y., Manasreh, M. O., Davis, E. A., and Weaver, B. D. (2006). Temperature dependence of the band gap of colloidal CdSe/ZnS core/shell nanocrystals embedded into an ultraviolet curable resin, *Appl. Phys. Lett.*, **89**, 131907.

Karakoti, A. S., Sanghavi, S., Nachimuthu, P., Yang, P., and Thevuthasan, S. (2011). Probing the size-and environment-induced phase transformation in CdSe quantum dots, *J. Phys. Chem. Lett.*, **2**, 2925–2929.

Kilina, S., Ivanov, S., and Tretiak, S. (2009). Effect of surface ligands on optical and electronic spectra of semiconductor nanoclusters, *J. Am. Chem. Soc.*, **131**, 7717–7726.

Kelley, A. M. (2010). Electron–phonon coupling in CdSe nanocrystals, *J. Phys. Chem. Lett.*, **1**, 1296–1300.

Kelley, A. M. (2011). Electron–phonon coupling in CdSe nanocrystals from an atomistic phonon model, *ACS Nano*, **5**, 5254–5261.

Klimov, V. I. (2003). Charge carrier dynamics and optical gain in nanocrystal quantum dots: from fundamental photophysics to quantum dot lasing, in *Semiconductor and Metal Nanocrystals: Synthesis and Electronic and Optical Properties* (Klimov, V. I. ed.), Marcel Dekker, New York, pp. 159–214.

Kowerko, D., Schuster, J., Amecke, N., Abdel-Mottaleb, M., Dobrawa, R., Würthner, F., and Borczyskowski, C. von (2010). FRET and ligand related non-FRET processes in single quantum dot-perylene bisimide assemblies, *Phys. Chem. Chem. Phys.*, **12**, 4112–4123.

Krause, S., Hartmann, M., Kahle, I., Neuman, M., Heidernätsch, M., Spange, S., and Borczyskowski, C. von (2013). Optical tracking of single Ag clusters in nanostructured water films, *J. Phys. Chem. C*, **117**, 24822–24829.

Krause, S., Heidernätsch, M., Hartmann, M., and Borczyskowski, C. von (2014). Optical investigation of diffusion of single Ag markers in confined water films, *J. Nanopart. Res.*, **16**, 2700.

Lee, J., Koteles, E. S., and Vassell, M. O. (1986). Luminescence linewidths of excitons in GaAs quantum wells below 150 K, *Phys. Rev. B*, **33**, 5512.

Lin, C., Gong, K., Kelley, D. F., and Kelley, A. M. (2015). Electron–phonon coupling in CdSe/CdS core/shell quantum dots, *ACS Nano*, **8**, 8131–8141.

Liptay, T. J., Marshall, L. F., Rao, P. S., Ram, R. J., and Bawendi, M. G. (2007). Anomalous Stokes shift in CdSe nanocrystals, *Phys. Rev. B*, **76**, 155314.

Lo, S. S., Khan Y., Jones, M., and Scholes, G. D. (2009). Temperature and solvent dependence of CdSe/CdTe heterostructure nanorod spectra, *J. Phys. Chem.*, **131**, 084714.

Martin, J., Cichos, F., Chan, I. Y. F. Huisken, F., and Borczyskowski, C. von (2004). Photoinduced processes in silicon nanoparticles, *Israel J. Chem.*, **44**, 341–351.

Martin, J., Cichos, F., Huisken, F., and Borczyskowski, C. von (2008). Electron-phonon coupling and localization of excitons in single silicon nanocrystals, *Nano Lett.*, **8**, 656–660.

Martin, J., Cichos, F., and Borczyskowski, C. von (2012). Spectral diffusion of quasi localized excitons in single silicon nanocrystals, *J. Lumin.*, **132**, 2161–2165.

Melnikov, D. V., and Fowler, W. B. (2001). Electron-phonon interaction in a spherical quantum dot with finite potential barriers: the Fröhlich Hamiltonian, *Phys. Rev. B*, **64**, 245320.

Mello Donega, C. de, Bode, M., and Meijerink, A. (2006). Size- and temperature-dependence of exciton lifetimes in CdSe quantum dots, *Phys. Rev. B*, **74**, 085320.

Miranda, R. P., Vasilevskiy, M. I., and Trallero-Giner, T. (2006). Nonperturbative approach to the calculation of multiphonon Raman scattering in semiconductor quantum dots: polaron effect, *Phys. Rev. B*, **74**, 115317.

Morello, G., De Giorgi, M., Kudera, S., Cingolani, R., and Anni, M. (2007). Temperature and size dependence of nonradiative relaxation and exciton-phonon coupling in colloidal CdTe quantum dots, *J. Phys. Chem. C*, **111**, 5846–5849.

Narayanaswamy, N., Feiner, L. F., Meijerink, A., and van der Zaag, P. J. (2009). The effect of temperature and dot size on the spectral properties of colloidal inp/zns core–shell quantum dots, *ACS Nano*, **3**, 2539.

Nomura, S., and Kobayashi, T. (1992). Exciton–LO-phonon couplings in spherical semiconductor microcrystallites, *Phys. Rev. B*, **45**, 1305–1315.

Nonoguchi, Y., Nakashima, T., and Kawai, T. (2008). Temperature-dependent exciton recombination dynamics of CdTe nanocrystals, *J. Phys. Chem. C*, **112**, 19263–19267.

O'Donnell, K. P., and Chen, X. (1991). Temperature dependence of semiconductor band gaps, *Appl. Phys. Lett.*, **58**, 2924.

Paessler, R. (1999). Parameter sets due to fittings of the temperature dependencies of fundamental bandgaps in semiconductors, *Phys. Stat. Sol. B*, **216**, 975.

Paessler, R. (2001). Temperature dependence of fundamental band gaps in group IV, III–V, and II–VI materials via a two-oscillator model, *J. Appl. Phys.*, **89**, 6235.

Sagar, D. M., Cooney, R. R., Sewall, S. L., Dias, E. A., Barsan, M. M., Butler, J. S., and Kambhampati, P. (2008). Size dependent, state-resolved studies of exciton-phonon couplings in strongly confined semiconductor quantum dots, *Phys. Rev. B*, **77**, 235321.

Schmidt, T., Chizhik, A. I., Chizhik, A. M., Potrick, K. Meixner, A. J., and Huisken, F. (2012). Radiative exciton recombination and defect luminescence observed in single silicon nanocrystals, *Phys. Rev. B*, **86**, 125302.

Schmidt, R., Krasselt, C., Göhler, G., and Borczyskowski, C. von (2014). The fluorescence intermittency for quantum dot is not power-law distributed: a luminescence intensity resolved approach, *ACS Nano*, **8**, 3506–3521.

Salvador, M. R., Graham, M. W., and Scholes, G. D. (2006). Exciton-phonon coupling and disorder in the excited states of CdSe colloidal quantum dots, *J. Chem. Phys.*, **125**, 184709.

Son, D. H., Wittenberg, J. H., Banin, U., and Alivisatos, A. P. (2006). Second harmonic generation and confined acoustic phonons in highly excited semiconductor nanocrystals, *J. Phys. Chem. B*, **110**, 19884–19893.

Takagahara, T. (1993). Electron–phonon interactions and excitonic dephasing in semiconductor nanocrystals, *Phys. Rev. Lett.*, **71**, 3577.

Takagahara, T. (1996). Electron–phonon interactions in semiconductor nanocrystals, *J. Lumin.*, **70**, 129–143.

Takagahara, T. (2002). *Semiconductor Quantum Dots: Physics, Spectroscopy and Applications* (Masumoto, Y., and Takagahara, T., eds.), Springer, New York.

Varshni, Y. P. (1967). Temperature dependence of the energy gap in semiconductors, *Physica*, **34**, 149.

Valerini, D., Creti, A., Lomascolo, M., Manna, L., Cingolani, R., and Anni, M. (2005). Temperature dependence of the photoluminescence properties of colloidal Cd Se/Zn S core/shell quantum dots embedded in a polystyrene matrix, *Phys. Rev. B*, **71**, 235409.

Wuister, S. F., Mello Donega, C. de, Bode, M., and Meijerink, A. (2004). Luminescence temperature antiquenching of water-soluble CdTe quantum dots: role of the solvent, *J. Am. Chem. Soc.*, **126**, 10397.

Yaremko, A. M., Yakimchuk, V. O., Dzhagan, V. M., Valakh, M. Ya., Azhniuk, Yu. M., Baran, J., Ratajczak, H., and Drozd, M. (2007). *Semicond. Phys. Quantum Electron. Optoelectron.*, **10**, 1–5.

Zenkevich, E. I., Stupak, A., Kowerko, D., and Borczyskowski, C. von (2012). Influence of single dye molecules on temperature and time dependent optical properties of CdSe/ZnS quantum dots: ensemble and single nanoassembly detection, *Chem. Phys.*, **406**, 21–29.

Zenkevich, E. I., Stupak, A., Goehler, C., Krasselt, C., and Borczyskowski, C. von (2015). Tuning electronic states of a CdSe/ZnS quantum dot by only one functional dye molecule, *ACS Nano*, **9**, 2886–2907.

Zhang, L.-G., Shen, D.-Z., Fan, X.W., and Lu, S.-Z. (2002). Exciton-phonon scattering in CdSe/ZnSe quantum dots, *Chin. Phys. Lett.*, **19**, 578–580.

Chapter 4

Optical Detection of Single Quantum Dots

Christian von Borczyskowski
Institute of Physics, University of TechnologyChemnitz, Reichenhainer Str. 70, D-09107 Chemnitz, Germany
borczyskowski@physik.tu-chemnitz.de

4.1 Introduction

Nanostructured materials such as semiconducting or metallic quantum dots (QDs) easily exhibit due to their size-dependent electronic wavefunctions (particle-in-a-box) tunable optical properties. Details of this tunability are of principal interest, both with respect to fundamental scientific aspects and with respect to applications as biological markers, in sensorics, and in photovoltaics. Especially at the level of nanosized objects, devices, and environments related to applications in biology, medicine, nanosensorics, and quantum optics, only very few QDs or even single ones are desirable to monitor (discrete) events at a high level of sensitivity. Optical detection of, respectively, tailored single quantum objects (atoms, ions, molecules, QDs, quantum well structures) was

Tuning Semiconducting and Metallic Quantum Dots: Spectroscopy and Dynamics
Edited by Christian von Borczyskowski and Eduard Zenkevich

ISBN 978-981-4745-24-6 (Hardcover), 978-981-4745-25-3 (eBook)
www.panstanford.com

an exciting breakthrough during the last few decades and is still dramatically developing, though many aspects are nowadays already standard. The main handle for optical tuning of QDs relies on the size-dependent particle-in-a-box concept (see Chapter 1). However, optical properties depend also on the embedding environment, both with respect to optical energies and even more dramatically with respect to the quantum yield (QY) of photon emission. These dependencies "outside particle-in-a-box" may at first glance be considered as a drawback, but on a second view they may turn out as extremely suitable sensors on the nanoscale, for example, for super-resolution microscopy.

The extremely high sensitivity inherent to single-quantum-object detection is not the only advantage. Additionally, single quantum object detection offers at least two more advantages, namely the response to single and/or rare events in the near environment, for example, of a QD, and the application as nanoprobe in heterogeneous materials. In the latter case each individual QD might respond selectively to static and dynamic properties of its specific environment, resulting in a nanoscopic "mapping" of energy, time, and space. Such detailed information is buried in any kind of ensemble spectroscopy (or microscopy). Thus single nanoprobes serve as a kind of nanoscope with a tunable sensitivity with respect to various physical, chemical, and biological properties. Keeping this in mind it is not surprising that the power of single-quantum-object detection is right from its "invention" closely related with the combination with optical microscopy techniques. The power of this approach has been acknowledged with the 2014 Nobel Prize in Chemistry for super-resolution microscopy.

In this respect the most advanced techniques rely up to now on the use of single (organic) dye molecules. Basically two reasons are responsible for this development. First, dye molecules can be most readily functionalized to be attached, for example, to biological nano- and microstructures. The second reason is related to the fact that dye molecules are to a minor degree sensitive to the embedding environment compared to (colloidal) QDs, thus providing a probe with fixed properties. The first reason can be overcome by functionalizing QD surfaces especially via self-aggregation processes [Borczyskowski and Zenkevich, 2014]. While the second reason seems to be of disadvantage with respect to QD applications, it might be turned around by using QDs exactly as sensitive probes of the nanoscopic environment. Obviously this will need a more thorough understanding of the basic principles of tuning QD properties via environmental influences. By combining this understanding with the high photostability of QDs, its optical narrow-band tunability will open a broad scenario for QDs as nanoprobes.

This chapter will focus on the experimental principles of optical single-QD detection and those properties of single QDs which can be tuned across the interface to the nanoscopic environment. The high sensitivity beyond the core of a QD becomes effective—opposite to dye molecules—since the exciton wavefunction is only partly confined in a box and leaks (tunnels) out of it. Though we will concentrate upon aspects relevant for materials science, we will also emphasize that single-QD detection offers an exciting view on the "individuality" of quantum objects giving direct access in time and space to essential microscopically relevant spectroscopic properties than those explored via ensemble experiments for centuries.

4.2 Single Quantum Object Detection: From Holes to Dots

As we have mentioned in the introduction, detection of single quantum objects became more and more challenge during the last 40 years. Detection of high-energy particles and photons became a standard technique. Emission in the visible range of single atoms and ions became detectable via laser or ion traps. The implementation of scanning tunneling microscopy (STM) and atomic force microscopy (starting in 1981) was a breakthrough in discovering single atoms and molecules at solid state surfaces with atomic resolution. However, at that time the detection of the optical absorption and emission of a single emitter at surfaces or, even more challenging, within condensed matter had not yet been achieved. The impetus to overcome this limitation came from two completely different approaches, namely the development of luminescence line narrowing [Szabo, 1970; Personov et al., 1972] followed by spectral hole-burning techniques [Kharlamov et al., 1974; Gorokhovskii et al., 1974] and scanning near-field optical microscopy (SNOM) [Synge, 1928; Pohl et al., 1984]. We will start with the description of luminescence line narrowing techniques and will shortly describe the SNOM approach in the beginning of Section 4.3.

4.2.1 Spectral Line Width

Basically four features are essential in identifying electric dipole transitions in the manifold of quantum states via steady-state optical spectroscopy: transition energy, spectral width, and (transition) probability, including polarization. Energies are, in principle, well defined for a given quantum system. However, since we concentrate in this book on semiconductor QDs embedded in a condensed phase, influences of the matrix but also intrinsic material properties, including defects, impurities, and the strongly heterogeneous surface, influence the Eigen energies of the respective quantum system noticeably. Even if a QD is crystalline, interactions within the crystal will vary among different positions due to strain and defects. Moreover, there may be depending on the system many internal degrees of freedom such as a manifold of electronic energies, phonons, and vibrational modes. As a consequence, optical spectroscopy on quantum systems in the condensed phase is most often faced with a broad distribution of parameters. It is a key issue of optical spectroscopy to unravel as many contributions as possible or at least as necessary. However, ensembles of colloidal QDs—which are in the focus of this book—add further constraints to an optimal spectral resolution, for example, due to the remaining size distribution, variations in crystal structure, and surface morphology, as has already been outlined in detail in Chapters 1 and 2. A decrease of spectral width is always obtained at low temperatures when most of the dynamic processes (besides tunneling) are frozen in. This will decrease the spectral width also for single QDs dramatically [Cui et al., 2014].

Despite spectrally narrow lasers and detection sources being nowadays easily at hand, spectral resolution is still intrinsically limited in the case of a broad distribution of all contributing energies. Whether these distributions can be unraveled experimentally depends critically on the intrinsic spectral width of a specific transition with respect to the energies which we apt to separate from each other.

The (spectral) line width is constituted by two basically different contributions, homogeneous and inhomogeneous. The homogeneous part is given by a characteristic time T_2, which is in a two-level system in principle described by an energy dissipation time T_1 and a dephasing term T^*_2 according to $1/T_2 = 1/2T_1 + 1/T^*_2$. The homogeneous contribution to an apparent line width cannot be circumvented, since it is the manifestation of the Heisenberg uncertainty principle. T_1 and T^*_2 depend specifically on the quantum system and the inherent interactions. A major contribution to T_1 is the lifetime (including, for example, energy transfer processes) of the excited state, while T^*_2 might be controlled by, for example, electron–phonon interactions. The shape of a homogeneous transition is characterized by a Lorentzian line. The theoretical treatment of the homogeneous part is most conveniently followed by the Feynman–Vernon–Hellwarth (FVH) two-level model [Feynman et al., 1957] resembling a spin S = ½ quantum system. The terms T_1 and T^*_2 can be separated by a set of coherent spectroscopic techniques, which is, however, beyond the scope of this contribution.

In almost all cases of spectroscopy in condensed matter the spectral resolution for an ensemble of quantum systems is not limited by the homogeneous width but the inhomogeneous one. Fortunately, this can be circumvented via varies adequate techniques such as double-resonance techniques or effect modulation. Here, we will concentrate on those approaches, which are the basis for the spectroscopic detection of single QDs.

To proceed, we remember that the first steps to such an approach go back to fluorescence line narrowing and spectroscopic site selection techniques on molecules, which rapidly developed following the availability of spectral narrow laser excitation. A major step to drive spectral resolution as far as the homogeneous width was the implementation of "hole burning" techniques, especially in the case of molecules [Kharlamov et al., 1974; Gorokhovskii et al., 1974]. By burning (photophysically or photochemically) a spectral hole into an inhomogeneous line width (of an ensemble) of molecules, the homogeneous line width became at temperatures below a few Kelvin as narrow as that given by the inverse of the luminescence decay time of a few nanoseconds. Experimentally such a narrow width was achieved by irradiating the sample for a while at a fixed energy with a spectrally very narrow laser. Then the excitation intensity was reduced and the narrow band laser was scanned across the absorption, while the fluorescence was monitored at very moderate spectral resolution. In this kind of excitation spectroscopy burnt holes appear on the background of the otherwise inhomogeneous emission. Though hole widths were as narrow as the expected line width of a single molecule the width was constituted by all molecules in the sample which have—within the homogeneous width—identical transition energy. Moreover, at elevated temperatures dephasing times (T^*_2) decrease rapidly, thus burring all the achieved spectral resolution.

Nevertheless, low-temperature hole-burning experiments (on organic molecules) can be considered as one of the most important stimuli to approach finally the detection of a single molecule (or QD), namely to detect only the holes without burning and the inhomogeneous background. To succeed, it was necessary to reduce the number of molecules to only one within the interaction volume with the scanning excitation laser beam at selected excitation energies.

In 1989 Moerner and Kador [1989] were the first to achieve via a complex technique the absorption of a single pentacene molecule in a *p*-terphenyl molecular crystal at low temperature. More closely related to the already well-established hole-burning technique at that time was the approach of Orrit and Bernard [1990]. They also reduced the concentration of emitters down to femtomol and detected the absorption via broadband fluorescence excitation spectroscopy (see Fig. 4.1). This approach can be considered as one important breakthrough on the path to single quantum object detection and the following Nobel Prize awarded in 2014 for optical microscopy below the classical resolution limit defined by Abbe in 1873.

At the applied low temperatures extremely small lifetime-limited line widths have been obtained. At that time several approaches revealed by quantum mechanical means the true nature of a single quantum object, for example, photon bunching [Orrit and Bernard, 1990], antibunching, and photon correlation [Orrit and Bernard, 1990]. Photon bunching occurs in case dark states of any kind are involved in the excitation cycle, which is explored by the excited quantum object (molecule) many times. Progress of single-molecule detection, which is in some respect the “delivery nurse” of single-QD detection, has been published in several monographs and reviews, for example, Moerner et al. (2008), Sauer et al. (2011), Makarov (2015), and Rigler et al. (2012).

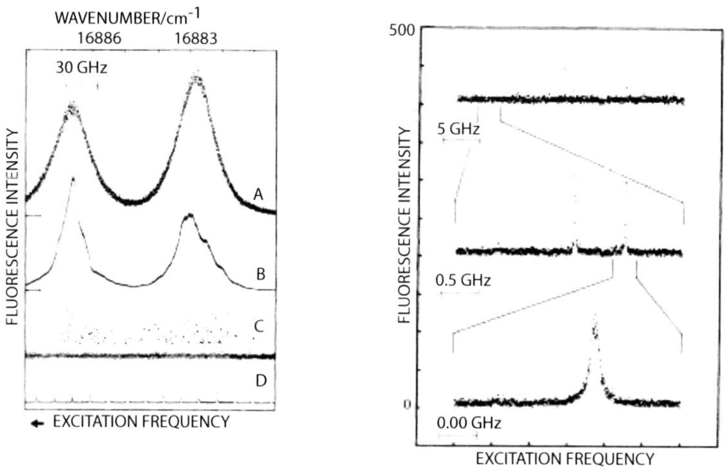

Figure 4.1 Fluorescence excitation spectra of pentacene (P) doped at various concentrations into *p*-terphenyl molecular crystals (1.5 K). Concentration of P is decreased from A to C (left) and (right). Finally a homogeneous Lorentzian line shape is obtained with a lifetime-limited width of 12 MHz. Reprinted (figure) with permission from [Orrit, M., and Bernard, J. (1990). Single pentacene molecules detected by fluorescence excitation in a p-terphenyl crystal, *Phys. Rev. Lett.*, **65**, 2716–2719.] Copyright (1990) by the American Physical Society.

This early approach of spectral selection of single emitters via spectral hole burning is limited to temperatures below a few Kelvin since spectral widths and thus selectivity are lost with increasing dephasing processes already at only moderately increased temperatures. However, we have already learned that the reduction of emitter concentration is an inevitable prerequisite of single-molecule (and single-QD) detection. We will come to this point in the experimental sections and will show that the combination with microscopic techniques resulted in a breakthrough of single quantum object detection. Before that we will spend some time considering aspects of sensitivity, which is a special demand in the case of single quantum object detection.

4.2.2 Optical Sensitivity

In the development of single quantum object detection in condensed matter the driving impetus from a historical point of view was the realizing of the ultimate line width. Naturally, approaching the single quantum object limit sensitivity matters has been the new crucial issue. A successful optical excitation cycle is shown in Fig. 4.2. Basically at least three electronic levels have to be considered: Population from the ground state | 1> is transferred resonant or nonresonant to the spectator state | 2> ("bright" state), from which emitted photons are detected. A competitive state is the "dark" state | 3>. In the first fluorescence-detected single-molecule experiments a narrow-band laser has been scanned across the resonant absorption range| 1>↔| 2> of the individual molecules with slightly different absorption energies [Orrit and Bernard, 1990], followed by a broadband detection of photons emitted at lower energies than the resonant emission to vibronically coupled states. Narrow-band laser excitation allows for a high excitation rate. Thus the classical ensemble average is replaced by a time average: photons (up to 10^6/s) are collected from just one molecule.

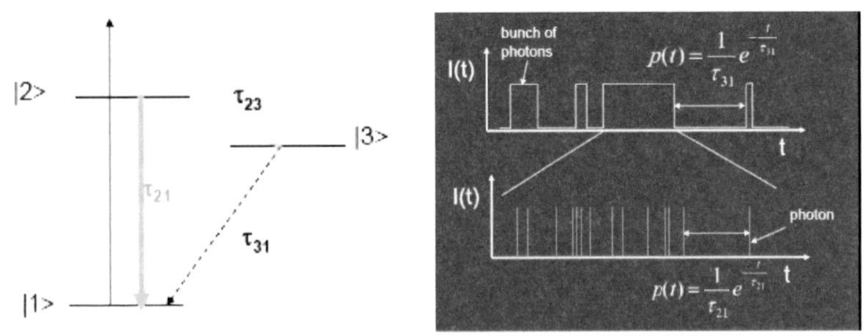

Figure 4.2 Left: Principal-level scheme for detection of single quantum objects via luminescence detection of the |2>→|1> optical transition. |3> is a dark-state interrupting emission (photon bunching). Right: Scheme of bunching, including the exponentially distributed probability of bunch separations according to dynamics in state | 3> (top) and the distribution of photons emitted from state | 2> (bottom).

4.2.3 Specific Nature of Single Quantum Objects

In optical single quantum object experiments emitted photons can be clearly related to each other with respect to their relative arrival times. This offers new possibilities for the analysis of interphoton arrival probabilities with respect to photoexcitation, as has been exploited in many experiments and is described in several recent publications [Verberk and Orrit, 2006; Lippitz et al., 2005; Cui et al., 2014]. In this chapter we will concentrate on phenomena like photon bunching, photon antibunching, and photon correlation.

4.2.3.1 Photon bunching

As we have shown in Fig. 4.2 (left) a single quantum object is driven by the exciting laser between ground and excited states, while the emitted photons are collected. In almost all cases photons are emitted from the lowest state(s). In case the population of dark states (e.g., |3> in Fig. 4.2) is competing with the population of the spectator state |2>, no photons are emitted (at the selected) detection wavelength during the time the quantum object is in that dark state. As a consequence, photons are emitted in bunches separated by a time typical for the residence (life) time in that dark state. Photon bunching is independent of the nature of the dark state as long as the timescale of this state is comparable or longer than that of the spectator state. Figure 4.2 (right) shows the bunching scheme with the involved time decay distribution in states |2> and |3>, respectively.

The general case of a dark state is for organic molecules the intrinsic (paramagnetic) electric dipole forbidden triplet state. Since the involved interruption of photon emission is also a typical fingerprint of QD luminescence, we show in Fig. 4.3 a typical experimental example of photon bunching for a terrylene dye molecule in a *p*-terphenyl molecular crystal at low temperatures [Vogel et al., 1995]. A detailed analysis, including the influence of magnetic resonance transitions (in the dark triplet state [Wrachtrup et al., 1993]), has been reported by Brown et al. (1994). The phenomenon of photon bunching has been termed "quantum jumps," "luminescence intermittency," or simply "blinking." Blinking can only be observed in the case of a single (or very few) quantum object(s). With only a few exceptions all up to now investigated single quantum objects show blinking effects in condensed matter environments [Cichos et al., 2005]. This can be used to investigate either intrinsic electronic states, which are in principle dark, or (slow) fluctuations of electronic-state properties. The latter are often caused by conformational changes of the quantum object or the embedding matrix. Such access to subtle properties is not possible by experiments on ensembles of quantum objects. We will discuss the blinking approach in more detail in Section 4.4.1 and intensively in Chapter 6.

Photon bunching is a first but not definite proof that we are dealing with a single quantum object. Electronically strongly coupled quantum objects might also show bunching effects. Moreover, in case not clear "on" and "off" intensities are observed this might indicate the excitation (observation) of several quantum systems.

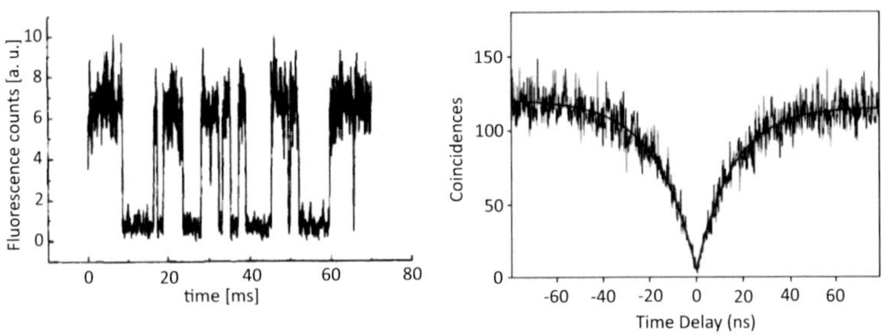

Figure 4.3 Left: Fluorescence counts (intensity) of a single terrylene molecule in *p*-terphenyl at 1.5 K as a function of observation time. Quantum jumps are directly visible as an interruption of the fluorescence. (A part of a trace of 200 s length is presented.) These interruptions are due to the population of the dark triplet state. Reprinted with permission from Vogel, M., Gruber, A., Wrachtup, J., et al. (1995). Determination of intersystem crossing parameters via observation of quantum jumps on single molecules, *J. Phys. Chem.*, **99**, 14915–14917. Copyright (1995) American Chemical Society. Right: Histogram of time delays between consecutive photon pairs (which is proportional to the intensity correlation function) detected from the fluorescence of a single CdSe QD. The histogram shows a dip at zero time delay, which is a clear signature of photon antibunching. The solid line represents a single exponential with a rise time of 16 ns, which corresponds to the PL decay time of the QD. Reprinted from Lounis, B., Bechtel, H., Gerion, D., Alivisatos, P., and Moerner, W. E. (2000). Photon antibunching in single CdSe/ZnS quantum dot fluorescence, *Chem. Phys. Lett.*, **329**, 399–404, Copyright (2000), with permission from Elsevier.

4.2.3.2 Photon correlation

Photon correlation analyzes the time spacing between pairs of arriving photons in the form of the intensity correlation function $g^{(2)}$ (t) for intensities $I(t)$ according to

$$g^{(2)}(\tau)=\frac{<I(t)I(t+\tau)>}{<I(t)><I(t+\tau)>} \tag{4.1}$$

where τ is a temporal delay. If treated as analog signals $g^{(2)}(t)$ gives the probability that $I(t)$ has changed after the delay time τ. In case $I(t)$ is described as a digital (single-photon) signal, $g^{(2)}(t)$ is related to the probability for a photon to be detected after $(t + \tau)$ with respect to the photon detected at t. In the case of Poisson statistics we arrive at $g^{(2)}(t) = 1$. Deviations from 1 indicate that photons are not independent from each another but are related by some physical mechanism such as luminescence decay of a single emitter.

Intensity correlation analysis has been successfully applied to the detection schemes of single molecules in the case that only a few blinking processes are present (such as those caused by the dark triplet state), which can altogether be described by a few exponential functions [Orrit and Bernard, 1990; Brown et al., 1994; Vogel et al., 1995]. However, in the case of single QDs, blinking dynamics (on long timescales) follow complex functions such as power laws and an analysis is extremely irreproducible since blinking undergoes often changes on long timescales [Messin et al., 2001], though a few successful reports on correlation analysis have been reported also for QDs [Verberk and Orrit, 2003].

4.2.3.3 Photon antibunching

While photon correlation failed in most cases of analysis of QDs, there is one exception for a two-level quantum emitter (molecule), namely in the case that $\tau \rightarrow 0$. In that case we arrive at $g^{(2)} = 0$ [Lounis et al., 2000; Lounis and Orrit, 2005]. The consequence of this is that detecting two photons simultaneously vanishes. This effect is called antibunching. Using the two detectors in a cross-correlation setup a photon will only show up in one of the detectors for $\tau \rightarrow 0$. Under continuous-wave excitation antibunching results in a dip in the cross-correlation of the two detectors. Corresponding experiments revealing photon antibunching are considered as a definite proof of the presence of a single emitter. A typical example of antibunching is shown in Fig. 4.3 (right).

4.2.3.4 Ergodicity

In the previous sections we have outlined some specific features like bunching, antibunching, and photon correlation inherent to single quantum object detection. It became obvious that antibunching is a feature clearly demonstrating the presence of only one quantum object. However, is this also true for photon bunching? Naturally, the presence of this cannot be detected via photon correlation experiments ($g^{(2)}(t) = 1$ for an ensemble of emitting quantum objects) since photon emission from several emitters is not correlated. Nevertheless, there are experiments on ensembles of emitters which closely resemble photon-bunching experiments. In the field of molecular spectroscopy the presence of a dark triplet state shows up as the so-called *fluorescence fading* (see Fischer et al. [1983] and references therein). Practically all those features are covered by this approach which can be determined from the intensity autocorrelation function at long times, including excitation power and magnetic resonance transitions [Brown et al., 1994]. Both blinking and photobleaching show up as a decrease of the ensemble luminescence during the time after turning excitation "on," as has been shown for Si and CdSe QDs [Cichos et al., 2005; Chung and Bawendi, 2004]. It could be shown that there is a close mapping between QD blinking of single QDs and photobleaching. However, the manifestation of such bleaching effects becomes only visible at reduced emitter concentrations in small sample volumes, a reason why these effects have escaped clear identification prior to single quantum object experiments. In case the excitation is turned "off" again, luminescence recovers (partly) in the dark. Such kinds of experiments allow for a convenient discrimination between reversible (blinking-type) and irreversible (permanent photobleaching) processes [Schuster et al., 2007].

These findings might suggest that single quantum object and ensemble experiments are at least for some of the observables identical (under the same external conditions). However, one has to realize that single quantum object detection is based on time-averaging over many excitation cycles contrary to averaging features of an ensemble. In other words, the equivalence of an ensemble and time average is mandatory which is fulfilled if ergodicity holds. Naturally, ensemble averaging is a superposition of an inhomogeneous distribution of properties of many objects. However, the same conditions could in principle be realized by summing up results from many single quantum object experiments, which shows practically in many cases a close relationship. However, identical results can only be expected in the case that ergodicity holds strongly. One reason for a possible failure of such a comparison is based on the fact that during a measurement, properties might irreversibly change, which can be related, for example, to aging processes.

(Weak) ergodicity breaking has been predicted [Stefani et al., 2009] and identified [Kuno et al., 2003; Margolin et al., 2006] in the case of single QDs. Though extreme care has to be taken when comparing single- and ensemble-QD experiments, at least qualitatively, in many practical cases a qualitative equivalence can be found from these at first glance very different experiments. In fact, transferring concepts from findings for single- to ensemble-QD experiments allows a conceptional understanding of (inhomogeneously distributed) features found in ensemble experiments [Zenkevich et al., 2015]. Naturally, ensemble experiments are a convolution of inhomogeneities among individual QDs and during observation of time-varying properties of a single QD (causing blinking). On the other hand, not all QDs might be identified in single-QD experiments, since only weakly emitting ones often escape identification, though they would (weakly) contribute in an ensemble experiment. In any case, a combination of ensemble- and single-QD experiments enables the most accurate determination of QD properties.

4.3 Experimental Concepts

In the following section we will discuss special experimental aspects with respect to single-QD detection. The majority of experiments are nowadays performed making use of a reduction of the detection volume which is doped with ultralow (ato- to femtomole) concentrations of QDs. Consequently, QDs are incorporated into thin films or deposited onto surfaces. Photoluminescence (PL) is detected via microscope setups or fluorescence correlation spectroscopy (FCS). The latter technique is the method of choice when investigating QDs in liquids. The detection volume is in this case determined by the confocal volume of the exciting laser, which is limited in plane by the classical optical resolution limit of a few hundred nanometers and about 1 μm in the perpendicular direction. During the last decade spatial resolution has been scaled down below the classical diffraction limit, which has been acknowledged by awarding the Nobel Prize in Chemistry in 2014 to Betzig, Hell, and Moerner [Betzig et al., 2006; Rittweger et al., 2009; Moerner, 2012]. These achievements are to a large extent based on single-quantum-object techniques.

A different approach to increasing optical resolution beyond the diffraction limit is based on scanning near-field optical microscopy (SNOM) or evanescent optical fields [Synge, 1928; Pohl et al., 1984; Lewis et al., 2003; Heinzelmann et al., 2011]. In principle this approach is based on the decrease in the aperture in the near field. The principal schemes of such detection principles are shown in Fig. 4.4. It has the advantage that besides SNOM, manipulation of the quantum objects, for example, by applying external fields to a cantilever, becomes easily possible. However, its application is limited to basically quantum objects on a surface.

Many more advanced optical techniques have been reported to increase both sensitivity and spatial resolution [Weisenburger and Sandoghdar, 2015], including surface plasmons, optical cavities and waveguides [Henessy et al., 2007], and thermal imaging [Naccache et al., 2010]. Spectral resolution could be considerably enhanced by photon correlation Fourier spectroscopy [Cui et al., 2014].

An interesting setup to directly detect charge fluctuations in QDs has recently been reported by Zbydniewski et al. (2015). This setup consists of a carbon nanotube field-effect transistor (FET), which directly monitors charge fluctuations and is associated with single trap charging, which exhibits power-law behavior. The charging energy is larger than 200 meV. Krauss and Brus (1999) have implemented an electric force microscope (FM) to investigate via scanning microscopy charging of single QDs in combination with laser excitation [Krauss et al., 2001].

Before entering the general description of the principal detection schemes of single-QD detection, we would like to make a few remarks with respect to sample preparation. As already addressed, very low concentrations and small detection volumes are necessary to reach the goal of single-QD detection. To proceed, in most cases QDs are diluted at very low concentrations in appropriate solvents and then

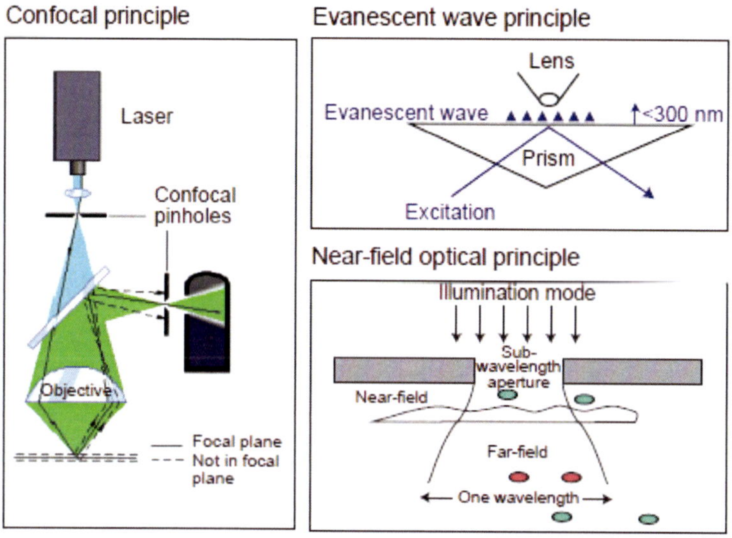

Figure 4.4 Comparison of microscopic schemes in the wide field (left) via evanescent waves (top right) and near field (bottom right). Instead of an illuminated aperture, nanosized optical waveguides or other types of cantilevers may be used. Reprinted by permission from Macmillan Publishers Ltd: [*Nature Biotechnology*] (Lewis, A., Taha, H., Strinkovski, A., Manevitch, A., Khatchatouriants, A., Dekhter, R., and Ammann, E. (2003). Near-field optics: from subwavelength illumination to nanometric shadowing, *Nat. Biotechnol.*, **21**, 1378–1386), copyright (2003)

spin-coated onto substrates such as quartz plates, mica, or Si-SiO_x substrates. The SiO_x layer must be thick enough (~100 nm) to avoid excitation energy transfer to the underlying Si wafer. Substrates and solvents have to be extremely clean to avoid spurious luminescence of any kind. Also all materials used in the preparation process have to be carefully cleaned. The solvent evaporates after spin coating, leaving QDs spatially well separated (>optical resolution) on the substrate. In case QDs are diluted, for example, in polymers, a certain amount of polymeric material has to be added to the solution to be spin-coated. The thickness of finally obtained films is in the order of 10–100 nm.

In the following section, we will concentrate on classical microscopy-based techniques. This description follows closely the work of Kowerko (2010).

4.3.1 Wide-Field Microscopy

Wide-field microscopy is used in case several tens to hundred QDs should be investigated simultaneously. A typical wide-field microscope is shown in Fig. 4.5. Excitation is achieved via lasers (e.g., argon ion, dye, or diode lasers), which are coupled into a single-mode polarization-maintaining fiber. An electro-optic modulator (EOM) is used to control and stabilize the excitation intensity. In the given example the sample is scanned below the objective to illuminate sample areas of typically 10 × 10 μm^2. PL is detected via the same objective with a high numerical aperture (NA). A low-pass filter is needed to separate laser light from the PL. Finally the PL is focused onto a charge-coupled device (CCD) camera.

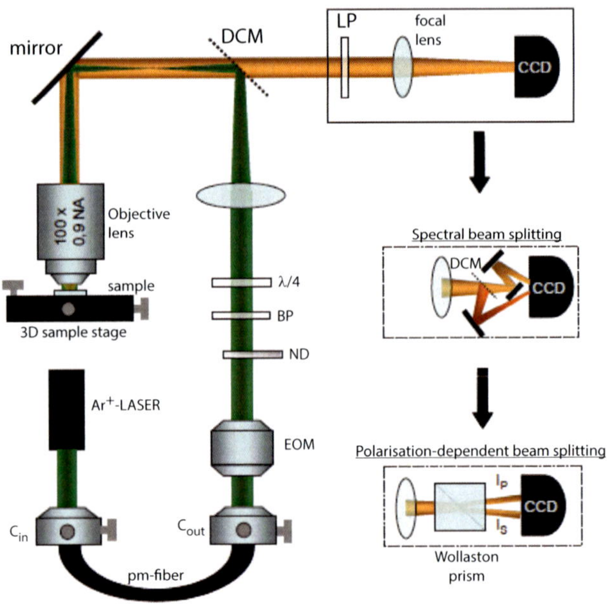

Figure 4.5 Scheme of an epifluorescence wide-field microscope. C, fiber coupler; EOM, electro-optic modulator; ND, neutral density filter (allows for continuous adjustment of excitation intensity); BP, band-pass filter (transmitting only the excitation wavelength); $\lambda/4$, quarter waveplate; DCM, dichroic mirror (beam splitter); LP, low-pass filter. Two alternative detection schemes to investigate optical polarization and/or to achieve spectral resolution are shown on the right. From Kowerko, D. (2010). *Dynamic Processes in Functionalised Perylene Bisimide Molecules, Semiconductor Nanocrystals and Assemblies*, PhD Thesis, TU Chemnitz, Germany, with kind permission from Danny Kowerko.

The most essential parts in a wide-field microscope with respect to single-emitter detection are the objective and the detector. A 2D detector with low background noise, high quantum efficiency, and fast read-out mode is mandatory. CCD cameras specialized for single-molecule applications meet all these demands. Peltier-cooled CCD cameras have frame rates up to 20 frames per second (fps) for an image size of 20 × 20 μm^2. An appropriate objective requires a high NA to enable an efficient collection of photons in parallel with a diffraction-limited resolution. Oil immersion lenses fulfill these conditions and additionally allow the detection of light emitted into angles diverging from the optical axis determination of 3D orientation of the optical transition dipole via defocused imaging techniques [Boehmer et al., 2003 Schuster et al., 2005]. This is of advantage since wide-field microscopes detect under conventional conditions only the in-plane orientation of the optical transition moment.

To achieve a homogeneous excitation the linearly polarized laser light is converted to circular polarized light by a quarter waveplate. Wide-field illumination of the sample is realized when the laser beam enters the objective divergently, which is achieved by an achromatic lens (with a typical focal length of f = 45 mm) at a distance d (with $f < d < 2f$).

A Wollaston prism or a dichroic beam splitter can be placed behind the last focal lens (see Fig. 4.5, right) in order to realize spectral resolution or to detect the two components of the polarization. A typical wide-field image is shown in Fig. 4.6.

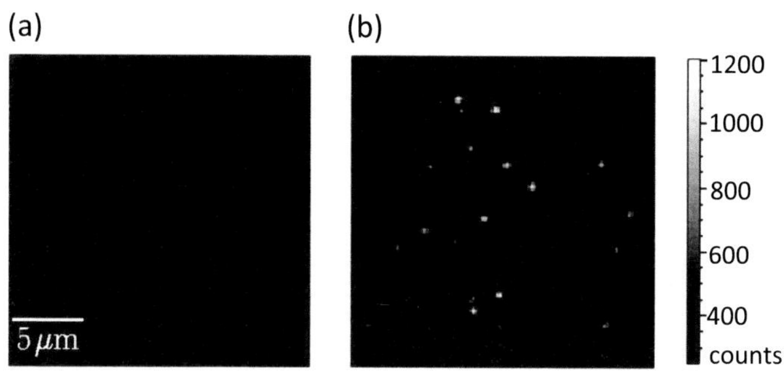

Figure 4.6 Wide-field-detected PL emission of CdSe/ZnS QDs on a Si-SiO_x substrate: (a) substrate only and (b) PL emission of QDs. From Krasselt, C. (2009). *Der Einfluss von Grenzflächen auf photoinduzierte Ladungslokalisierung,* Diploma Thesis, TU Chemnitz, Germany, with kind permission from Cornelius Krasselt.

4.3.2 Confocal Microscopy

In a confocal detection scheme the laser focus is scanned across the sample, while the PL is only detected from the confocal spot. Figure 4.7 shows a confocal microscope which allows for a multiparameter single-emitter detection. The confocal concept is realized via an intermediate image (via two lenses with focal lengths 12.5 cm and 25 cm) at the same position both for the exciting and for the emitted light. The shown setup allows for a simultaneous measurement of PL spectra, PL intensity, and PL lifetime. For excitation a pulsed laser diode (75 ps pulse width with a repetition rate up to 80 MHz at a wavelength of 465 nm) has been used. The confocal microscope depicted schematically in Fig. 4.7 uses a beam-scanning instead of a sample-scanning scheme. This allows for faster scanning and can also be used, for example, under conditions for which scanning of the sample is not easily achieved (e.g., in a cryostat). The laser beam is scanned in the sample plane by a mirror mounted on a piezo scanner, whereas scanning perpendicular to the plane is realized by a galvano scanner.

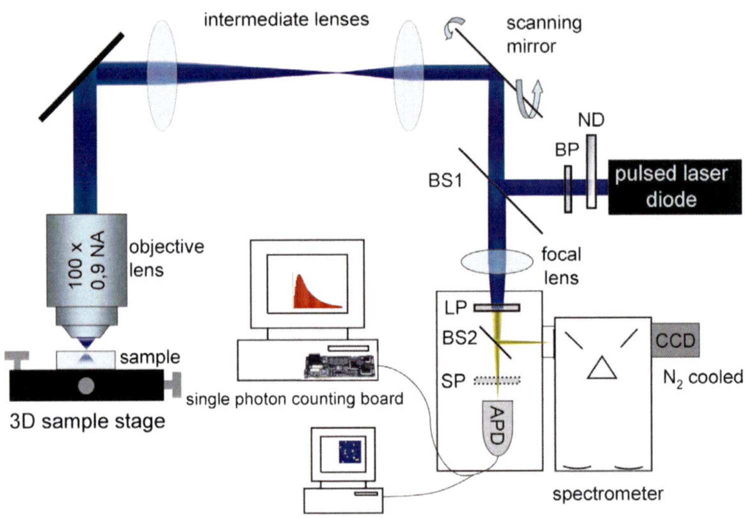

Figure 4.7 Scheme of a confocal microscope. ND, neutral density filter; BP, band-pass filter; BS, neutral beam splitter; LP, long-pass filter; SP, short-pass filter. From Kowerko, D. (2010). *Dynamic Processes in Functionalised Perylene Bisimide Molecules, Semiconductor Nanocrystals and Assemblies,* PhD Thesis, TU Chemnitz, Germany, with kind permission from Danny Kowerko.

Basic optical components are similar to those shown in the wide-field setup (Fig. 4.5). Contrary to a dichroic beam splitter, the neutral beam (90:10) splitter allows for PL excitation spectroscopy without realignment of the variable excitation wavelengths.

The emitted light is separated by a neutral beam splitter (3:1). This allows for a spectral and time-resolving detection channel in parallel. One part is focused by an achromatic lens to the entrance slit of a monochromator equipped with a CCD at the exit and the other one to a single-photon-counting avalanche photodiode (APD). Sequences of single-emitter spectra are typically recorded with a time resolution of seconds to several hundreds of milliseconds.

In the second part slow (>1 ms; macrotimescale) and fast (<1 μs; microtimescale) luminescence dynamics can be detected. The APD signal is split to a single-photon-counting board (time resolution) and to an integration unit. The initial TTL signal (in the fast channel) has to be converted to the fast signal processing standard NIM by a time-correlated single-photon-counting (TCSPC) router. While the excitation pulse width allows a TCSPC time resolution down to 75 ps, the jitter of the APD reduces time resolution to about 700 ps. Eventually, filters have to be used in the detection path to separate different spectrally selectable emitters. The setup allows for time tagging of each single photon with its microtime (arrival time related to the excitation pulse) and its macrotime (arrival time related to the start of the measurement). Such a time-tagged time-correlated single-photon-counting (tt-TCSPC) mode permits the simultaneous detection of PL intensity and PL lifetime on long experimental timescales. Furthermore, autocorrelations and interphoton statistics derived from tt-TCSPC provide insight into dynamics of single emitters [Xu et al., 2007]. Time binning allows a correlation with spectral features even after the experiment.

Figure 4.8 demonstrates the identification of three different spectroscopic emitters via a confocal microscope. Part (a) shows a resolution limited (5 × 5 μm) scan of the (integrated luminescence intensity (100 ms binning [integration] time). Note that some of the spots show intermittent intensities, which will be discussed in Section 4.4.1. Parts (b), (c), and (d) show examples of the recorded luminescence spectra of a dye molecule (perylene diimide [PP]), a CdSe/ZnS QD and a self-assembled PP/QD nanoassembly, respectively. Note that part (d) reveals a superposition of similar spectra, as shown in parts (b) and (c).

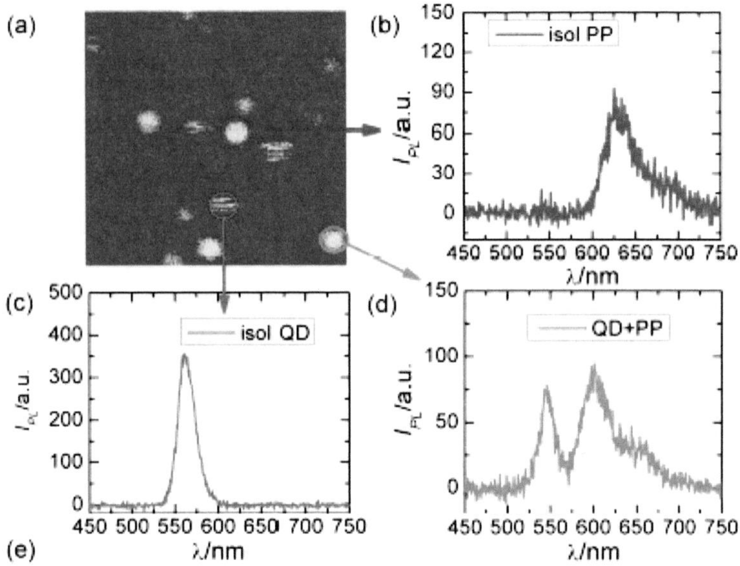

Figure 4.8 Confocal 5 × 5 μm^2 scan (a) of the luminescence intensities of three different emitters. The arrows point to the related spectra of a single perylene diimide (PP) dye molecule (b), a single CdSe/ZnS QD (c), and a self-assembled PP+QD aggregate (d). The initial relative concentration of the two compounds was 1 before spin-coating a toluene solution onto a SiO_2 substrate. Note the intensity intermittence in some of the confocal images, which is due to luminescence intermittency (see also Section 4.4.1). Integration (binning time) of the spectra is 1 s. The spectral resolution is 5 nm. From Kowerko, D. (2010). *Dynamic Processes in Functionalised Perylene Bisimide Molecules, Semiconductor Nanocrystals and Assemblies,* PhD Thesis, TU Chemnitz, Germany, with kind permission from Danny Kowerko.

4.3.3 Data Analysis

In this section (which has been written by D. Kowerko), we will discuss the challenges of data analysis in single-molecule fluorescence-based methods [Kowerko, 2010]. It is noteworthy to mention that comprehensive evaluation software packages are commercially not available or of limited functionality with respect to the demands single-molecule methods usually are confronted with. Since data analysis is complex and often requires a sequence of (data) processing steps, often home-built solutions are developed via programming and script languages like C++, Matlab (Mathworks), Python (open source), Igor Pro (Wavemetrics), and OriginPro (Originlab) but unfortunately often are not published or are accessible only on demand.

The basic difference with bulk methods is often the large number of datasets generally stemming from collecting several tens to hundreds or even thousands of molecules/datasets. Some problems are common to both, wide-field and confocal microscopy, while some are method specific and different. We give in the following paragraphs some typical examples of the latter.

In wide-field fluorescence, intensity time trace generation requires the summation over several pixels around the localized molecule of interest through the whole sequence. Image transformation and co-localization [Comeau, 2006] of subimages are necessary in the case of two or multichannel applications, like shown in Fig. 4.7. Techniques like defocused imaging necessitate subimage pattern analysis, freely available as the Matlab package developed by the group of Enderlein (Uji-i2006). Data acquisition of tt-TCSPC data in confocal microscopy requires the definition of time binning, which must not necessarily be fixed. Along with the fluorescence intensity time trace generation, fluorescence lifetime histograms have to be built for each time bin and deconvoluted from the excitation pulse. The deconvolution together with fast (multi) exponential fitting of fluorescence decay has also been published by the group of Enderlein and is available as a Matlab function on demand [Enderlein and Erdmann, 1997].

Some typical questions and problems encountered in both microscopic methods are:

- Especially for methods that are linked with large data sizes classically arising from videos and tt-TCSPC data, the data are stored in binary format for the sake of memory and computation speed. Hence, special import filters reading instrument-specific binary, ASCII, or image data are required. C++-based scripts are often used along with Matlab-based functions and are available on demand.
- Other challenges in both methods are single-molecule detection/localization, signal correction, for example, for camera offset, unwanted sample background, spectral bleed-through, filter transmission/absorption, and direct

excitation of acceptor. Since single-molecule methods often operate at small signal-to-noise values, denoising of the intensity time trace can help to improve data evaluation by changing the binning size, wavelet analysis, sliding average, or nonlinear filters [Taylor, 2010, 2011; Haran, 2004]. To determine kinetic rates from intensity time traces, they need to be discretisized followed by a (cumulative) histograms analysis. The discretization of such traces by hidden Markov models and available software tools has been reviewed by Walther [Blanco and Walter, 2010]. Accuracy and artifacts of threshold methods are discussed in Greenfield (2012) and Amecke et al. (2014). Depending on the underlying model, the dwell time histograms derived after discretization are subsequently fitted to exponential [Zhuang, 2000], gamma [Floyd, 2010], or power-law functions [Krasselt, 2011, Issac et al., 2012]. Nonlinear regression or curve fitting is the standard procedure in (single-molecule) spectroscopy and (tt)-TCSPC-based methods. Contrary to their bulk counterparts, large datasets of hundreds to several thousands have to be quickly processed. Henceforth automation via scripting and programming is necessary. Typical fit equations of multi-Gaussian and multiexponential functions are presented here:

$$I(x) = \sum_{n=1}^{n_{\max}} a_n e^{-(x-b_n)^2 / c_n} + \text{offset} \tag{4.2}$$

$$I(t) = \sum_{n=1}^{n_{\max}} a_n e^{-\frac{t}{\tau_n}} + \text{offset} \tag{4.3}$$

The offset is required to account for specific background noise and arbitrary offset values of light detectors. Important for exact PL lifetime extraction from a total decay time histogram using Eq. 4.3 is the knowledge of the correct offset, in the following equation being referred to as I_0. The latter is determined by the dark count rate I_{dark} (cts/s) of APD multiplied by the observation time t_{obs} divided by the number n_{micro} of time channels in the microtime histogram (Fig. 4.9) according to

$$I_0 = \frac{I_{\text{dak}} \cdot t_{\text{obs}}}{n_{\text{micro}}} \tag{4.4}$$

Alternatively, the stretched exponential function Eq. 4.5 is sometimes applied instead of Eq. 4.3, for example, if $n_{max} \geq 2$ or in case the number of τ_n is distributed rather continuous than discrete.

$$I(t) = A \cdot \exp(-\left(t/\tau\right)^{\beta}) + \text{offset} \tag{4.5}$$

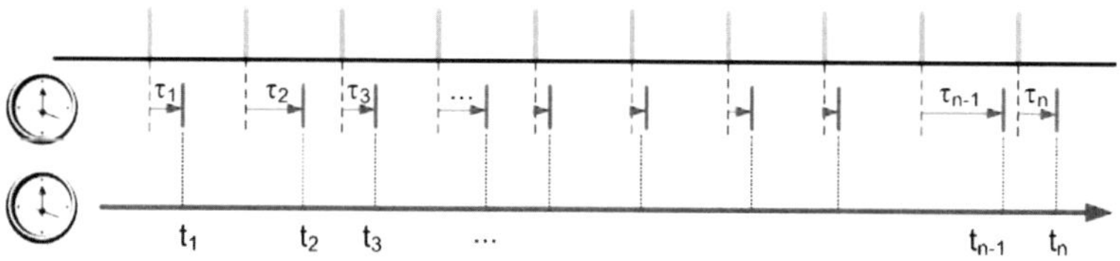

Figure 4.9 Data collection for tt-TCSPC. The excitation pulses act as clock pulses. Emitted photons are labeled twice: (i) time relative to the clock pulse (microtime, τ_j) and (ii) time relative to the start of the experiment (macrotime, t_j). From Blaudeck, T. (2007). *Self-Assembly of Functionalized Porphyrin Molecules on Semiconductor Nanocrystal Surfaces,* PhD Thesis, TU Chemnitz, Germany, with kind permission from Thomas Blaudeck.

Software tools for deconvolution of the experimental PL lifetime histogram from the instrument response function (IRF) and fitting of exponential functions to TCSPC data to derive the PL lifetime/heterogeneity have been published elsewhere [Enderlein and Erdmann, 1997] (www.fluortools.com). A straightforward way of fitting the series of fluorescence spectra by multiple Gaussians is provided by Matlab's graphical user interface-based curve-fitting toolbox. It offers the functionality of fitting sample data and exporting source code necessary and helpful to include it in automated fitting routines.

The advantage of the wide-field technique lies in fast recording of several tens of single emitters. Thus it is a convenient method for applications using single emitters as probes for the local environment. Such studies often require repetition of experiments under variation of a physical parameter, for example, temperature for the examination of polymers close to glass transition. Studying systems of high complexity usually requires a combination of different analytical methods. Here the confocal microscopy technique benefits from its variety of expendabilities allowing synchronous access to the PL properties, such as spectrum, lifetime, intensity, polarization, and dipole orientation. In addition the presented tt-TCSPC method offers a variety of elaborated analysis techniques used to disentangle dynamic processes on a submillisecond timescale, for example, autocorrelation [Verberk and Orrit, 2002] and change point analysis (CPA) [Watkins and Tang, 2005].

4.4 Optical Features of Single Quantum Dots

In the following section we address characteristic features of single QDs and their application in materials science. Especially, we will discuss those aspects which are related to tuning of the relevant optical parameters as they can only be detected by single-QD spectroscopy. More general features of tuning optical properties have been discussed in Chapter 1. The following content is not intended to be a review, since the field is too broad to be covered by a comprehensive section. Moreover, we will exemplarily concentrate on selected classes of QDs, that is, CdSe colloidal QDs and Si nanocrystals (NCs). However, similar features apply to other colloidal QDs [Kuno et al., 2000; Wang et al., 2006; Peterson and Krauss, 2006], quantum rods [Wang et al., 2006; Glennon et al., 2007], Si nanostructures [Bruhn et al., 2011], epitaxially grown quantum structures [Pistol et al., 1999; Sugisaki et al., 2001], and diamond NCs [Bradac et al., 2010].

4.4.1 Luminescence Intermittency

Luminescence intermittency (also often called blinking) is, as has been discussed already in Section 4.2.3.1, typical for many quantum systems [Cichos et al., 2007]. An illustrative snapshot of PL intensities is shown in Fig. 4.10. Blinking has been reported for colloidal QDs for the first time by Nirmal et al. (1996) and has been attributed to be caused by Auger recombination following photoinduced charging of the QD core. Blinking decreases the PL QY and varies for different individual QDs under otherwise the same conditions. The latter indicates a considerable influence of crystal imperfections, inhomogeneous interfaces (including the impact of ligands or surfactants), and/or the local (inhomogeneous) environment [Cichos et al., 2007]. It is difficult to assign whether blinking is also present in ensemble experiments. It has been shown, however, that blinking shows up as (reversible) photobleaching in small ensembles of CdSe/ZnS [Chung and Bawendi, 2004] and Si

QDs [Cichos et al., 2004] or molecules [Schuster et al., 2007]. A close relationship has been pointed out recently [Qin et al., 2013]. Even blinking of QDs in liquid solution has been detected by FCS [Verberk and Orrit, 2003].

In Chapter 6 we will discuss the principal relationship between single- and ensemble-QD spectroscopy since blinking is a widespread phenomenon, which has been observed already two decades ago and has been intensively investigated both experimentally and theoretically. Nevertheless, many aspects are still not yet understood. Several blinking-related review papers have been published [Cichos et al., 2007; Marcus, Stefani et al. 2009; Riley et al., 2012, Cordones and Leone, 2013] also comparing blinking in very different quantum systems. For this reason we will in the following section only shortly outline the basic principles but will discuss in more detail how blinking parameters can be tuned by intrinsic or extrinsic conditions in Chapter 6.

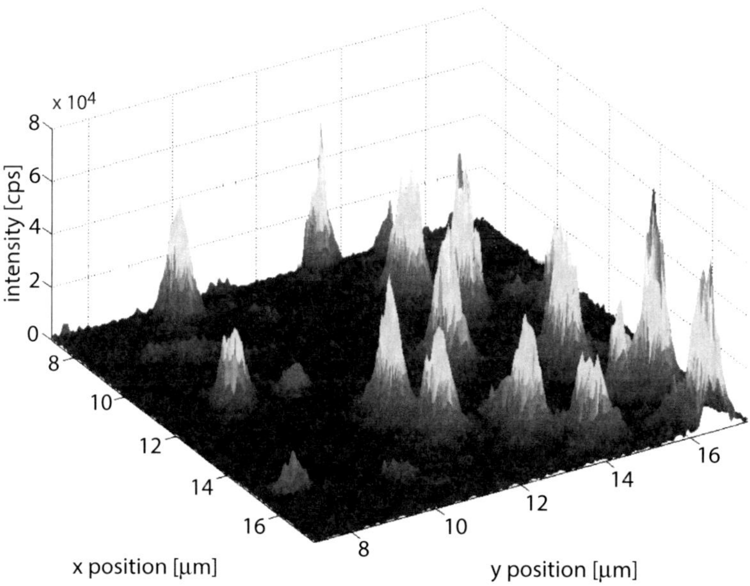

Figure 4.10 Three-dimensional presentation of PL intensities of single CdSe/ZnS QDs on a SiO_2 substrate. Integration time is 1 ms. Appearance of different PL intensities is due to PL blinking. From Schmidt, R. (2010). *Zeitaufgelöste Konfokalmikroskopie an einzelnen CdSe/ZnS-Halbleiternanokristallen,* Diploma Thesis, TU Chemnitz, Germany, with kind permission from Robert Schmidt.

Figure 4.11 shows a typical PL intensity intermittency of a single CdSe/ZnS QD in polystyrene (PS) on two different scales of the time trace. Two observations emerge immediately:

(i) The timescale of fluctuations is between 50 ms (binning time) and tens of seconds. Timescales down to microseconds have been detected by FCS techniques [Verberk et al., 2003].

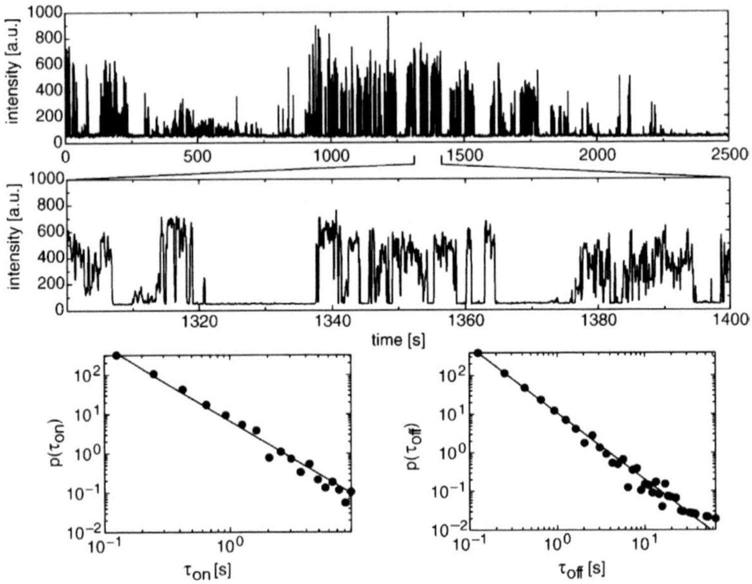

Figure 4.11 Top: Blinking trace of a single QD (CdSe/ZnS in PS) showing random telegraph patterns. Note the self-similarity of the fluctuations by comparing with the magnified trace (middle). Bottom: Distributions of "on" and "off" times follow power laws over, at least, respectively two to three decades in time and probability (see Issac et al., 2005). Direct detection of "on" and "off" times on the PL time trace limits the resolution to about 1 ms. Reprinted from Cichos, F., Borczyskowski, C. von, and Orrit, M. (2007). Power-law intermittency of single emitters, *Curr. Opin. Colloid Interface Sci.*, **12**, 272–284, Copyright (2007), with permission from Elsevier.

(ii) Fluctuations refer to "on", "off," and intermediate ("gray" or "dim") intensities. There are reports in the literature which show only "on" and "off" states. When setting an intensity threshold to discriminate between "on" and "off" intensities somewhat above the noise level, it is possible to measure the distribution of "on" and "off" times, as is shown in the lower part of Fig. 4.11 on double-logarithmic scales. It is obvious that distributions follow apparently a power law over several orders of magnitude in probability and time. However, one has to take into account that when setting a low-intensity threshold "on" times do not discriminate between "on" and "dim" intensities. Moreover, closer inspection of blinking times reveals that deviations from a power law are typical for most cases.

Figure 4.12 shows the influence of three different embedding environments on the blinking characteristics of single Si QDs [El-Kork et al., 2011]. The PL emission is probably due to single defects in the Si/SiO_2 interface [Schmidt et al., 2012]. Besides blinking it seems that (permanent) bleaching also occurs. However, PL recovery experiments show that up to 90% of the apparent bleaching can be recovered at long waiting times in the dark [Chung and Bawendi, 2004; Cichos et al., 2004; Martin et al., 2004]. A strong and immediately obvious influence of the embedding matrix on blinking characteristics as has been reported in the literature by several authors and also for molecules [Schuster et al., 2005] is shown in Fig. 4.12. These and other effects on blinking properties will be discussed after inspection of the principal concepts of blinking analysis.

4.4.1.1 Threshold-based analysis

The assignment of "on" and "off" times is achieved by setting an intensity threshold above the background noise level. A typical example is shown in Fig. 4.13. Since the lengths of bright and dark periods are defined by the choice of the threshold, it is important to check the influence of the threshold level on the observed statistics. The influence of threshold settings and the possibly related pitfalls have been recently described [Crouch et al., 2010; Amecke et al., 2014]. An alternative approach will be discussed in Section 4.4.1.2.

Additionally, from Fig. 4.13 it is evident that the number of photons averaged over an experimentally given integration time (resulting in a time resolution described by a so-called binning time typically in the range of milliseconds to seconds) varies over a broad range, as can be seen in Fig. 4.14, which shows a histogram of intensities from the blinking QD shown in Fig. 4.13. The maximum occurrence at an intensity level of around 55 is originating from the background. The long asymmetric tail is the signature of fluorescence from the QD. Assuming a Gaussian distribution, the contribution from the background reduces to 0 at an intensity level of about 75. In the threshold-based analysis the presence of intermediate PL intensities (dim intensities) is neglected, which can be avoided, as will be shown in Section 4.4.1.2.

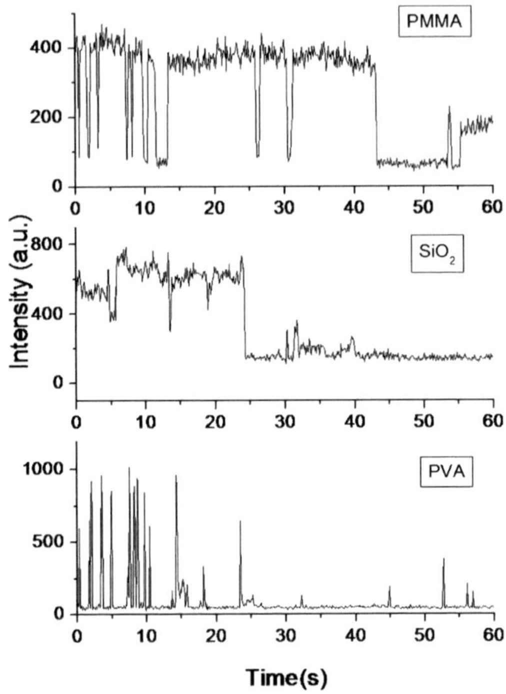

Figure 4.12 Typical intensity time traces of single Si/SiO_2 QD embedded in matrices with different dielectric constants. PMMA, poly(methyl methacrylate): $\varepsilon = 3.4$; SiO_2 (substrate): $\varepsilon = 3.8$; PVA, poly(vinyl alcohol): $\varepsilon = 14$. Reprinted from El-Kork, N., Huisken, F., and Borczyskowski, C. von (2011). Dielectric effects on the optical properties of single silicon nanocrystals, *J. Appl. Phys.*, **110**, 074312, with the permission of AIP Publishing.

In the next step histograms for the duration of bright or dark periods are generated either for a single QD or by summing up histograms for a collection of QDs. A typical histogram for the duration of dark periods is shown Fig. 4.15a both for a collection of 100 QDs and for a single QD (inset).

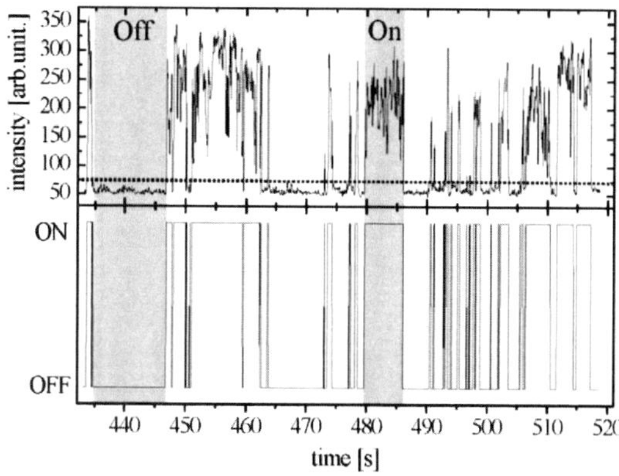

Figure 4.13 Top: Part of a 42-minute time trace which shows emission intermittency. "On" and "off" states are separated by a predefined intensity threshold (dotted line). Bottom: Equivalent binary nature of intermittency with respect to the chosen threshold. QD is either "on" or "off" at a given time. From Issac, A. (2006). *Photoluminescence Intermittency of Semiconductor Quantum Dots in Dielectric Environments,* PhD Thesis, TU Chemnitz, Germany, with kind permission of Abey Issac.

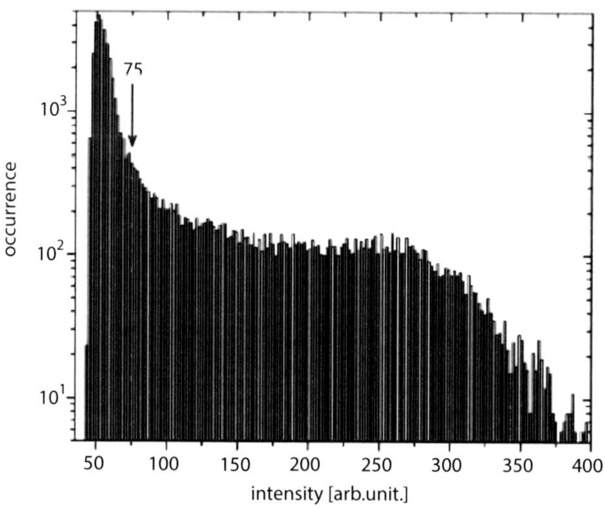

Figure 4.14 Distribution of PL intensities from the single CdSe/ZnS QD shown in Fig. 4.13 within an observation time of 42 min. The arrow represents the intensity level at which contribution to blinking statistics from the background reduces to zero. From Issac, A. (2006). *Photoluminescence Intermittency of Semiconductor Quantum Dots in Dielectric Environments,* PhD Thesis, TU Chemnitz, Germany, with kind permission of Abey Issac.

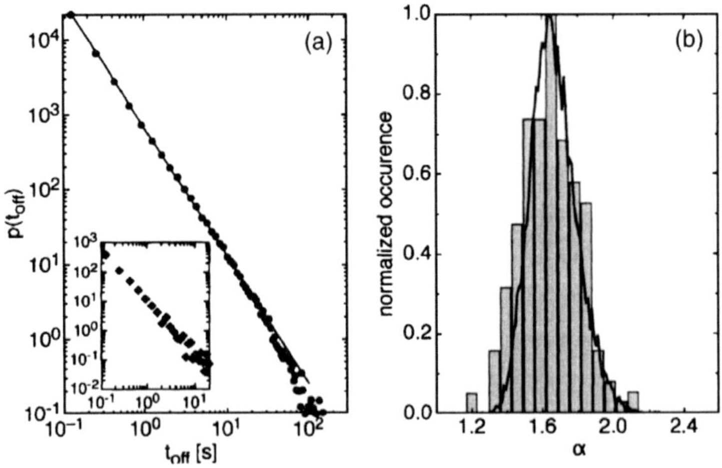

Figure 4.15 (a) Histogram of "off" times of CdSe/ZnS QDs embedded in PMMA. The histogram was constructed from 100 individual PL intensity time traces, measured at 1.3 mW excitation power, each with 5104 frames and 50 ms time resolution. The inset shows the corresponding "off" time statistics of a single CdSe/ZnS QD in PMMA. (b) Histogram of power-law exponents determined from 100 individual PL intensity time traces as well as a simulated power-law exponent distribution for time traces with 1000 "off" time events per trace. Reprinted (figure) with permission from [Issac, A., Borczyskowski, C. von, and Cichos, F. (2005). Correlation between photoluminescence intermittency of CdSe quantum dots and self-trapped states in dielectric media, *Phys. Rev. B*, **71**, 161302.] Copyright (2005) by the American Physical Society.

The line in the double-logarithmic presentation of Fig. 4.15a shows that the "off" time distribution clearly deviates from a single exponential distribution, which would be the characteristic of a single rate process. As has been shown for the first time by Shimizu et al. (2001), "off" time distributions fit well to a power law (Eq. 4.6) over the total range of "off" times, which covers more than 4 orders of magnitude in $P(t)$ and more than 3 orders in time. The "off" time distribution can be characterized by

$$P(t_{\text{off}}) = P_0 t_{\text{off}}^{-\alpha_{\text{off}}} \tag{4.6}$$

with P_0 as a constant and α_{off} as the power-law exponent. A slight deviation from the power-law behavior is observed for very long "off" times. However, one has be careful that the overall length of the time trace is long enough to still provide adequate statistics.

The same procedure of data analysis is performed for "on" times. Naturally, this evaluation is inherently more questionable since "on" intensities are due to the presence of gray states not uniquely defined. Figure 4.16 shows a comparison of "on" and "off" times both for a single QD and for a collection of 100 QDs. It is immediately evident that "on" times deviate from a power-law behavior more pronounced and much earlier in waiting time on the experimental timescale than "off" times do. This is a general observation reported throughout the literature. In the case of "off" times noticeable deviations from a power law are often not observed at all.

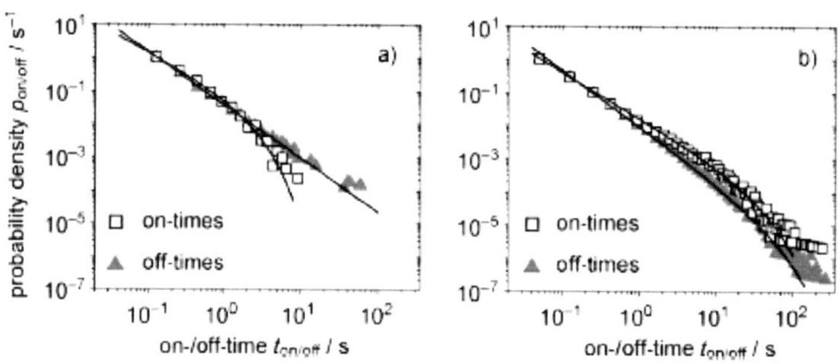

Figure 4.16 Comparison of "on" and "off" time distributions of CdSe/ZnS QDs on a SiO_x substrate for (a) a single QD and (b) 100 assembled individual QDs. Full lines correspond to fits by a truncated power law. Reproduced from Issac, A., Krasselt, C., Cichos, F., and Borczyskowski, C. von (2012). Influence of the dielectric environment on the photoluminescence intermittency of CdSe quantum dots, *ChemPhysChem*, **13**, 3223–3230, with permission from John Wiley & Sons, Inc.

Deviations from a power law can be described by a truncation rate $\Gamma_{on/off}$, resulting in a truncated power law:

$$P(t_{on/off}) \approx t^{-\alpha(on/off)}\exp(\Gamma_{on/off}t) \quad (4.7)$$

It has been generally observed and will be discussed in detail in Chapter 6 that truncation rates are more sensitive to various parameters of the experiment, such as temperature, excitation power, excitation wavelength, capping shell, ligands, and embedding matrix [Issac et al., 2012; Schmidt et al., 2014].

The appearance of a power law implies that transitions from the bright to the dark state and vice versa are not governed by simple exponential kinetics. While this observation may be easily explained for "off" times as being due to a large distribution of after photoexcitation accessible dark states accompanied by the presence of many different back relaxation rates, it is not so readily understood in the case of "on" time statistics. The reason is that even if many accessible dark states exist in parallel the QD state should decay monoexponentially since all decay rates will sum up to one effective rate. Verberk et al. (2002) have suggested that this contradiction can be solved by assuming that reexcitation is not taking place into exactly the same QD state compared to the previous one. This implies that the nature of the complex system of QDs and its environment is changing on slow timescales during continuously driven excitation cycles.

The description by a power law and thus the presence of a broad distribution of states have been ascribed to tunneling processes of electrons (or holes) [Cichos et al., 2007] since already very small variations of the tunneling parameters (distance, energy barrier) will cause considerable differences of relaxation rates. The magnitude of power-law exponents, α, has been found to be $1 < \alpha < 2$ in the majority of experiments with a high probability to be close to 1.5. This apparent universality has stimulated the implementation of universal models [Tang and Marcus, 2005].

With respect to experimental precision it has to be noticed that absolute values of α have to be taken with care since they depend (see above) on many (sometimes unknown) parameters, the kind of analysis (truncated power law or not), and the threshold setting (and thus the signal-to-noise ratio). Because of this it is difficult to compare results from different labs or even different sample charges (though expected to be identical). This implies that parameter variations have to be carefully performed under comparable conditions and optimum sensitivity over many orders in probability and time.

Even deviations from a truncated power law have been observed under optimal conditions, as is shown in Fig. 4.17. Such deviations are also evident in other reported QD data without further notion or even in the case of blinking of single molecules [Schuster et al., 2005]. In the simplest approach such deviations can be accounted for by adding a well-defined and specific exponential relaxation channel (for "on" times). According to [Krasselt et al., 2011]

$$P(t_{\text{on}}) = P_{\text{A}} t^{-\alpha(\text{on})} \exp(\Gamma_{\text{on}} t) + P_{\text{B}} \exp(\gamma_{\text{on}} t) \tag{4.8}$$

describes by the first term a truncated power law superimposed by a second exponential term with decay rate γ_{on}. Such a superposition can be assigned to a broad distribution (power law) of deactivation channels and one well-defined deactivation channel populated via a rate γ_{on}.

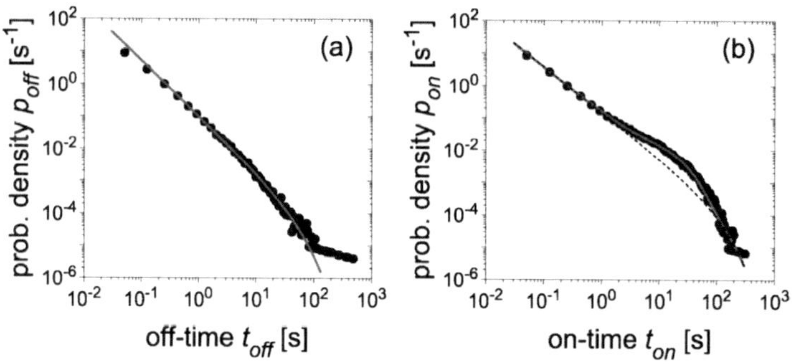

Figure 4.17 Distributions for (a) "off" and (b) "on" times of 100 CdSe/ZnS QDs on highly hydroxylated silicon oxide. The data in (a) and (b) are fitted (red lines) by a truncated power law according to Eq. 4.7 in (a) and its extended version with an additional superimposed exponential function defined in Eq. 4.8 in (b). The dashed line in (b) denotes the best approximation according to Eq. 4.7. Reproduced from Krasselt, C., Schuster, J., and Borczyskowski, C. von (2011). Photoinduced hole trapping in single semiconductor quantum dots at specific sites at silicon oxide interfaces, *Phys. Chem. Chem. Phys.*, **13**, 17084, with permission from the PCCP Owner Societies.

Figure 4.18 shows exemplarily how the evaluation of complex blinking data depends on the threshold settings. Recent publications discuss in detail the pitfalls and limitations of the threshold settings, especially in the case of "on" times [Crouch et al., 2010; Amecke et al., 2014]. The larger sensitive of "on" times on threshold settings compared to "off" times is readily understood in the presence of dim or gray PL intensities during a blinking time trace. In Section 4.4.1.2 we demonstrate how such dependencies can be avoided by applying the so-called CPA.

4.4.1.2 Change point analysis

Figure 4.18A shows for a typical example of QD PL blinking that intermediate PL intensities occasionally occur during a blinking time trace. Such a situation is often observed, though not often taken care of. There are reports which clearly do not report intermediate intensities and assign that this is due to an excellent quality of QDs without any defects, neither in the core nor in the shell material. Additionally, some reports also argue that intermediate intensities are caused by a limited time resolution (too long binning times), averaging thus over fast cycling processes between "on" and "off" states [Amecke and Cichos, 2011; Galland et al., 2011]. All these arguments might apply under certain conditions and have to be checked carefully. However, as has been shown recently it can be shown systematically that dim states are a reality [Schmidt et al., 2014]. How can this be taken adequately into account?

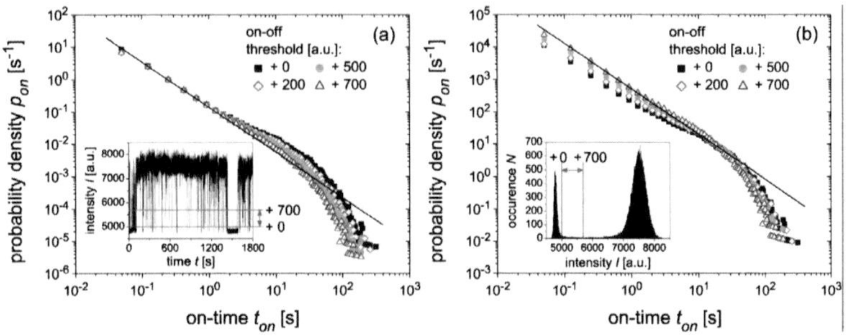

Figure 4.18 "On" time distributions of 100 CdSe/ZnS QDs on highly hydroxylated silicon oxide as a function of different threshold settings separating "on" and "off" events. The graphs compare the respective (a) normalized/scaled and (b) absolute probability densities of the "on" events. The black lines denote an untruncated power law according to Eq. 4.8 with α_{on} = 1.4. The insets illustrate the range of thresholds with respect to a typical blinking time trace and its related intensity histogram for a single QD. Reproduced from Krasselt, C., Schuster, J., and Borczyskowski, C. von (2011). Photoinduced hole trapping in single semiconductor quantum dots at specific sites at silicon oxide interfaces, Phys. Chem. Chem. Phys., 13, 17084, with permission from the PCCP Owner Societies.

Watkins et al. have introduced the so-called CPA [Watkins and Yang, 2005]. This approach makes it possible to identify beyond the conventionally used time binning PL emission of one and the same intensity with, in principle, a time resolution given by a TCSPC of about 50 ns. The related analysis is complex and its adequate description is beyond the scope of this presentation. Qualitatively this approach is visualized in Fig. 4.19A, in which we mark (in black) with respect to the CPA method a significant change in PL intensity in an otherwise "noisy" blinking time trace recorded with a 1 ms binning time. Following an additional proposal by Watkins et al., the obtained distribution of PL intensities can be "clustered" [Watkins et al., 2005]. The result is shown in Fig. 4.19B as a probability for a given PL intensity in comparison to binning the intensities. The clustered PL intensities are almost equally distributed in probability besides an increased density of intensities at very low ones close to the noise level and a considerably reduced density at the highest intensities.

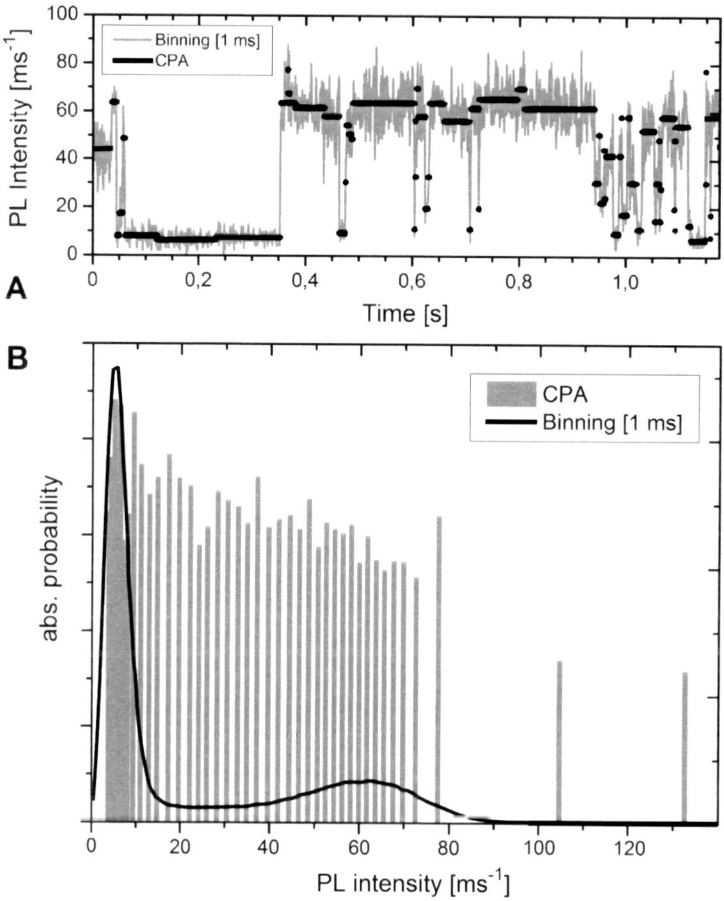

Figure 4.19 (A) PL blinking time trace of a single CdSe/ZnS on SiO_x with 1 ms binning time. CPA-identified PL intensities are shown by a kind of step function shown in black. Data have been obtained by confocal single-QD detection. (B) Probability to find a certain PL intensity (given in counts/ms) after CPA and clustering (see text). The black curve shows the intensity distribution obtained by the conventional threshold-based approach with 1 ms binning time. From Göhler, C. (2014). *Schnelle Spektralanalyse des Blinkens von Halbleiternanokristallen,* M.Sc. Thesis, TU Chemnitz, Germany, with kind permission from Clemens Göhler.

The CPA approach allows for a sophisticated analysis. Figure 4.20d demonstrates that also the distribution of intensity jumps ΔI can be identified. Moreover, the probability (and jump time) of jumping among different intensities can be determined [Schmidt et al., 2014]. In the given example, side bands show up in the distribution of ΔI (Fig. 4.20d). These side bands are correlated with a deviation from a threshold-based truncated power law (see Fig. 4.20a) and an increased density of PL intensities I in the range before probabilities of (high) intensities (e.g., at 27 kcps) are considerably reduced (Fig. 4.20c).

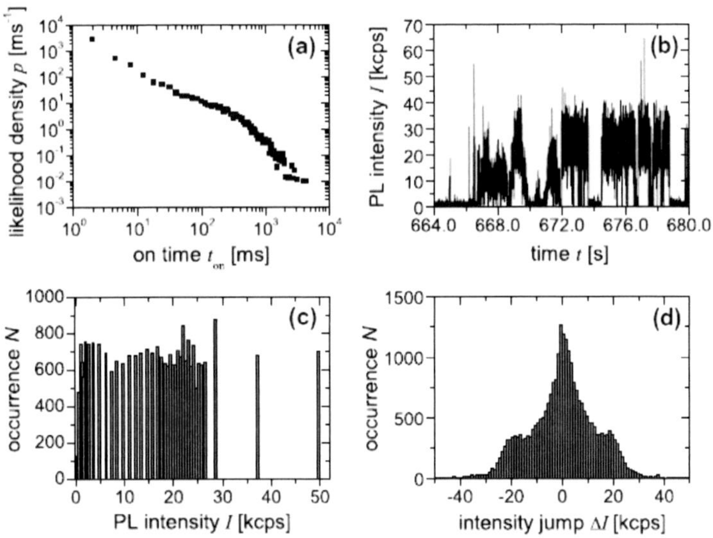

Figure 4.20 Blinking and change point analysis (CPA) of a single CdSe/ZnS QD in PS with <1 ms binning time detected via confocal microscopy. (a) "On" time distribution, (b) blinking time trace with varying PL intensities, (c) clustered distribution of PL intensities, and (d) distribution of PL intensity jumps ΔI between subsequent intensities. Reprinted from Schmidt, R., Krasselt, C., and Borczyskowski, C. von (2012). Change point analysis of matrix dependent photoluminescence intermittency of single CdSe/ZnS quantum dots with intermediate intensity levels, *Chem. Phys.*, **406**, 9–14, Copyright (2012), with permission from Elsevier.

Figure 4.21 (left) shows a power-law-type presentation of blinking data analyzed by CPA (top) for "off" times. In comparison, data analyzed via threshold setting are shown on the bottom. "Off" times follow a power law for three decades in time and five in probability. The dynamic ranges and the power-law exponent are in reasonable agreement with the threshold-based analysis (bottom). However, below about 5 ms the probability flattens in CPA with decreasing "off" times. "On" time analysis (right) shows that this flattening is not due to limited time resolution. However, presently we do not have an explanation for the observation that the probability for very short "off" times is decreasing. A decrease in probability would imply that "off" times approach a cutoff at short times. However, "off" times shorter than the time needed for reexcitation of the QD cannot be observed.

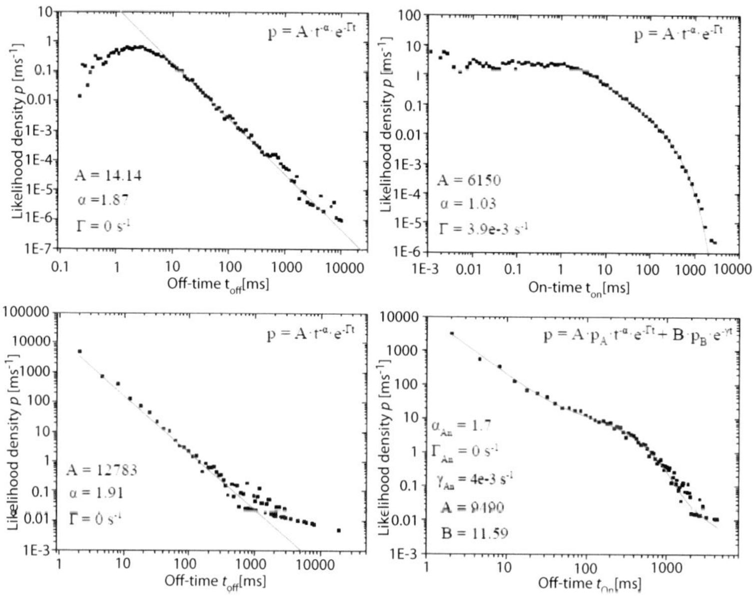

Figure 4.21 Left: Power-law-type presentation of "off" times obtained by CPA (top) and threshold setting (bottom) for a single CdSe/ZnS QD in PS. Note the different timescales. Gray lines show fits to a truncated power law. Effective time resolution of CPA is about 500 μs. Right: Power-law-type presentation of "on" times obtained by CPA (top) and threshold setting (bottom). Data have been fitted according to Eq. 4.8. Fitting parameters are included in the respective figures in all cases. From Schmidt, R. (2010). *Zeitaufgelöste Konfokalmikroskopie an einzelnen CdSe/ZnS-Halbleiternanokristallen,* Diploma Thesis, TU Chemnitz, Germany, with kind permission from Robert Schmidt.

With respect to "on" times data have been analyzed for each intensity separately. The limiting condition of reexcitation does not apply to "on" times. Figure 4.21 (top right) shows that "on" times do not show power-law behavior over a broad dynamic range, as has also been discussed recently [Schmidt et al., 2014]. The findings are different from those obtained by a threshold-based analysis (bottom right). Though CPA has no intrinsic limitations on time resolution (>50 ns), the effective one will depend on the detectable photon rate. Thus, time resolution results in the given experiment in about 500 μs. The time resolution can be improved by increasing the number of collected photons, for example, via increasing of the excitation intensity.

4.4.1.3 Autocorrelation analysis

Houel et al. (2015) implemented an intensity autocorrelation analysis for the unbiased determination of power-law exponents. To proceed they detected simultaneously 450 blinking time traces of a CdSe/CdS QD at a frame rate of 250 Hz with single-photon sensitivity. Though "on" and "off" power-law functions are intermixed in a correlation-type experiment, they could determine in an elaborated way the respective power-law exponents with high precision from a comparison to numerical simulations. They identified (truncated) power-law behavior on a timescale from 2 ms to 1 s and aimed at a final timescale down to 10 μs. This approach seems to be promising, though it is experimentally demanding to find very exact power-law coefficients. Certainly this recent work is a breakthrough with respect to analytic precision, though it has not become clear yet whether it can also handle multiple dim states. However, as we have discussed above it is obvious that not all blinking quantum systems show (truncated) power laws. The question arises, Is the understanding of blinking merely a matter of precision? Or what other parameters or modifications are needed to understand more precisely the physical background of luminescence intermittency?

4.4.1.4 PL intensity-lifetime correlation

Besides the timescale of blinking in the range of >500 μs (which we term "macrotimescale") it is also possible to determine the intrinsic PL decay which is typically between hundreds of femtoseconds and microseconds (microtimescale) when making use of tt-TCSPC according to Section 4.3.3. Typical time sequencing is shown in Fig. 4.22.

Figure 4.23 collects a summary of data which can be obtained via CPA. Part (b) shows (top) a threshold-based blinking time trace with 1 ms binning time and how it compares to a reconstructed (bottom) time trace obtained by CPA. It is evident that all relevant features are present in both approaches, especially also high-intensity spikes with a short "on" time. Due to the applied binning times limiting resolution, such spikes have not been reported in previous experiments. In the present experiments, the macro-time resolution that is the timescale of blinking events (for definitions of timescales see also Fig. 4.22) amounts to about 500 μs. Part (c) shows the probability distribution of PL intensities. After clustering, intensities are sorted into characteristic ranges (Fig. 4.23d) [Schmidt et al., 2014]. The intrinsic decay of an individual PL intensity can be investigated on the microtimescale τ in the range of nanoseconds, which is typical for the radiative decay of QDs (see Chapter 1). The results following a monoexponential data fitting are shown in part (e) as a function of PL intensities determined by CPA.

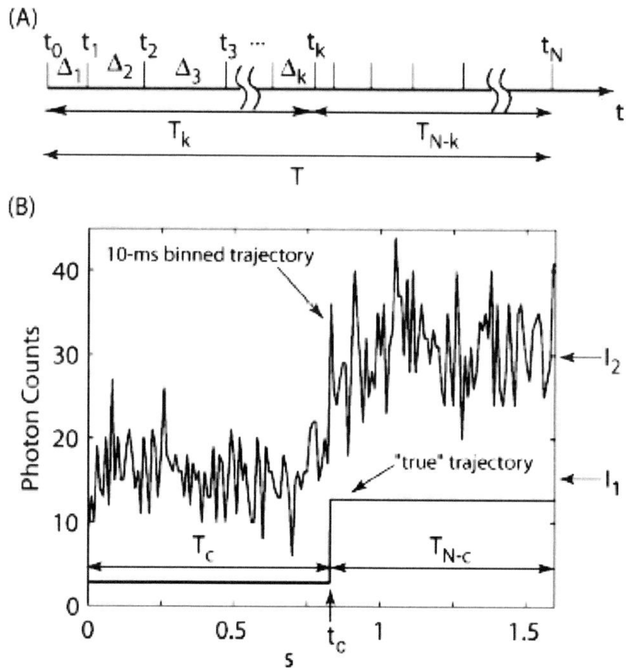

Figure 4.22 Time sequencing of CPA (A) and change of intensity at T_c (B) [Watkins and Yang, 2005].

It is obvious that the PL decay time τ decreases with decreasing PL intensity across the intensity ranges *N–I* and becomes slightly longer again in the range *L*. Remarkably, the decay time remains constant across the large total intensity range *S*. To understand this behavior we have to discuss how the PL decay time τ is related to the PL intensity I.

In the simplest approximation, luminescence intensity is proportional to the decay time τ according to

$$I \approx k^{\mathrm{r}}\tau \tag{4.9}$$

where $1/\tau = k^{\mathrm{r}} + k^{\mathrm{nr}}$. k^{r} is the intrinsic radiative decay rate and k^{nr} is the sum of all nonradiative decay rates. Concluding, the PL intensity should follow linearly the PL decay time. An explanation of the observed violation of this approximation has recently been published [Schmidt et al., 2014].

Making additional use of TCSPC, time-resolved PL decay down to a subnanosecond range is also as shown in Fig. 4.24, accessible for each PL intensity along a blinking time trace, as will be discussed in Section 4.4.1.4. Figure 4.24 shows a direct comparison of PL intensities and (monoexponentially evaluated) PL decay times during a (threshold-based) blinking time trace with 20 ms binning time. It can be seen that $I \approx \tau$. However, as already evident from Fig. 4.23 close inspection of the PL reveals that the PL does not decay monoexponentially across the total intensity range. At first glance this is surprising since experiments are performed on a single QD. However, we have to keep in mind (see also Section 4.4.2) that single-QD detection occurs over an extended sampling time, exploring various different excitonic states. As soon as fluctuations in the overall quantum system occur, PL properties might change during the observation time accompanied by "lifetime blinking."

Figure 4.25 shows the valuation of the PL decay as a function of threshold setting. Similar to recently reported data [Schlegel et al., 2002; Fisher et al., 2004; Zhang et al., 2010] the PL decay has been fitted by a stretched exponential function according to

$$I \approx \exp(-t/\tau_0)^{\beta} \tag{4.10}$$

with $0 < \beta < 1$. It can be seen that the fitted decay time τ_0 and β decrease with low threshold settings. Above a threshold of about 150 (a.u.) PL decays monoexponentially ($\beta = 1$), while decreasing linearly with PL intensity and becoming multiexponential below this threshold. The bending of the linear relationship with intensity has already been previously observed [Zhang et al., 2010]. It has been explained recently [Schmidt et al., 2014] as being due to a sequential superposition of a few well-defined QD configurations in course of the exploration time and corresponds to a kind of lifetime blinking.

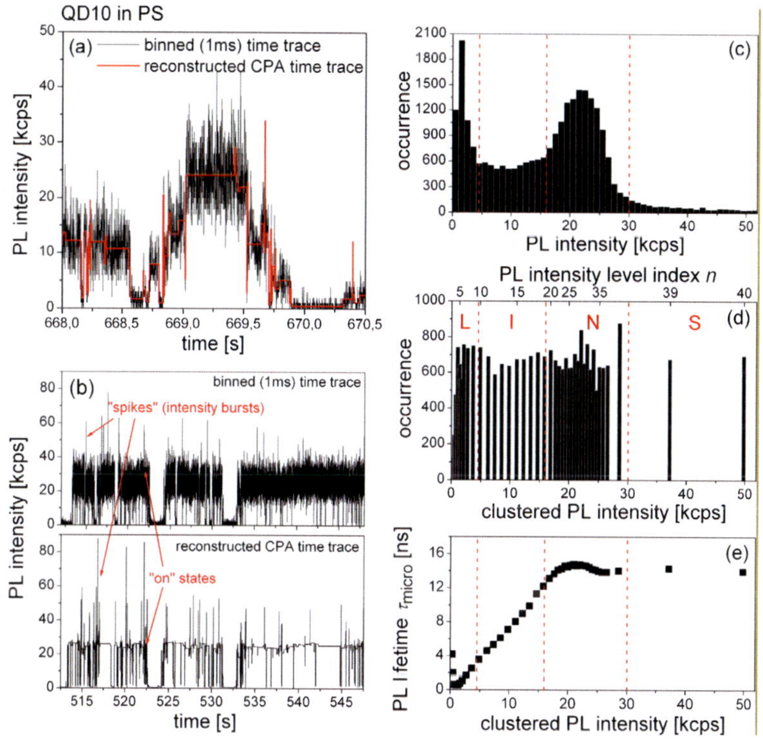

Figure 4.23 PL data of an (exemplary) single CdSe/ZnS QD10 in PS. (a) Subset of the PL intensity time trace with black and red data presenting the binned (integration time 1 ms) and CPA-reconstructed time trace, respectively. The multi-intensity blinking as well as the reproduction of the binned time trace by CPA are clearly visible. (b) Long PL time trace binned at 1 ms (top) and CPA reconstructed (bottom). Strong intensity spikes are visible in both datasets for this QD in PS. (c, d) CPA histogram of intensities before (c) and after (d) PL intensity clustering. Besides the intensity given in kcps, we also indicate an intensity level index *n*. In this example of CdSe/ZnS on PVA, a high density of intensities is visible in the *N* range. (e) PL lifetime as a function of clustered PL intensity of CdSe/ZnS on PVA. Data evaluation is approximated by a monoexponential fit for simplicity. Deviations from a monoexponential fit are detectable across the intensity range *N*–*L*, especially below $I \approx 15$ kcps. Reprinted with permission from Schmidt, R., Krasselt, C., Göhler, G., and Borczyskowski, C. von (2014). The fluorescence intermittency for quantum dot is not power-law distributed: a luminescence intensity resolved approach, *ACS Nano,* **8**, 3506–3521. Copyright (2014) American Chemical Society.

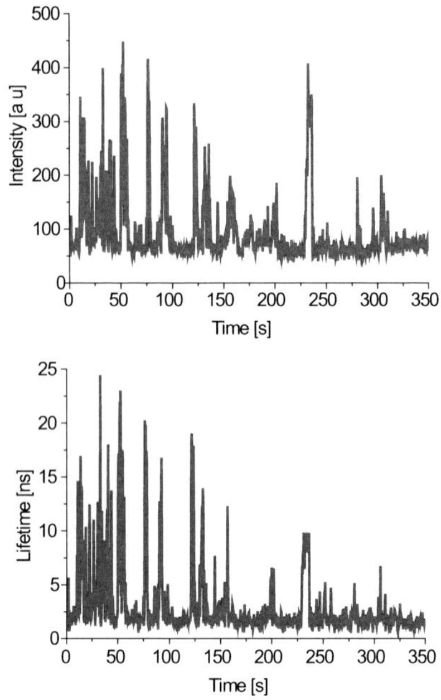

Figure 4.24 PL intensity during a blinking time trace (top) and monoexponentially evaluated PL decay (life) time of CdSe/ZnS in PS. Binning time is 20 ms. Reproduced with kind permission from Thomas Blaudeck.

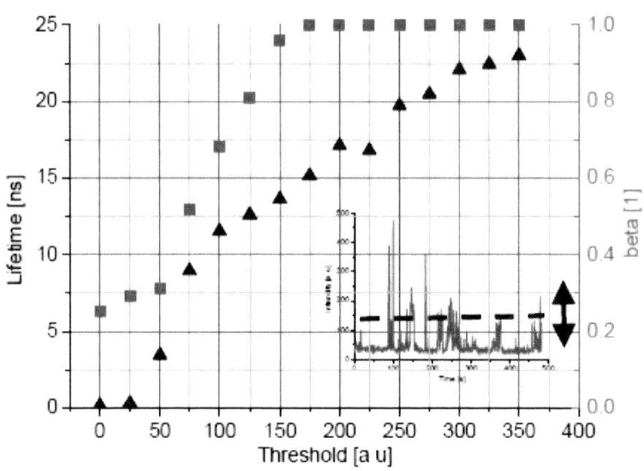

Figure 4.25 PL decay time parameter τ (left scale) and stretched exponential β (right scale) of CdSe/ZnS in PS as a function of the threshold setting, which is schematically shown in the inset, during a blinking time. From Blaudeck, T. (2007). *Self-Assembly of Functionalized Porphyrin Molecules on Semiconductor Nanocrystal Surfaces*, PhD Thesis, TU Chemnitz, Germany, with kind permission from Thomas Blaudeck.

4.4.1.5 Correlation of PL energy with blinking and PL decay times

In the previous discussion it became obvious that blinking reveals that not only PL intensities are fluctuating but also PL decay times since both are inherently related to each other. If decay times are fluctuating this might indicate that emission might stem from different electronic states which are inherently related to different PL energies. Though energy differences might be small and hidden in the line width even for a single QD they might be detected as several observations of spectral diffusion indicate (see Section 4.4.2). The ultimate goal thus is to find a correlation of all three observables during a blinking time trace revealing intensity, decay time, and energy blinking, which will be shown in the following section to be feasible.

Experimental limitations based on sensitivity and time resolution arguments will not allow determining all the three observables simultaneously on timescales down to milliseconds with high spectral resolution. For this reason a modified scheme has been suggested by Biebricher et al. (2006). This approach is based on the incorporation of a dichroit centered within the PL band. A typical experimental setup is shown in Fig. 4.26.

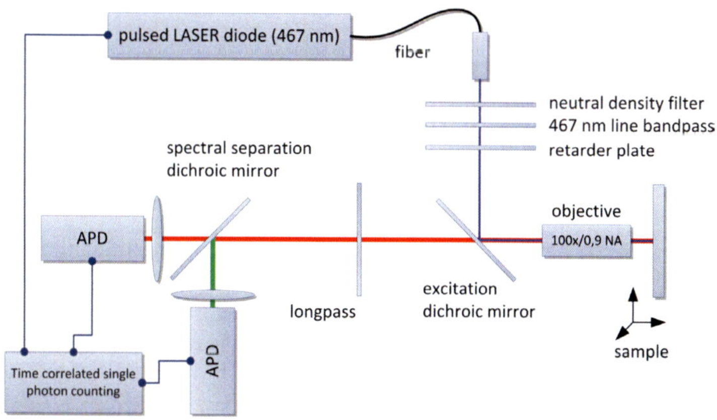

Figure 4.26 Scheme of a confocal microscope with pulsed laser excitation. A dichroit centered within the PL band splits the PL emission into a blue and red avalanche photodiode (APD)-based detection channel. In case that the PL energy of the QD fluctuates (spectral diffusion), the relative PL intensity in the detection channels changes. Making use of CPA, this can be related with PL intensities and decay times. From Göhler, C. (2014). *Schnelle Spektralanalyse des Blinkens von Halbleiternanokristallen,* M.Sc. Thesis, TU Chemnitz, Germany, with kind permission from Clemens Göhler.

Figure 4.27 shows the distributions of energy fluctuations *dE* (between the two detection channels) for a single CdSe/ZnS QD on SiO_x (left) and CdSe/CdS on SiO_x as a function of an additional threshold setting (discriminating between "on" and "off" intensities) and absolute duration length *T* of the respective intensity. Data have been obtained by CPA. The results show that fluctuations depend critically on the absolute PL intensity and *T* during the blinking time trace. Details will be discussed in Chapter 6 with respect to a complete analysis of all blinking observables.

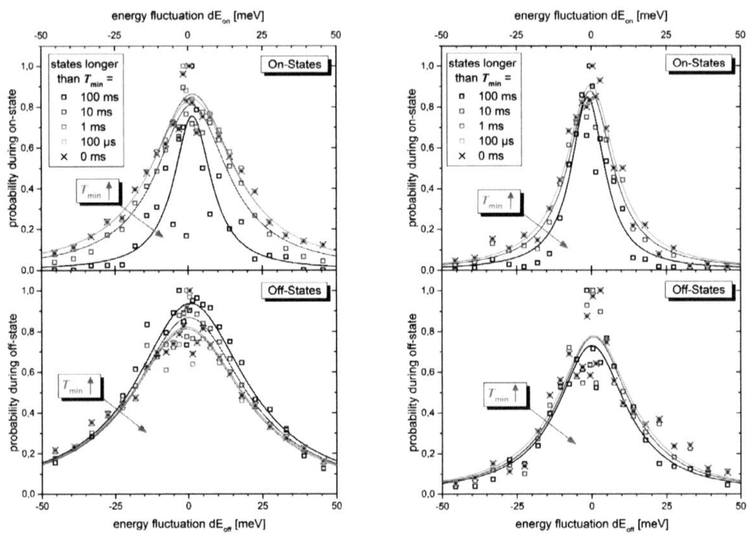

Figure 4.27 Distributions of energy fluctuations *dE* of "on" intensities determined by CPA for different duration times *T* of a selected intensity and at two (superimposed) threshold settings ("on," top, and "off," bottom). Data are collected for CdSe/ZnS on SiO_x (left) and CdSe/CdS in PMMA (right). Fluctuations are broader for "off" compared to "on" intensities. Distributions of *dE* do not depend on the length of *T* after being in an "off" intensity but become narrower with increasing *T* during an "on" intensity. This implies that short "on" times are correlated with a broader manifold of PL energies. The respective widths of the distributions are somewhat broader for CdSe/ZnS compared to CdSe/CdS. Lorentzian fits have been applied for simplicity to the data (but see also Chapter 6). From Göhler, C. (2014). *Schnelle Spektralanalyse des Blinkens von Halbleiternanokristallen,* M.Sc. Thesis, TU Chemnitz, Germany, with kind permission from Clemens Göhler.

4.4.1.6 Suppression of blinking

Though blinking is an intriguing phenomenon as such there are applications where suppression of it may be of advantage [Hollingsworth, 2013]. These are increased QYs for labeling biosystems or to create single-photon sources or QD lasers. Efficiency in photovoltaic applications will also profit from blinking suppression. Passivation of surface traps [Hohng and Ha, 2004; Jeong et al., 2005; Hammer et al., 2006; Ko et al., 2010; Nadeau et al., 2012; Antelmann et al., 2012] via appropriate ligands (especially of electron donor type) and coupling to TiO_2 particles [Hamada et al., 2010] will decrease blinking efficiencies. Other approaches are based on increasing the radiative rates close to metal surfaces [Shimizu et al., 2002; Yuan et al., 2009a; Yuan et al., 2009b; Ma et al., 2010] or in buffer solutions [Fomenko and Nesbitt, 2008]. Recently blinking of CdSe QDs was suppressed efficiently via thick (giant) CdS shells [Chen et al., 2008; Mahler et al., 2008; Vela et al., 2009; Galland et al., 2012; Chen et al., 2013] or alloyed CdZnSe/ZnSe QDs [Wang et al., 2009]. A completely different approach compared to material manufacturing was realized (see also Section 4.6.1.2) by applying external electric fields to charge QDs in electrolytic cells [Galland et al., 2011; Qin et al., 2012; Qin and Guyot-Sionnest, 2012].

Though considerable success controlling blinking has been achieved in the past the mechanisms are still under debate. Additionally, throughout this book we show what features tune blinking processes. When understanding the microscopic origin, blinking might turn into an advantage since it might be in turn implemented as a sensitive tool of processes on a nanoscale close to a single QD. Moreover, recent advantages in optical super-resolution microscopy might also profit from blinking by having a switchable (and very photostable) marker of various structures at hand [Wang et al., 2013].

4.4.1.7 Blinking models

4.4.1.7.1 *Summary of experimental findings*

Recently Cichos et al. (2007) have summarized the main experimental facts related to power-law blinking. Since then several new review articles have been published updating some of the findings

[Frantsuzov et al., 2008; Stefani et al., 2009; Lee and Osborne, 2009; Riley et al., 2012; Cordones and Leone, 2013; Cui et al., 2014]. In the meantime new experimental findings have to be added or modified. The updated list is as follows:

- Blinking is observed for many different classes of emitters and for different chemical compositions within each class.
- Blinking is only found in disordered materials (polymers, complex interfaces, silicate glasses, surfactant shells).
- A number of phenomena are correlated to intensity blinking (fluctuations of PL decay times, spectral diffusion) or its natural consequences (weak ergodicity breaking, statistical aging).
- Statistics of "on" and "off" times have in many cases been described by an exponentially truncated power law.
- Power-law exponents are found between −1.1 and −2.2 for both the "on" and "off" times of NCs and at least for the "off" times of molecules.
- Dielectric properties of the environment (matrix, substrate, shell, ligands) are important and lead to systematic changes of the power-law exponent and/or of the cutoff times (often correlated with the dielectric constant of the environment or the ligands).
- Deviations from a power law often appear as exponential cutoffs at long "on" times and also at long "off" times. In the latter case they are more difficult to detect since they occur at much longer times than for "on" times with a very low probability.
- A superposition of a (truncated) power law and an exponential decay has been identified for "on" times of QDs in a few cases, but re-inspecting the literature reveals that such a superposition has been overlooked in several previous publications.
- The "off" time exponent is independent of or only weakly dependent on temperature and excitation intensity or wavelength. For "on" times, cutoff times are reported to depend weakly on temperature, laser intensity, and dielectric constant of the matrix. If cutoffs for "off" times are observed (which is rare), these depend similarly on these parameters.

- In the case of some molecules but also for Si nanostructures, NV centers in diamond NCs and metallic nanoparticles monoexponential blinking statistics are observed.
- Combination of blinking with time-resolved experiments and implementation of CPA question the power-law behavior for "on" times.
- Both holes and electrons can be trapped.
- Recent experiments reveal for one and the same QD several different types of blinking, which might be controlled via external charging/external electric fields or laser intensity.

4.4.1.7.2 *General considerations*

Since luminescence intermittency (blinking) is an intriguing, inherent feature of many single quantum systems (molecules, semiconducting and metallic QDs, NV centers in diamond NCs), a huge number of experimental investigations have been reported. Typical examples for blinking in quite different systems are compared in Fig. 4.28 for a dye molecule (top) and a colloidal QD (bottom) together with a threshold-based analysis of blinking intensities (right scale). Generally it is found that blinking of molecules is more closely related to true "on-off" intensities, whereas QD blinking is in many cases related to a broad distribution of intensities during a blinking time trace [Cichos et al., 2007].

Despite the fact that blinking is known since more than two decades investigations are still continued since the phenomenon has currently not yet been completely understood. Several review and feature articles are devoted to this subject, for example, Cichos et al. (2007), Frantsuzov et al. (2008), Stefani et al. (2009), Riley et al. (2012), and Cordones and Leone (2013). Due to the related general aspects several models have been proposed. However, we will only concentrate on the most challenging ones. Before proceeding, we would like to make three general statements.

Firstly, in this context we will neglect those blinking features which are caused by intrinsic photon bunching (see Section 4.2.3.1) in the sense that during the experimental observation time, several electronic states with different emission properties are explored (e.g., singlet and triplet states in the case of molecules). In that case blinking dynamics correspond to the inherent dynamics of the quantum system [Brown et al., 1994]. As long as such a quantum system is completely decoupled from its environment, each single quantum system will show identical blinking features, depending on the specific configuration (e.g., the molecular conformation) [Cichos et al., 2007].

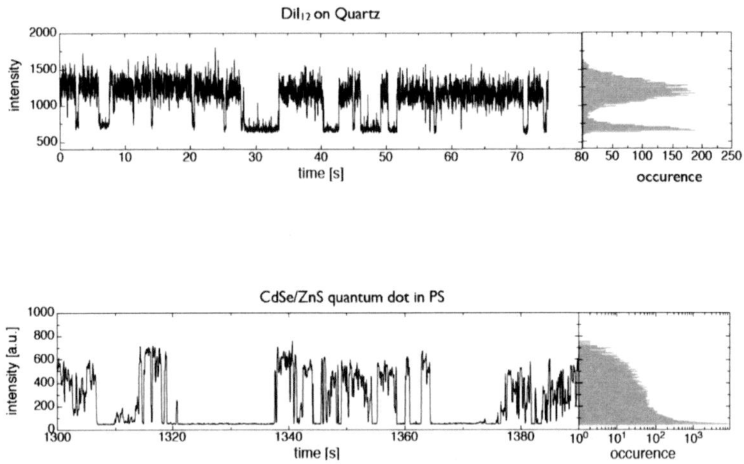

Figure 4.28 Time traces of blinking of a single di-iodide molecule on a quartz substrate (top) and a single CdSe/ZnS QD in PS (bottom). The occurrence of the PL intensities determined by the threshold setting is shown on the right side. While the blinking of the dye is very closely related to an "on-off" situation, PL intensities are broadly distributed in the QD case.

Secondly, in most reported experiments blinking depends, independent of whether charging models apply or not, on coupling of the quantum system to a bath (beyond the particle-in-a-box approach: environment, surrounding). The environment might be thought of as the embedding matrix, surfactants (e.g., ligands in the case of QDs), or an interface (e.g., core/shell systems in the case of QDs). Cichos et al. have claimed that disorder in the form of heterogeneous coupling to a bath is an essential ingredient of blinking processes [Cichos et al., 2007]. At least two observations are in favor of this statement: (i) Single-color (NV) centers of diamond have been since the first experiments [Gruber et al., 1997] intensively investigated without any identification of blinking features. Only recently, NV centers in very small NCs are subject to blinking [Bradac et al., 2010]. It has been argued that these NV centers couple to the heterogeneous surface of such nanodiamonds. (ii) One and the same dye molecule shows only blinking if it is embedded into a polymer matrix instead of a (rigid) molecular crystal [Pfab et al., 2004]. However, as soon as the molecular crystal is disordered, due to a distribution of orientations of the phenyl rings in *p*-terphenyl, single crystals' orientation fluctuations of the phenyl rings will give rise to spectral jumps possibly accompanied by fluctuations of the intensities of a single pentacene molecule embedded in those crystals [Reilly and Skinner, 1995]. This will also show up as luminescence intensity fluctuations, though it has not been monitored in such a manner. The general view on these findings is that the electronic wavefunction extends far beyond the quantum system itself (in principle to infinity) and thus "feels" the features of the embedding bath, including its fluctuations. It is generally known that heterogeneous systems (like polymers) are not in a thermal equilibrium but that the related energy landscapes can be described by a broad distribution of two-level systems fluctuating (tunneling) on broad timescales, which is, vice versa, felt by the embedded quantum system.

Thirdly, one intriguing observation from the very first blinking investigations [Nirmal et al., 1996] was that blinking dynamics can in many cases be described by a power law (very often with power-law exponents close to 1.5) over a very broad timescale. This stimulated universal blinking models. However, the present status is that though a power law is in many cases an appropriate description of the observed phenomena, it is not universal for molecules [Haase et al., 2004], NV centers [Bradac et al., 2010], Si nanostructures [Bruhn et al., 2011], or epitaxial QDs [Cichos et al., 2007]. Even in colloidal QDs deviations from a (truncated) power law have been observed [Krasselt et al., 2011]. Recently it has been shown experimentally [Schmidt et al., 2014] that "on" times are not power-law-distributed for typical colloidal II–VI QDs, as has also been confirmed by numerical simulations [Amecke et al., 2014]. In any case, identification of a power law needs investigations over a broad range both in probability and in time, which is often difficult to realize. Moreover, absolute parameters for nominally one and the same quantum system/matrix vary often considerably from experiment to experiment or from lab to lab. In fact, this is not surprising since one of the basic parameters of blinking is the coupling to a (heterogeneous) environment which may not be in its thermal equilibrium. Consequently, investigated dynamics may differ from experiment to experiment. Conclusions with respect to parameter variation (as we will discuss in Section 4.5) have to be made with care as long as experiments have not been performed under conditions as identical as possible. Moreover, as we have outlined in the previous sections the kind of analysis enters critically the quantitative determination of blinking dynamics. Additionally, excitation conditions like excitation intensity and wavelength, detection wavelength, detection scheme (wide-field or confocal microscopy), and the kind of laser (continuous wave or pulsed) influence experimental findings.

A comprehensive summary of blinking models has been published in 2007 [Cichos et al., 2007]. Since that time a few new concepts have emerged which question the general validity of the power-law behavior for QD blinking, such as the multiple recombination center (MCR) model [Frantsuzov et al., 2009; Frantsuzov et al., 2013]. In the following paragraphs we will present a critically updated version following our previous review [Cichos et al., 2007] and will include new developments.

Most models assign blinking of semiconductor NCs to photoejection of a charge. Previously it has been argued that the ejected charge is most likely an electron because of its spatially more extended wavefunction (see Chapter 1). However, hole surface trapping as been predicted and observed experimentally [Li et al., 2014]. Dark states are associated to the presence of a charge (say, a hole) in the dot's core, which efficiently quenches the fluorescence by Auger processes [Efros and Rosen, 1997]. Looking for a unified picture, most researchers naturally interpret blinking as the emitter's switching between a neutral (bright) state and a charged (dark) state. Models disagree, however, as to the nature and dynamics of the charge-separated state. In trap models, the ejected charge is assumed to tunnel back and forth between the emitter and immobile trap sites, whereas the random-walk models assume the ejected charge to diffuse (in space and/or energy) in the environment of the ionized species.

4.4.1.7.3 *Trap models*

Power-law distributions of blinking times emerge when an exponentially large value of a quantity (e.g., time) is obtained with an exponentially small probability. The static trap model is based on this property. The model [Kuno et al., 2000; Verberk et al., 2002] assumes trapping of the ejected charge in the emitter's environment, including shell or ligand capping. Photoionization and recombination rates are supposed to be strongly correlated and broadly distributed. At low enough temperatures, recombination can only proceed by tunneling. For a particle with mass m_e, the Wentzel–Kramers–Brillouin theory predicts a tunneling rate k, depending exponentially on barrier width d and height V according to

$$k \approx \exp(-2d\sqrt{m_e V})\cdot \qquad (4.11)$$

When the exponent is large and distributed, the tunneling rates span many orders of magnitude. Tunneling in charge transfer involves an electronic coordinate and a nuclear reorganization coordinate [Marcus, 1960; Dorfmann et al., 1991]. A dominant nuclear coordinate would imply activation at higher temperatures, and rate changes by orders of magnitude, which are not observed.

One is thus led to assume that trapping and recovery rates are governed by the exponential decay of electronic wavefunctions with the emitter–trap distance [Dorfmann et al., 1991]. Further assuming a continuous distribution of possible trapping sites, we obtain a 1D version of the model with an ionization rate toward the trap at distance x

$$p_T(x) = A\exp(-\alpha x) \qquad (4.12a)$$

and a recovery rate from this trap

$$p_R(x) = B\exp(-\beta x) \qquad (4.12b)$$

Expressing the trap distance x as a function of the (average) recovery time τ, we find a power law [Verberk and Orrit, 2002; Verberk and Orrit, 2003] for the probability distribution of recovery time

$$p(t) \approx \tau^{-\mu} \text{ with } \mu = 1 + \alpha/\beta. \qquad (4.13)$$

This model is closely related to McWhorter's model [Surya and Hsiang, 1986] of the 1/f-noise in certain field-effect transistor (FET) structures, where trapping of charge carriers at different depths in the insulator material leads to a broad range of times in current fluctuations. A similar derivation was proposed 40 years ago by Tachiya and Mozumder [1975] to explain the kinetics of geminate ion recombination in rigid glasses.

The static trap model predicts the correct range of exponents (between 1 and 2), only with the natural assumption that the trapped state has lower energy than the ionizing state, that is, $\alpha \leq \beta$. Tunneling of electrons or holes explains the absence of a strong temperature dependence and suggests that the rates should depend on the substrate material in or on which the emitters are placed. Another virtue of the model is its prediction of a wide distribution of rates already for a modest range of accessible distances (electronic wavefunctions decrease by about a factor of *e* for 0.1 nm in vacuum, which easily leads to a spread of rates by 10 decades or more within 1–2 nm). In this model, however, the emitter spends most of its time in an ionized dark state, which means that the corresponding dots or molecules would not be the ones generally picked up in single-molecule experiments. Such dark dots are best studied in ensembles, either by fluorescence correlation methods (Tittel et al., 1997] or by TCSPC [Smith et al., 2006]. Nonetheless, this static trap model has been found to reproduce the properties of "off" time distributions and correlation functions of certain uncapped CdS QDs, down to the single-dot level [Verberk et al., 2002].

The interesting single emitters (molecules or NCs) are the bright ones, that is, those with relatively long "on" times. In the static traps model, the rate of decay of "on" times is the total trapping rate, largely determined by transfer to the nearest-by traps. The distribution of "on" times is a single exponential instead of a power law. This defect of the static traps model persists even with a low QY for trapping, or when experimental imperfections such as a weak detection yield, dark count rates, improper choice of threshold, etc., are considered [Verberk and Orrit, 2003].

Long "on" times require that the ionization or trapping mechanism be turned off for long periods, although the emitter is constantly excited and spends a significant fraction of its time in the excited state. Trapping may be turned off by making the traps inaccessible, for example, energetically, if the excited state's energy becomes lower than the trap's. If this configuration is permanent, the "on" time never ends, resulting in nonblinking dots. Therefore, the trap model can be made dynamic by switching between two configurations. Verberk et al. (2002) proposed that a first charging event changes the emitter's potential and that Coulomb blockade prevents further charge ejection because it costs too much energy to extract a second charge. If the dot is to remain bright, this first remaining charge (say, a hole) should be stabilized in the close environment (we call it the shell). The bright "on" state will persist for as long as a net positive charge remains on the emitter.

An appealing feature of this modified trapping model is its natural explanation of the power-law distribution of the "on" times (the "on" and "off" time exponents may differ if the potentials for trapping and recovery of the trapped electron are modified by the trapped hole). Yet, several points remain open. The energy of the ionizing state determines the electronic leakage coefficient α and thereby the exponent. Is the ionization mechanism a one- or multiphoton process? Higher excited states are much shorter lived but may ionize much more efficiently. By which mechanism is a hole stabilized in the shell, and what could prevent it from accepting an electron from the excited emitter? A possible mechanism would be an electrochemical reaction which could turn the QD into a kind of type II heterostructure [Kim et al., 2003] and switch back upon return of the trapped electron. Very recently electrochemical experiments of blinking have been reported [Jha and Guyot-Sionnest, 2010; Galland, et al., 2011; Qin et al., 2012; Qin and Guyot Sionnest; 2012; Amecke, 2015], as we will discuss further. Since blinking is in a certain sense a general feature of single quantum objects, how is it possible to imagine such mechanisms for molecules whose charged states have very different optical properties?

In the general trap model, the microscopic nature of the traps remains unaddressed. We will discuss, however, a few related proposals in Section 4.4.1.7.9. From their experiments on different dielectric substrates, Cichos et al. (2004) have suggested that the ejected charge can self-trap in the surrounding dielectric. This hypothesis solves two problems at once. The distribution of traps is nearly continuous (even on nanometer scales), and no special trapping sites are required. The trap model requires, however, that the self-trapped charge be nearly immobile in the matrix so as to keep the memory of the trapping position before recombination and, with it, the critical correlation between trapping and recombination rates. However, the MCR model (see Section 4.4.17.5) does not need a continuous distribution of trap states. Additionally, recent experiments question the power-law model (see Section 4.4.7.8). Nevertheless, self-trapping remains an interesting approach since blinking depends critically on dielectric properties of the embedding environment (see also Chapter 6).

Self-trapping of charges in crystals and amorphous solids such as polymers has been studied for a long time. Landau first recognized that the transport of charges in crystals is affected by the local polarization the charge creates inside the crystal. Strong interactions (including lattice deformation) lead to a trapping of the charge [Sluger and Stoneham, 1993]. These considerations underlie the concept of a polaron as a charge surrounded by a local lattice distortion or a phonon cloud [Fröhlich, 1954, Holstein, 1959]. Under strong electron–phonon interaction, a polaron will only move by hopping [Holstein, 1959]. A polymer chain, for example, which may be the embedding matrix of a QD, can be viewed as a chain of identical units, with a band structure and large band gaps of typically 8 eV. Any disturbance in the periodicity, for example, due to kinks in the chain conformation, side groups, or chain ends, causes a spread of the band edge and possibly localized states [Lowel and Rose Innes, 1980]. This spread is not large enough to create states in the semiconductor conduction band, about 4 eV below vacuum level. However, a charge injected into the system will be stabilized upon polymer backbone relaxation by the formation energy of the polaron. This energy is nothing else than the reorganization energy λ, a key factor in electron transfer and solvation theory, extensively discussed by Marcus (1960). This reorganization energy can be quite large. Values of 1 eV and larger are suggested for many systems [Lewis, 1989]. Thus dielectric reorganization and local disorder can stabilize deep electronic states within the band gap of insulating materials. The large spread of these electronic states is well known, for example, from transient photocurrent or charge release [Watson et al., 1992] measurements in amorphous materials, which show a power-law current transient as analyzed in great detail by Scher and Montroll (1975). They related the current transient to a power-law distribution of hopping times for the charges in the amorphous material, where the long tail corresponds to isolated traps in the material. We suppose these to be relevant, too, for charge trapping and emitter blinking in amorphous materials. Interestingly, these current transients also follow power laws in close-packed ensembles of colloidal semiconductor QDs [Novikov et al., 2005].

A theoretical analysis of deep trap states in polymers has been carried out by Duke and Fabish (1976), who consider side groups of polymers as one further key ingredient for the stabilization of charges. Even though a negative charge on a benzene side group of PS is not stable, the side group interacts with the polymer backbone, shifting the energy levels via intra- and intermolecular distortions. Intramolecular contributions are related to nuclear and electronic rearrangements in the accepting molecule or side group and have been estimated to be on the order of 2 eV [Duke and Fabish, 1976]. Intermolecular contributions arise from dielectric relaxation, as described above for polarons and solvation. These contributions are also on the order of 2 eV and, from simple electrostatic theory, found to be proportional to

$$f(\varepsilon) = 1 - 1/\varepsilon \tag{4.14}$$

where ε is the dielectric constant of the material [Blaise and Sarjeant, 1998]. Thus states at 4 eV below vacuum level are possible. These are well in the range of the conduction band of many metals and semiconductors and have been experimentally investigated by contact charging experiments [Gady et al., 1998]. Furthermore, they are intrinsic to the polymer, which can explain the large charge densities of up to 10^{26} m^{-3} estimated from many experiments [Lowell and Rose-Innes, 1980]. Finally, it is because of the large energy spread of these states (about 1 eV [Lowell and Rose-Innes, 1980; Gady et al., 1998]) that the charges remain well localized in the polymer, as energy differences between neighboring states are much larger than their coupling [Lowell and Rose-Innes, 1980]. Thus, these charges can keep their position within the polymer for a long time [Watson et al., 1992]. The above discussion based on polymers is generally valid for other amorphous systems [Scher and Montroll, 1975; Williard, 1973] and may also hold for the (amorphous) shell of surfactant molecules. The main requisite ingredients for self-trapping are disorder, which provides a spread of localized electronic states, and the presence of polarizable degrees of freedom, which stabilizes the charge by dielectric relaxation.

Furthermore, the self-trapping model may help understand the apparent requirement of disorder for power-law intermittency. We propose here (see Fig. 4.29) that a smooth Coulomb energy funnel, as can be expected in ordered systems, acts as an efficient conducting path for recombination, kinetically preventing self-trapping. By roughening the funnel potential surface, disorder can create local potential minima and lengthen the residence time sufficiently for the system to cross the barrier toward a self-trapped state. Even if rare, these events can then compete efficiently with recombination and lead to the observed long trapping events. Tachiya and Mozumder [1974] proposed a similar scenario for electrons in low-temperature glasses: an initial shallow trapping event was required for deep solvation to take place. Recently Orrit [2010] discussed this scheme with respect to chemical and physical aspects of charge transfer in the intermittency of single molecules and QDs.

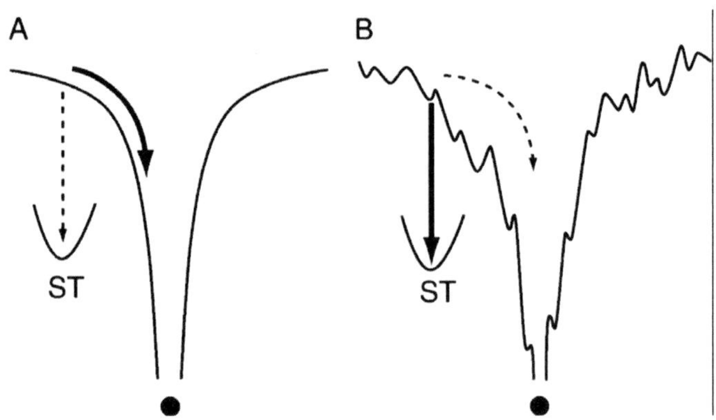

Figure 4.29 Proposed pretrapping mechanism leading to a self-trapped charge (ST) in a disordered medium. The potential landscape shown is that of the ejected charge around the remaining counterion (solid circle). (A) In an ordered material, the smooth Coulomb potential quickly funnels the ejected charge back to the center, avoiding self-trapping. No dark state is observed. (B) In a disordered material or on a surface, local inhomogeneities can transiently confine the carrier in a shallow well, from which tunneling or activation to the self-trapped state has a higher branching ratio. Once in the self-trapped state, the system enters a long-lived dark period, which ends only upon tunneling of the charge toward the counterion through a high and thick barrier. Reprinted from Cichos, F., Borczyskowski, C. von, and Orrit, M. (2007). Power-law intermittency of single emitters, *Curr. Opin. Colloid Interface Sci.*, **12**, 272–284, Copyright (2007), with permission from Elsevier.

4.4.1.7.4 *Random-walk models*

As the experimental distributions of "off" and "on" times of single emitters often follow power laws with exponents μ around 3/2, it is tempting to relate intermittency to a random walk involving a first-passage problem. The time of first passage, required for the random walker to return to its starting point, is distributed as a power law with an exponent of −3/2 in the simplest case of a 1D unbiased random walk. A 1D random walk therefore intrinsically includes the power-law distribution of times observed experimentally. Assuming an Auger quenching mechanism for the excited state emission, this model just requires a random walk in the dark or bright state on any coordinate (energy or space) to provide power-law statistics for the distribution of "on" and "off" times. The random walk, however, has to be restricted to one or two dimensions, since, according to Polya's theorem, only a fraction of random walkers return to their origin in three dimensions. An emitter would thus turn dark forever after one or a few blinking events, contrary to observations. Further, the first-passage time distribution for a 2D random walk obeys a power law with an exponent of −1, which is also not observed in the experiments.

Since the blinking mechanism requires a charge separation, the random walk is presumably influenced by the Coulomb interaction between the two charges. You should thus consider the first-passage time distribution in a biased random walk. It is expected that the general $\tau^{-3/2}$ power law is modified by Coulomb forces. This issue has been discussed by Hong et al. [1981] with respect to diffusion in space modeling the geminate electron–hole recombination in amorphous semiconductors. Although no exact solution to this problem exists so far, one would expect the local dielectric properties to be important, since they determine the strength of the Coulomb interaction. Because the first-passage statistics result from a chain of independent steps, the timescale, the temperature dependence of the steps, and other physical details of the individual events are difficult to unravel.

Statistics of "on" and "off" times were first discussed by Shimizu et al. [2001] based on a random walk along an energy coordinate. A simple realization could be a spectrally diffusing (or wandering) excited state of the QD, which may or may not hit the trap's energy.

Whenever the NC level comes into resonance with the trap, the QD would switch from bright to dark or vice versa. The spectral diffusion, due to charge rearrangements, could be photoinduced or thermally activated.

While the random-walk or diffusion models often only aim at explaining the experimentally observed power laws, Tang and Marcus [2005] have developed a model of diffusion-controlled electron transfer (DCET), explaining also other observations, including exponential cutoffs in the "on" and "off" time statistics and spectral diffusion. The diffusion process itself is assumed to be light-induced and closely related to spectral diffusion with a certain diffusion constant D, that is, to local charge rearrangements [Neuhauser et al., 2000]. Tang and Marcus reduced these rearrangements to a single reaction coordinate. Spectral diffusion thus results from diffusion on a parabolic free-energy surface over this reaction coordinate. The emission intermittency is then related to a projection of this parabolic free-energy surface on a different reaction coordinate, connecting a charged dark state and a neutral bright state of the QD. Both states are subject to spectral diffusion and thus represented by parabolas along the reaction coordinate.

Each "on"/"off" period then starts and ends at the crossing point of the bright- and dark-state parabolas. By solving the diffusion equation, Tang and Marcus [2005] find three different time regimes for this first-passage time distribution. The first regime is determined by a change between bright and dark states due to the dynamics in the close vicinity of the crossing point. This results in power-law statistics with an exponent of $\mu = 1/2$. At longer times, the system reveals the common first-passage time distribution of a random walk with an exponent of $\mu = 3/2$. At long time, the "on" and "off" time distribution cannot grow to infinity due to the parabolic shape of the free-energy surface. Thus an exponential cutoff, $\sim\exp(-\Gamma t)$, to the power law appears. This cutoff time is related to the relative energetic positions of the bright- and dark-state parabolas as well as to the spectral diffusion dynamics in both states [Tang and Marcus, 2005]. At still longer times, charge transfer dynamics as described in photoinduced charge transfer appear, that is, dependencies on electronic overlap integrals and redox potentials. While there is direct or indirect experimental evidence for a cutoff in the "on" and "off" time distributions [Shimizu et al., 2001; Chung and Bawendi,

2004], there is, however, no indication for a turnover from $\tau^{-1/2}$ to $\tau^{-3/2}$ at short times. The observations of Smith et al. even suggest that such a turnover does not exist, since they find a power-law statistics with an exponent of roughly −3/2 down to a nanosecond timescale [Smith et al., 2006].

Like all random-walk models, the DCET has to incorporate anomalous diffusion to explain power-law exponents different from −3/2. This is done by introducing a time-dependent diffusion constant $D(t)$, which is related to the frequency-dependent dielectric relaxation function [Tang and Marcus, 2005]. The dielectric relaxation can be represented by a Cole–Davidson dielectric response function $\chi(\omega)=(1+i\omega\tau_{\mathrm{D}})^{-\beta_{\mathrm{CD}}}$. If $\beta_{\mathrm{CD}} = 1$, normal Debye relaxation is obtained and the diffusion constant is time independent. For $\beta_{\mathrm{CD}} < 1$, the surrounding material includes a spread of relaxation times, which leads to anomalous diffusion. The DCET theory then predicts a power-law exponent of $\mu = 2 - \beta_{\mathrm{CD}}/2$. Thus there is a direct relation between the dielectric relaxation time distribution and the power-law intermittency exponent of single emitters. However, with available relaxation data on polymers, the model does not predict the correct trends for the power-law exponents.

4.4.1.7.5 *Multiple recombination center model*

Many of the previous discussions are based on PL quenching caused by Auger-type processes after reexciting an already charged QD. However, recent experiments, especially on CdSe/CdS QDs, questioned the long-time charging model, since the underlying Auger effects are obviously too small for those QDs to account for effective quenching. Additionally, states with very low PL intensities but fluctuating (blinking) PL decay times have been observed [Jha and Guyot-Sionnest, 2009; Zhang et al., 2010; Zhao et al., 2010; Rosen et al., 2010; Cordonnes et al., 2011]. Amecke and Cichos (2011) found that also large CdSe/ZnS QDs show evidence for the presence of low-intensity states.

While we have already recently favored trapping-type models [Cichos et al., 2007] since they most conveniently account for many observations, Frantsuzov et al. (2009, 2013) recently presented a model called the *multiple recombination center* (MRC) model. In this model the initial step of a blinking sequence is a hole trapping to one of N recombination centers (RCs), followed by nonradiative recombination of the trapped hole with the remaining electron. In that approach the nature of an RC is a priori not predetermined but is thought to be related to the QD surface or to the core/shell interface. Naturally, these RCs will be sensitive to properties of the embedding environment, for example, its dielectric properties. To be a universal model and to extend its applicability also to molecules for which clearly a surface or interface cannot easily be defined, RCs should also be present in the surrounding environment.

Now the essential idea of this model is that each RC has two conformations corresponding to a high and low trapping rate. This way the RC acts like a switch turning PL "on" and "off." We suggest that these switches can be considered to be two-level systems. This analogy is charming in a sense that many properties of heterogeneous materials have been described by an energy landscape of stochastic two-level systems. In that sense, the optical active quantum system (QD, nanostructure, molecule) couples to a bath of two-level systems. At a more complex level, there may be hierarchies of such levels as has, for example, been identified in the case of proteins. We find such a concept intriguing since, as outlined in Section 4.4.1.73, blinking seems to take place only under conditions of disorder, either because the quantum system itself is heterogeneous or at least the environment is.

According to Frantsuzov et al. (2012) the total nonradiative quenching or recombination rate k^{r} is given by the sum of individually fluctuating trapping rates according to

$$k_t(t) = \sum_{n=1}^{N} k_n \sigma_n(t) + k_0 \tag{4.15}$$

$\sigma_n(t)$ becomes 1 when the respective RC is in the active (hole-trapping) conformation and 0 otherwise, while k_0 applies in case that all RC are in the "off" state. Each RC performs independently telegraph noise-transitions between the two conformations with transition rates $\gamma_n^{\pm}$. The assumption now is that these switching rates are broadly distributed, for example, due to tunneling processes in slightly different (asymmetric) two-level states. The distribution of switching rates controls the power spectral density (PSD). Under quasi-steady-state conditions [Frantsuzov et al., 2013] we can apply again Eq. 4.9 (see Section 4.4.1.4) with the only difference that k^{nr} is now converted into $k^{nr}(t)$, thus resulting in luminescence intermittency including a continuous range of PL intensities with the limiting values "on" and "off." The MRC model can explain unlike other models both the distribution of intensities and PL decay times and is independent of threshold settings (as CPA is). The model is based on the concept of f^{-1} noise, with r being a value close to 1. Parametric variables in the RC are N, the number of RCs, and the switching rate γ (including its distribution). Additionally, an interaction among the RCs has partly been taken into account.

Within this MRC framework experimental blinking data (analyzed by the evaluation of the power spectral density approach) could be modeled for very different quantum systems, as shown in Fig. 4.30. The number of reaction centers was between 1 and 12 with variously distributed trapping rates. In most cases an f^{-r} noise was observed with r ranging in $0.7 < r < 1.3$. The satisfying overall agreement for a large manifold of quantum systems within the $1/f$ approach (assuming a limited number of RCs) is a successful complex extension of the multitrap models proposed earlier. Also, the independence from a threshold setting (though there are also other methods to circumvent this drawback) can be considered as an advantage. However, the assumption of various trapping rate distributions remains as a very open fitting parameter. To compare, power-law evaluation needs at most six parameters to describe the experimental findings (see Section 4.4.1). Naturally, the RC model does not answer the question what the physical nature of the traps is. We feel that the main advantage of the MRC model is the close relationship with switching two-level systems as they are typical under heterogeneous conditions.

The previous discussion of the MRC model aimed at two goals, namely to implement a threshold-free analysis of blinking in the frame of $1/f^r$ ($r \approx 1$) noise and to combine this analysis with a possible scenario of coupling the quantum system to two-level (switching) RCs. We have also discussed in Section 4.4.1.2 that CPA is a threshold-free analysis. However, as has been argued recently [Houel et al., 2015], this method suffers though of advantage in comparison to threshold-based approaches from detecting only "real" state (intensity) changes (see also Section 4.4.1.3).

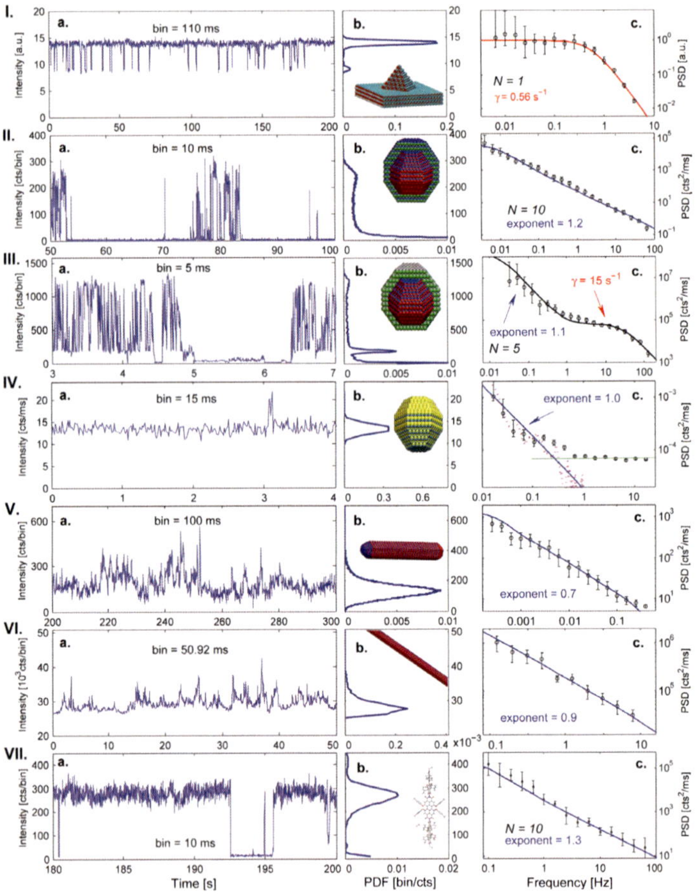

Figure 4.30 Luminescence intermittency in various nanoscale fluorophores. Experimental fluorescence trajectories (panels a), with the corresponding intensity histograms (panels b) and power spectral densities (panels c, circles, experiment; solid lines, theory) are shown for (I) a single self-assembled InP QD, (II) a colloidal CdSe/ZnS (core/shell) QD, (III) a colloidal CdSe/CdS (core/shell) QD, (IV) a CdZnSe/ZnSe nonblinking QD, (V) a colloidal CdSe/ZnSe nanorod, (VI) a CdSe nanowire, and (VII) a perylene diimide molecule. (Panels c) Solid blue lines show f^{-1}-type PSDs calculated from the MRC framework, whereas the black line (III.c) was calculated with specially chosen RCs. The red line (I.c) is a Lorentzian fit to the PSD corresponding to a single RC, while the green line and the dotted magenta line (IV.c) show shot noise and signal spectra after filtering, respectively. The insets in (panels b) show the schematic structure of each fluorescent nanoscale system. The PSD in (panels c) were obtained from a modified average Periodogram method. The error bars correspond to a 95% confidence interval. Reprinted with permission from Frantsuzov, P. A., Volkán-Kacscó, S., and Jankó, B. (2013). Universality of the fluorescence intermittency in nanoscale systems: experiment and theory, *Nano Lett.*, **13**, 402–408. Copyright (2013) American Chemical Society.

4.4.1.7.6 *Other models*

Besides the above-mentioned models to explain power-law intermittency, a number of other attempts and extensions have been made. Frantsuzov and Marcus (2005) have suggested a model which is also based on a diffusion process in energy space. However, it relies on a deep hole trap state as the source of large changes in the nonradiative rate of the QD [Frantsuzov and Marcus, 2005]. Consequently, most of the features found in the DCET model (power law $\tau^{-3/2}$, exponential cutoff) are also present in the model of Frantsuzov. Nevertheless, this model does not seem to apply to the blinking of single organic molecules, since it requires a quenching mechanism specific only for QDs.

On the other hand, most of the features of power-law intermittency coincide with the features of Lévy statistics. Thus emission intermittency of QDs and even molecules can be described in the framework of Lévy walks, which has been worked out in detail by Barkai and colleagues [Margolin and Barkai, 2005; Barkai et al., 2004; Margolin et al., 2006]. This model also addresses the issue of aging. However, the details of the Lévywalk deliver a mathematical description of the experimental phenomena but still lack a connection to physical quantities.

Xia and Silbey (2006) relate the emission intermittency of organic molecules and QDs to Anderson localization. Anderson localization arises from coherent scattering by disorder in a lattice with band structure, leading first to localized states at the edges of the energy band and finally to localization of all band states when the disorder exceeds a critical value. The model therefore includes disorder, which is also supposed to play an important role for the power-law blinking of single molecules. As a result, the model predicts a power law with an exponent of −2 for the "on" time probability distribution. The "off" time is recovered by quantum tunneling according to the model by Verberk et al. [2002].

Most stimulating are recent experiments on charging QDs in electrochemically cells (see for details Section 4.6.1). Such experiments will elucidate more and more details of charge-trapping processes. It has already been shown that blinking processes can be assigned to discernible processes. Galland et al. [2011] identified two types of blinking discriminating charge fluctuations on the

QD surface (B-type blinking) from charging processes of the QD. Schmidt et al. [2014] identified four different blinking processes via CPA. These findings are more deeply outlined in Chapter 6. Osad´ko et al. [2015] assigned two blinking mechanism, namely ionization/neutralization and diffusive fluctuations of atoms on the core-shell interface.

Finally, different blinking processes can be discriminated against each other by the variation of the optical pumping cycle. Evidence of both dark- and light-induced blinking has been shown [Baker et al., 2009]. Lifetime blinking in nonblinking QDs has been explained by a competition of biexciton generation and Auger-type PL quenching [Galland et al., 2012].

4.4.1.7.7 *Comparison of models*

Since the first comprehensive summary of blinking models in 2007 [Cichos et al., 2007] several new experimental insights into blinking dynamics, especially for QDs have been reported. The most important ones are firstly that the long-time charging models have been challenged [Zhao et al., 2010, Rosen et al., 2010; Cordonnes et al., 2011]; secondly, different types of blinking have been identified for one and the same QD [Galland et al., 2010; Schmidt et al., 2014; Osad´ko et al., 2015]; and thirdly, a non-power-law dependence has been found [Krasselt et al., 2011; Schmidt et al., 2014]. Taking these recent experimental findings into account (see also Chapter 6), we will now discuss the strengths and weaknesses of three main approaches explaining power-law intermittency: first the random-walk models, then the trap models, and finally the MRC models. It is amusing to note that a very similar debate arose several decennia ago in the analysis of geminate ion recombination in solid organic glasses, a problem whose physics now appears to be akin to that of blinking. Then, too, diffusion [Debye and Edwards, 1952] and trapping [Tachiya and Mozumder, 1974] models were proposed to explain power laws in the slow recombination kinetics.

The main strength of random-walk models is their natural introduction of a power law as a consequence of the many diffusing steps taken by a random walker. The exponent in the first-passage time distribution is 3/2 in 1D space and decreases to 1 when the dimension tends to 2. Several recent investigations have shown that power-law exponents are more broadly distributed than previously assumed. Moreover, systematic deviations from a power law have been observed also in the case of QDs [Krasselt et al., 2011; Schmidt et al., 2014]. Though being very general, specific findings cannot be readily explained. Moreover, most of the recently elaborated crucial questions (see a detailed summary in [Cichos et al., 2007]) have not been answered. Conclusively, trap or MRC models are superior compared to random-walk models.

The trap models, as mentioned before, are more specific than the random-walk models. The invoked trapping recovery tunnel mechanisms have been known for many years in the physics of solid state devices and in physical photochemistry. The combination of the exponential dependences of the trapping and recovery rates naturally leads to power laws with exponents in the right range, and observed dynamics over up to nine decades of times can easily be accommodated by tunneling rates within a few nanometers from the emitting center, be it an NC or an organic molecule. The weak temperature dependence is a general feature of electronic tunneling processes, in which large excesses of electronic energy are dissipated into vibrational heat. The puzzling question of the origin of a quasi-continuous distribution of trap distances is neatly answered by the hypothesis of self-trapping of the photoejected charge. Self-trapping requires no chemical impurity but presumably requires a local potential minimum (pretrapping), which can explain why power-law intermittency is observed in disordered media only. As the energy of the self-trapped state is very low [Duke and Fabish, 1974; Gady et al., 1998], the associated large distortion along the reaction coordinate entails very low mobilities, even in strong fields or in presence of thermal fluctuations. However, one has to be critical that the obviously broad distribution of trapping states is merely a convolution of various physically distinct blinking processes.

At this point we would like to strengthen that models which concentrate on universal power-law coefficients are probably not realistic, since experiments show fluctuations of the respective coefficients. Additionally, more and more deviations from power laws are nowadays explored and several different blinking mechanisms can be identified for a single QD which are related "on-off" charging processes (of Auger type) and/or fluctuations of charges, ligands, surface atoms, or polarons. Cordones and Leone (2013) have recently summarized the richness of charge-trapping processes in relation to blinking dynamics.

The present discussion is an attempt to convey our personal conviction that the main features of the trap model (or the MRC variant) can account for many, if not all, of the observations of power-law intermittency on single molecules and single NCs. New experiments, more directed to the mobility of the separated charges, are required to specify the different models. In the next section we will outline in detail that power-law behavior of blinking is only one (practical variant) to describe blinking dynamics.

Finally, studies of blinking may eventually give us access to the structure and dynamics of (self-) trapped charges in electric insulators and at semiconductor interfaces [Montanari and Morshuis; 2005].

4.4.1.7.8 *Questioning power-law models*

In the previous sections we have reviewed the most common models for luminescence intermittency for very different types of single quantum objects. All the concepts have in common that a broad distribution of blinking kinetics is involved, often described by power-law-type behavior. Though the power-law approach is simple and convenient, the question arises whether it is of intrinsic nature or merely a convenient approximation. Recent observations [Schmidt et al., 2014] indicate, however, that power laws do not always apply and are often only a rough approximation, especially for "on" times. The latter is not surprising since in a first approximation one would expect an exponential behavior in the case of "on" times. Most experiments have been performed using a threshold-based analysis which not only may suffer from artifacts [Crouch et al., 2010] but is only under certain conditions expected to be applicable [Amecke et al., 2014]. The MRC model predicts a more versatile approach than the power-law function on the base of $1/f$ noise [Frantsuzov et al., 2009, 2013]. Altogether we are convinced that multiple traps (or RCs) are the most convincing basic models to describe complex intermittency statistics. To proceed in questioning power laws we will introduce our new experimental findings and their consequences in the following.

One step further than evaluating the statistics is the assignment of the physical relevant identification and nature of the predicted trap or self-trapped states. To approach this goal two steps have to be undertaken:

(i) Variation of intrinsic or extrinsic parameters in order to tune blinking statistics. (We will turn to this point in Chapter 6).

(ii) Correlated observation of as many as possible different spectroscopic observables during a blinking time trace, that is, blinking statistics, PL intensities, QYs, PL decay times, and spectral fluctuations. In the following paragraphs, we will summarize such an approach based on CPA [Schmidt et al., 2012, 2014].

In the preceding discussion we concentrated on the question of how to determine adequately blinking statistics. However, equally important are the underlying physical models in terms of the kind of quenching process and the involved excitonic states. We have described recently a CPA-based investigation of blinking of CdSe/ZnS QDs in two different polymers [Schmidt et al., 2012; 2014] via confocal microscopy under pulsed laser excitation. Later spectral resolution has been added in correlation to blinking [Zenkevich et al., 2015; Göhler, 2014]. Time resolution covers the microtime of the PL decay with a resolution of 200 ps and the macrotime (blinking) with a corresponding resolution of (depending on PL intensity) typically 500 μs. Roughly speaking, microtimes do not depend noticeably on the embedding matrix, while macrotimes change considerably. The principal findings are the following:

- CPA-clustered PL intensities show different regimes of density (see Figs. 4.19B and 4.20c).
- Deviations from truncated (threshold-based) power laws for "on" times (according to Eq. 4.8 and Fig. 4.20a) corroborate with high densities in the range N (Fig. 4.20c) and an increased probability for large intensity jumps ΔI (Fig. 4.20d). This implies the presence of a specific state with a well-defined single (exponential) "on" time.
- The remaining PL intensities show only small intensity jumps ΔI, thus proceeding from "on" to "off" (and vice versa) in only small intensity jumps ΔI (see Fig. 4.31).

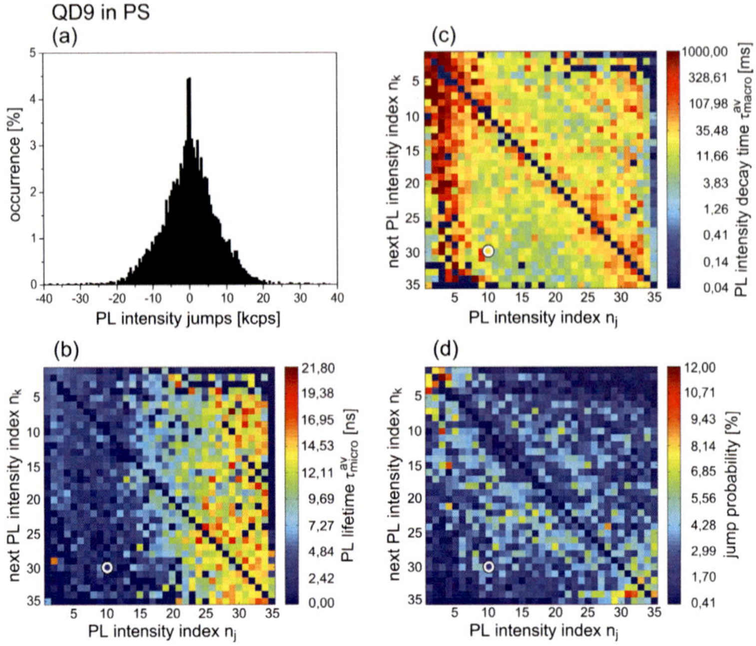

Figure 4.31 Jump dynamics between consecutive (clustered PL) intensities for the single CdSe/ZnS QD9 in PS. (a) Distribution of intensity jumps ΔI. (b) Averaged PL lifetimes τ_{microav} (color code) as a function of clustered intensities n_j before jumping to the next intensity n_k (c) Averaged intensity decay times τ_{macroav} (color code) as a function of clustered intensity n_j before an intensity jump to the next intensity n_k. Note the long time τ_{microav} for $n_j \approx 4$, independent of n_k. (d) Jump probability (color code) for an intensity jump from intensity n_j to n_k. The same notations apply as for (c). The white circles in (b–d) relate to an example of how the graphs should be interpreted as a relation between the starting intensity level $n_j = 10$ jumping to $n_k = 30$. We have chosen a linear scale n_j, n_k for clustered intensities I since the distribution of the very strong (spike) intensities (S) is much less dense compared to those for the other intensities (see Figs. 4.19B and 4.20c. Reprinted with permission from Schmidt, R., Krasselt, C., Göhler, G., and Borczyskowski, C. von (2014). The fluorescence intermittency for quantum dot is not power-law distributed: a luminescence intensity resolved approach, *ACS Nano,* **8**, 3506–3521. Copyright (2014) American Chemical Society.

- The intensity distribution is accompanied by a systematic variation of the PL decay time (Figs. 4.23e and 4.24), only roughly following Eq. 4.9 across part of the intensity range.
- However, for most of the PL intensities the PL decay is at least bi- and at very low intensities even triexponential with

different radiative rates. This leads to the conclusion that three rather well-defined states have to be taken into for each QD. We suggest, that these states are related to three different hole-trapping situations [Schmidt et al., 2014].

- At very high intensities (with very short, matrix-dependent, "on" times), PL decays monoexponentially without any dependence on PL intensity (Fig. 4.23e). This can be explained according to Fig. 4.32 by taking into account that the excitation into energetically high exciton states creates hot electrons and is obviously also subject to blinking processes, a finding which is supported by observations from PL excitation spectra discussed in Chapter 6 and more recently on single-QD data where blinking is influenced via external electric fields [Galland et al., 2011].

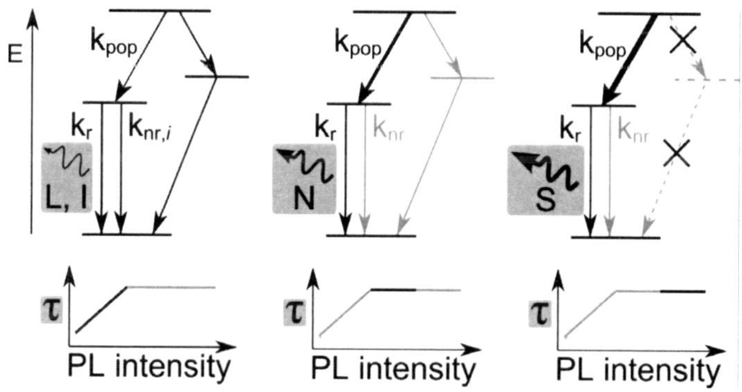

Figure 4.32 Population dynamics and PL intensity–dependent lifetime. PL intensity classes L, I, N, and S (see Fig. 4.23) are shown as a function of hot-electron relaxation pathways. Spikes S are observed in the case that the nonradiative bypass is blocked. Reprinted with permission from Schmidt, R., Krasselt, C., Göhler, G., and Borczyskowski, C. von (2014). The fluorescence intermittency for quantum dot is not power-law distributed: a luminescence intensity resolved approach, *ACS Nano,* **8**, 3506–3521. Copyright (2014) American Chemical Society.

- Finally, the decay of each PL intensity is on the macrotimescale characterized by a monoexponential decay different at various PL intensities (see Fig. 4.33a). It is clearly seen that "on" times are in this type of QD not power-law-distributed but that only a few macro-decay times (which correspond to on times) are identified with upper and lower limits, as shown schematically in Fig. 4.33a. A superposition of these decays reconstructs the apparent power law, as shown in Fig. 4.33b. This result resembles the finding of Frantsuzov et al. (2013) that only a few RCs are needed and that a power-law behavior applies for a given system only across a limited frequency (respectively time) range.

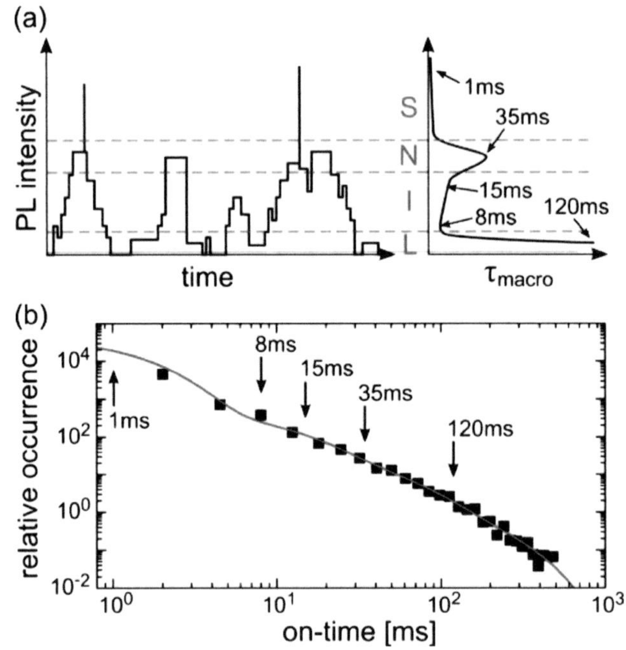

Figure 4.33 (a) Schematic presentation of the time-dependent PL intensity (left) and the corresponding decay times τ_{macro} as a function of clustered intensities I (right). (b) Fit of the "on" time histogram (see Fig. 4.20a) by five exponential decay functions with fixed decay times $\tau_{macro,i}$ = {1, 8, 15, 35, 120} ms (a, right) instead of a truncated power law. Reprinted with permission from Schmidt, R., Krasselt, C., Göhler, G., and Borczyskowski, C. von (2014). The fluorescence intermittency for quantum dot is not power-law distributed: a luminescence intensity resolved approach, *ACS Nano*, **8**, 3506–3521. Copyright (2014) American Chemical Society.

The overall model is shown in Fig. 4.34. From this scheme it becomes evident [Schmidt et al., 2014] that the experimentally observed PL is not necessarily related to PL from the exciton but may also be due to PL caused by recombination of a trapped hole and the electron. In that case excitonic emission can be only observed with a low intensity (slightly blue-shifted with respect to PL from trapped hole-electron recombination). We tentatively suggest that the eventually observed PL slightly above the noise level [Zhao et al., 2010; Amecke and Cichos, 2011; Cordonnes, 2011] might alternatively be due to (quenched) excitonic emission and not due to Auger-type PL. We would like to point out, however, that the scheme (Fig. 4.34) is only a very tentative model to strengthen the complexity of blinking processes. A more detailed analysis is presented in Chapter 6.

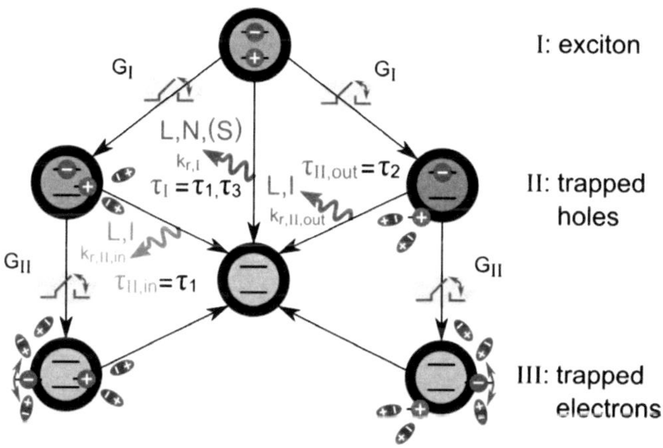

Figure 4.34 Schematic population and switching dynamics in a core/shell QD: hierarchy of exciton (I) relaxation via hole (II) and electron trapping (III). Wavy lines denote the three radiative transitions with corresponding PL intensity ranges (L, I, N, S), wavelength of the PL, and PL decay times ($\tau_{\rm I}$, $\tau_{\rm II,in}$, $\tau_{\rm II,out}$; indices "in" and "out" indicate trap positions of the hole within the QD shell) related with experimental PL lifetimes (τ_1, τ_2, τ_3; τ stands for $\tau_{\rm micro}$). We assume similar radiative rates $k_{\rm r,I} \approx k_{\rm r,II,in} \approx k_{\rm r,II,out}$. Gate symbols denoted by $G_{\rm I}$ and $GI_{\rm I}$ switch decay channels accessible (gate open) or blocked (gate closed; see also Fig. 4.10). In the case of open (closed) gates $G_{\rm I}$ the decay of the exciton (*I*) is characterized by a low PL intensity *L* (*N*) and a short (long) PL lifetime τ_3 (τ_1). We assume only two hole-trapping configurations visualized by trapping at the inner (left side, index in) or outer shell interface (right side, index out). Electron trapping is assumed to follow a broad distribution of processes indicated by the double arrow among schematic electric dipoles. Reprinted with permission from Schmidt, R., Krasselt, C., Göhler, G., and Borczyskowski, C. von (2014). The fluorescence intermittency for quantum dot is not power-law distributed: a luminescence intensity resolved approach, *ACS Nano*, **8**, 3506–3521. Copyright (2014) American Chemical Society.

We would like to demonstrate in this discussion and the one in Chapter 6 that a microscopic blinking model needs more than evaluating statistical properties but relies on additional input, for example, from spectroscopy (with time and spectral resolution), to unravel the complex interplay of electron and hole kinetics in combination with a high degree of heterogeneity. Moreover, variation of temperature, excitation intensity, and excitation energy is necessary to conceive a completely coherent model. Additionally, the embedding matrix and the kind of capping and ligand shell should be varied to tune blinking effects. Applying an electric field will also promote a thorough understanding of blinking dynamics. We will discuss these aspects later in this chapter and in Chapter 6 in more detail.

4.4.1.7.9 *Nature of traps*

Blinking models do not necessarily need a microscopic understanding of the nature of traps, though Issac et al. (2005, 2012) suggested self-trapped states in a sense of polarons in polymers or the surfactant shells. This is supported by the influence of AC and DC electric fields on blinking dynamics in nonconducting matrices [Krasselt, 2015], as also shown in Section 4.6.1.2. Krasselt et al. (2011) and Schmidt et al. (2012, 2014) explained a certain type of (exponential) blinking dynamics in CdSe/ZnS QDs by coupling to OH groups, which traps holes in the environment of the QDs. Interestingly, very recently Zbydniewska et al. (2015) found a (power-law-type) correlation between luminescence blinking and charge blinking via a carbon nanotube–based single-charge sensor in a field-effect transistor and identified a blinking responsible trap in the band gap of the CdSe/ZnS QD with a charging energy of 200 meV. Orfield et al. (2015) correlate the atomic structure (surface and internal effects determined by electron transmission microscopy) of an individual CdSe/CdS QD with its PL properties including blinking. They could, for example, relate certain structural defects with permanently dark QDs.

Trapping at chemical impurities seems excluded, because already in moderately pure substrate materials, the probability of having more than one impurity in the relevant volume around a single emitter (10–100 nm^3) is negligible. However, doping of CdSe QDs by copper (Cu^+) created electron traps shortening "on" times during a blinking time trace [Whitham et al., 2015]. Recently we have shown that Cu-porphyrin attached on a nearly 1:1 base to the surface of CdSe/ZnS QDs induces (at temperatures below 230 K) deep intraband states, resulting in PL quenching and longer "off" times [Zenkevich et al., 2015; Zenkevich and Borczyskowski, 2016].

Several attempts are reported to calculate the dynamics at surfaces which are responsible for fluctuating trapping centers for charges or relaxation (in spirit of the MRC model). Such processes are thought to be due to (structural or binding) fluctuations within the ligand shell [Voznyy, 2011] or even of surface atoms [Voznyy and Sargent, 2014]. Brawand et al. (2015) calculated that dangling bonds cause trap states at the surface of oxidized Si QDs resulting in exponential blinking dynamics.

4.4.2 Spectral Diffusion

As has been outlined in Chapter 1, analysis of QD spectra is the basic approach to characterize QDs with respect to excitonic absorption and PL emission energies. Naturally, in a first approach of optical spectroscopy on single QDs, all inhomogeneous contributions to the line width should vanish and the spectral width should be only limited by the PL dynamics. However, contrary to spectroscopy on single molecules (in molecular crystals) at low temperatures this has not been observed in the case of QDs.

One of the reasons is that similar to blinking dynamics the PL emission energy is not fixed in time but fluctuates on timescales similar to those observed for blinking. This phenomenon is called spectral diffusion (or spectral jitter) typically also for single molecules in polymers [Moerner and Orrit; 1999] and as detected much earlier in spectral hole-burning experiments [Fleury et al., 1993]. A typical example of spectral diffusion is shown in Fig. 4.35 for PL from a single Si NC [Martin et al., 2012]. The PL stems from a single defect center probably at the Si/SiO_2 interface. It can be clearly seen that all parameters (energies of the PL band maxima, line widths, and PL intensities vary on timescales of seconds.

Figure 4.36 demonstrates that time-dependent spectral shifts of the most intense band are intimately related with the integrated PL intensity. We have ascribed these observations to fluctuating charges close to the emission center, causing a fluctuating Stark effect which shifts spectra upon an increasing electric field to lower energies, thereby strengthening simultaneously radiationless transitions (followed by a decrease in PL intensity) due to enhanced exciton–phonon coupling. Obviously fluctuations of PL intensity and PL energy are caused by the same mechanism. These intensity fluctuations can be assigned to a kind of "blinking" of PL energies.

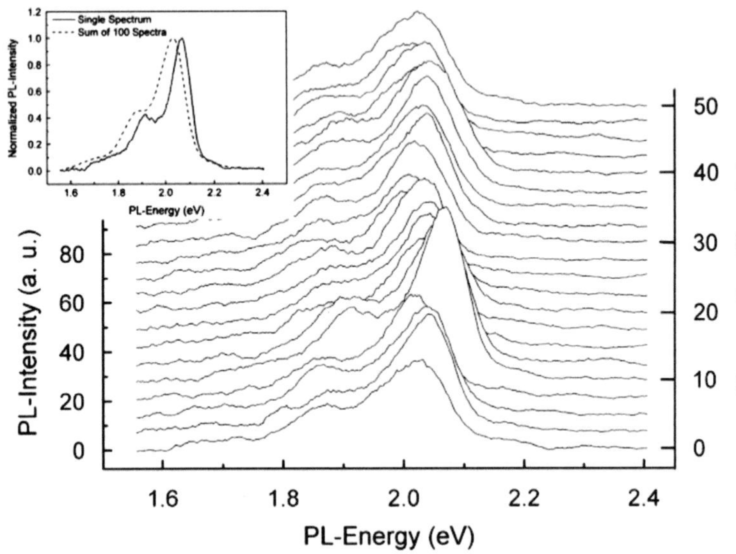

Figure 4.35 Series of subsequent PL spectra of a single Si NC with 0.5 s integration time over a time period of 50 s. Excitation occurs at 532 nm (2.33 eV) with an excitation intensity of 6×10^3 W/cm^2. The inset shows a comparison of PL spectra of a single spectrum with 1 s integration time (full line) and the sum of 100 subsequent spectra (broken line) of one typical Si NC. Reprinted from Martin, J., Cichos, F., and Borczyskowski, C. von (2012). Spectral diffusion of quasi localized excitons in single silicon nanocrystals, *J. Lumin.*, **132**, 2161–2165, Copyright (2012), with permission from Elsevier.

To investigate the close relation between spectral fluctuations (diffusion) and PL intensities, we have undertaken a more systematic investigation for two different types of blinking QDs, CdSe/ZnS and CdSe/CdS (in poly(methyl methacrylate) [PMMA]), on a SiO_x substrate to test whether such a close relation holds also in these cases. We have made use of the two-channel detection scheme outlined in Section 4.4.1.5. This experimental approach allows for a faster (1 ms) and more sensitive analysis compared to the one when recording complete spectra as in the case of Si NCs with about 1 s integration time. Intensities during the blinking time trace have been analyzed with CPA. This setup allows also the simultaneous detection of the PL decay and blinking dynamics.

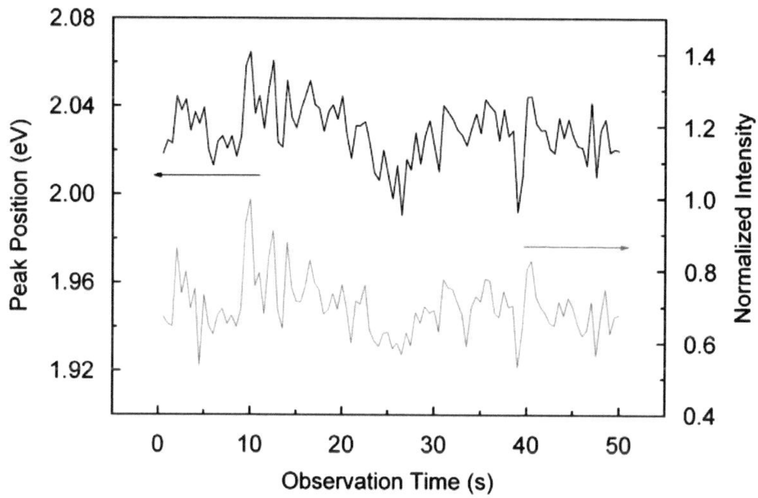

Figure 4.36 Correlation of the PL energy of the principal PL line of a single Si NC and the integrated PL intensity, both as a function of observation time at 300 K. PL intensity has been set to 1 at the highest intensity during observation time of 50 s. Reprinted from Martin, J., Cichos, F., and Borczyskowski, C. von (2012). Spectral diffusion of quasi localized excitons in single silicon nanocrystals, *J. Lumin.*, **132**, 2161–2165, Copyright (2012), with permission from Elsevier.

Figure 4.37 compiles a comparison of threshold-based "on" times (a), the shift of the spectral position (with respect to the separation wavelength of the dichroit) as a function of the PL intensity (b), and the probability of PL intensities during a blinking time trace. We observe a shift to the red with decreasing PL intensity. However, there are a few distinct observations. At intensity level $N \approx 170$ we observe a relatively high density of states which is accompanied by a slight spectral blue shift. We have shown that an increased density of intensities in the intensity range N is in all cases related to a deviation (see Fig. 4.37a, top) from a truncated power law [Schmidt et al., 2014].

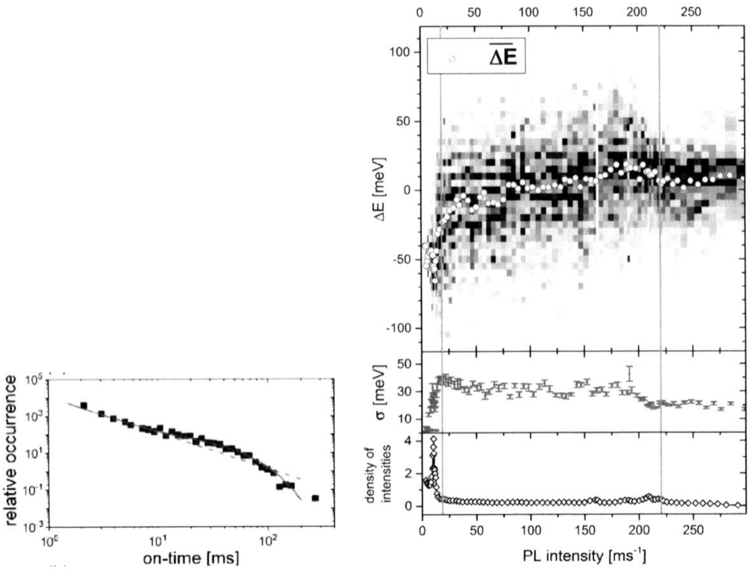

Figure 4.37 Left: Threshold-based (binned) "on" time histogram of the single CdSe/ZnS QD7 on a SiO_x substrate. The solid curve indicates a fit according to a superposition of a truncated power law and an additional exponential decay according to Eq. 4.6, whereas the dashed line depicts a power law without truncation for comparison. Right top: Spectral jumps (diffusion) ΔE with respect to the dichroic beam splitter at 567 nm during a blinking time trace (of typically 20 min) as a function of the PL intensity (determined by CPA) of a single CdSe/ZnS QD on SiO_x. Circles correspond to the algebraic average of fluctuating ΔE at the respective PL intensity. The two black bars indicate in the high-intensity range the maximum of the intensity distribution and at 1/10 of it. The latter corresponds to intensities still detectable in continuous-wave PL spectra. The spectral shift between these two marks is typically ~35 meV. Right bottom: σ (crosses) corresponds to the Gaussian width of the spectral diffusion at a given PL intensity and is typically 20–40 meV in the intensity range between the two black bars. Density of intensities (diamonds) is a function of PL intensity.

Down to $N \approx 50$, the average spectral shift follows linearly the PL intensity as in the case of Si NCs (see above). For $N < 50$ the relationship is more complex but a larger red shift is observed, indicating that in this range, states with considerably different electronic nature than those at higher intensities contribute to the PL intensity and PL energy.

It is not evident from results reported previously in the literature that spectral diffusion of QDs is proportional to the PL intensity during a blinking time trace. On the contrary, Neuhauser et al. [2000] report that spectral jumps occur only following an "off" time. However, there are not many experiments reported on spectral diffusion with respect to blinking as, for example, in Plakhotnik et al. (2010).

Basically three major influences might be thought of as contributing to spectral diffusion processes, as will be discussed in more detail in Chapter 6. An intrinsic one is related to photoinduced charging processes and/or generation of multiexcitonic states which all have different PL energies in the order of tens of millielectron volts [Blanton et al., 1996; Empedocles and Bawendi, 1999]. Extrinsic influences are related to fluctuations of the environment, such as diffusing charges [Neuhauser et al., 2000], or local fluctuations of dielectric properties of the embedding surrounding [Gomez et al., 2006]. This hypothesis has been tested applying an external electric field, resulting in quantum-confined Stark interactions [Empedocles and Bawendi, 1997]. The influence of charges or the dielectric environment on spectral fluctuations or spectral line width has been investigated by several groups [Müller et al., 2004; Fernée et al., 2012].

A third process somewhat in between intrinsic and extrinsic effects might be thought of as being due to (structural or binding) fluctuations within the ligand shell [Voznyy, 2011] or even of surface atoms [Voznyy and Sargent, 2014]. Naturally, most of these processes will be influenced, if not even controlled, by the embedding matrix and the interfaces. This applies especially if charge transfer or diffusion of charges are involved. The conceptional background for this coupling is the reorganization energy accompanied by charge reorganization, a concept which has been suggested by Tang et al. (2005) and implemented into the DCET mechanism to explain power-law blinking (see Section 4.4.1.7.4).

Though spectral diffusion is detected on timescales longer than milliseconds and spectral fluctuations larger than a few millielectron volts they can be detected by classical spectroscopic means. However, in case they are much faster on an even smaller energy scale (spectral jitter) they contribute to the line width of the PL transitions, due to which the homogeneous width cannot be directly approached [Müller et al., 2004]. To proceed [Fernée et al., 2010; Plakhotnik et al., 2010], it needs a spectral two-channel detection scheme, as described shortly in Section 4.4.1.5. An identification of very small spectral jitter has been successfully achieved via photon correlation Fourier spectroscopy, as has been described recently in a comprehensive review [Cui et al., 2014] to obtain an ultranarrow line width of 10 μeV at low temperatures. More details of this technique and evaluation of detailed information on spectral diffusion are also described in that review.

4.5 Quantum Dot Nanoprobes

The discussion in previous sections has shown that QDs are very sensitive to subtle changes with respect to both intrinsic and extrinsic properties. There are several reasons for this behavior. According to the particle-in-a-box model outlined in Chapter 1, the wavefunction of the exciton is in principle confined to the core of the QD (quantum confinement). However, like in every quantum system, confinement is not perfect and the corresponding wavefunction tunnels beyond the confining energy walls. Such tunneling is according to the concrete conditions different for electron and holes. In the case of colloidal QDs, tunneling is reduced via appropriate semiconducting shells (see also Chapter 1), for example, a CdSe core covered by a thin ZnS shell, which introduces energetic barriers both for hole and electron tunneling. On the other hand, CdS shells are efficient barriers mainly for the holes [Chen et al., 2013] which has, for example, consequences for Auger recombination processes since the overlap of electron–hole wavefunctions is reduced [Chen et al., 2008]. A schematic presentation of wavefunction tunneling of the radial part is shown in Fig. 4.38. A systematic investigation of such tunneling effects has been published by Dabbousie et al. (1997). In Chapter 7 we will discuss an application of this approach to explain non–Förster resonance energy transfer (non-FRET) PL quenching [Blaudeck et al., 2008; Kowerko et al., 2010; Blaudeck et al., 2012].

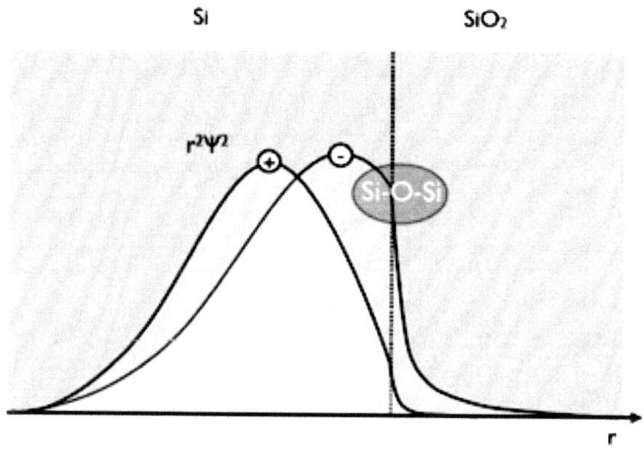

Figure 4.38 Scheme for the tunneling of the radial part of the electron and hole wavefunctions at the interface of a Si QD to its SiO_2 interface. Si-O-Si indicates the coupling of the (electron) wavefunction to Si-O-Si vibrations, as experimentally observed. Reprinted with permission from Martin, J., Cichos, F., Huisken, F., and Borczyskowski, C. von (2008). Electron-phonon coupling and localization of excitons in single silicon nanocrystals, *Nano Lett.*, **8**, 656–660. Copyright (2008) American Chemical Society.

Naturally, in nano-objects the core-to-surface-volume ratio is relatively small. Due to this surface properties are extremely important in the case of QDs. Intrinsic surface properties are to a large extent determined by unsaturated (dangling) bonds, which is not considered in the particle-in-a-box model (see also Chapters 1 and 2). Recent theoretical approaches are able to take surface properties into account and calculate the near-band-edge optical properties (e.g., Frenzel et al. [2007] and Fischer et al. [2012]).

Besides these, intrinsic quantum mechanical "sensitivity" to material properties beyond the QD core synthetic concepts for the preparation of colloidal QDs distinctively influences their PL properties. Basically two constituents play a major role in this respect, namely the morphology (crystal structure, shell thickness, amalgamation at the interface) of the semiconducting shell and the influence of organic ligands (surfactants). Though the major role of the latter is to prevent further growth and aggregation of QDs during wet-chemical synthesis and further on in solution, they considerably influence PL properties due to their chemical nature (in relation to the intrinsic electronic properties of the surface) and the degree of surface coverage [Kowerko et al., 2010; Blaudeck et al., 2012]. In solution, surface coverage is even dynamic equilibrium between solvated and surface-bound ligand molecules. On the other hand, semiconducting shells exhibit strain on the QD core material as long as lattice constants are not matching.

Taking all these features into account it is obvious that optical properties depend critically on the conditions at the surface, the interface(s), and the embedding environment. Obviously many applications of QDs will depend on their interface properties [Zenkevich and Borczyskowski, 2016]. Additionally, for the sake of these properties, single QDs are sensitive probes of their nanoenvironment, as will be demonstrated in the following examples.

4.5.1 Influence of Substrates and Interfaces

In all cases QDs have to be deposited onto a substrate or to be incorporated into a solid (polymer) or liquid solvent. In the following paragraphs we will discuss qualitatively some aspects related to the influence of the matrix (substrate and dielectric properties of polymers) on optical properties of single QDs. The impact of ligands will be presented more rigorously in Chapter 6 and Trenkmann et al. (2016). The influence of solvents on PL energies, QY, and PL decay times has been reported repeatedly for ensembles of QDs (see, for example, Leatherdale and Bawendi [2001]). There is ample evidence that the matrix modifies PL energies and radiative decay times the overall effects are relatively small. More dramatic is the influence on the QY. As the QY is intimately related with PL decay dynamics (ratio between radiative and nonradiative decay rates), it can be more directly investigated on a single-QD level, especially with respect to blinking dynamics. Moreover, single QDs offer the possibility to investigate the influence of local inhomogeneities in combination with microscopic approaches.

Several investigations take the influence of the substrate (for example, SiO_x, indium tin oxide [ITO]) into account. Also graphene influences blinking dynamics [Rogz et al., 2014]. A further illustrative example of the influence of a substrate on PL blinking has been reported by Fu et al. (2007) in Fig. 4.39 for streptavidin-conjugated CdSe/ZnS QDs immobilized on a biotinylated bovine serum albumin monolayer on a glass substrate or near–silver island films (SIFs) on that substrate. The arrangement of the overall nanostructure is shown in Fig. 4.40.

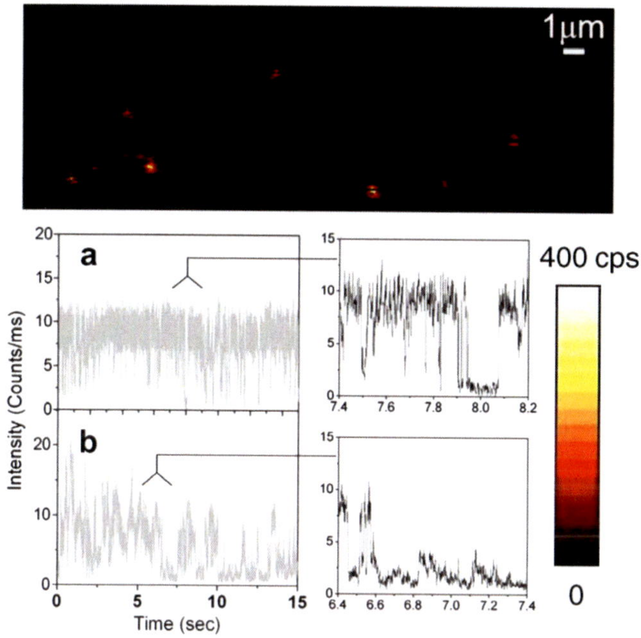

Figure 4.39 Left: Single QD images (top) and representative blinking time traces and their expanded sections recorded for single QDs immobilized on a glass substrate (bottom). Right: Single QD images (top) and representative blinking time traces and their expanded sections recorded for single QDs immobilized on a SIF substrate. Reprinted from Fu, Y., Zhang, J., and Lakowicz, J. R. (2007). Suppressed blinking in single quantum dots (QDs) immobilized near silver island films (SIFs), *Chem. Phys. Lett.*, **447**, 96–100, Copyright (2007), with permission from Elsevier.

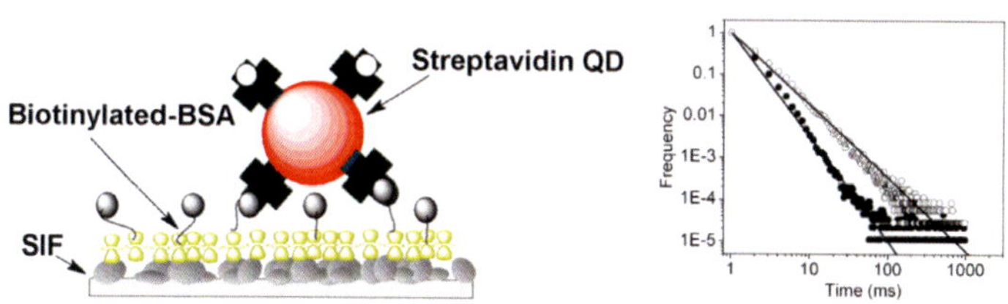

Figure 4.40 Scheme of the nanostructure including CdSe/ZnS QDs (left). "Off" time histograms compiled from around 70 single QDs. The lines are approximately inverse power-law fits to the histogram, giving rise to an exponent of −1.74 on glass (open circles) and −2.5 on an SIF (solid circles). Reprinted from Fu, Y., Zhang, J., and Lakowicz, J. R. (2007). Suppressed blinking in single quantum dots (QDs) immobilized near silver island films (SIFs), Chem. Phys. Lett., 447, 96–100, Copyright (2007), with permission from Elsevier.

It can be clearly seen that blinking is dramatically reduced (thus QY is enhanced) upon addition of an SIF. Blinking suppression has been ascribed to the modification of PL by the surface plasmon resonance from the metallic nanostructures. When the optical transition dipole of the QD is localized near the metal particle, the radiative rate of the QD is altered through coupling with the metal plasmons [Lakowicz, 2001].

Most probably the overall radiative rates have been changed in the example of QDs near to the SIF, demonstrating how properties of two completely different nanostructures can be coupled.

More closely related to local properties of a substrate—in the following a SiO_x surface—is the influence of the degree of hydroxylation on blinking of CdSe/ZnS QDs, as shown Fig. 4.41. This example is the first published one which investigated deviations from a (truncated) power-law behavior. In case that the concentration of substrate silanol (OH) groups is increased the truncated power law for "on" times is superimposed by an exponential contribution according Eq. 4.8 [Krasselt et al., 2011]. It turned out that this process is completely reversible but does not show up for "off" times. It results on average in an increase of the QY by a special coupling to the substrate since the contribution turns out to be single-exponential. It depends also on the kind of ligands. Details of this finding are more deeply discussed in Chapter 6 and Zenkevich and Borczyskowski (2016).

The third example deals with SiO_x surface layerson Si NCs. The impact of surface layers on Si NCs is different from the ones of II/VI compounds since they serve not as a barrier to more strongly confine the exciton but to saturate dangling bonds at the Si surface which otherwise inhibit PL. Figure 4.42 shows PL spectra of two single Si/SiO_x NCs on a SiO_x substrate. The PL emission stems from emission close to the interface (partly localized excitons [Schmidt et al., 2012]), as can be concluded from the strong interaction with Si-O-Si type phonons [Martin et al., 2008] showing up as side bands at lower energies of the PL.

As Fig. 4.42 shows, individual NCs emit at different energies and show different intensity ratios of phonon side bands with respect to the principal line at high energies. This behavior could be treated in terms of electron–phonon coupling, depending on the PL energy of the Si NC, as shown in Fig. 4.43. The Huang–Rhys factor varies across the range of PL energies nonmonotonously, which has been explained by distinguishable localization of the electron and hole wavefunction, respectively [Martin et al., 2008]. It should be noted that electron–phonon coupling depends in this case also on the dielectric function.

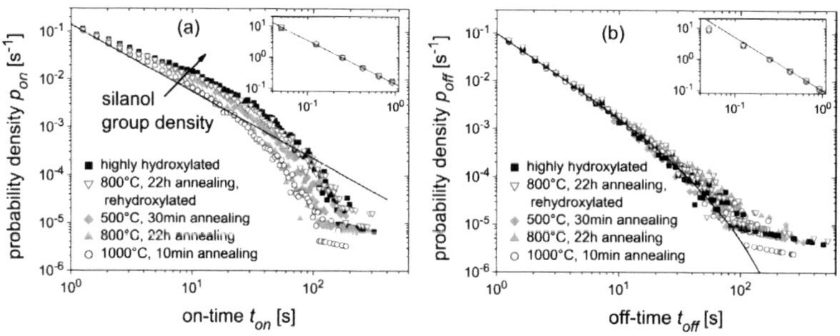

Figure 4.41 Distributions for (a) "on" and (b) "off" times of at least 100 CdSe/ZnS QDs on silicon oxide surfaces with varying degrees of hydroxylation and silanol group densities resulting from different preparation conditions. The insets display the timescale at very short "on" and "off" times. The black line in (a) corresponds to an untruncated power law (Eq. 4.6) with $\alpha_{\text{on}} = -1.4$, whereas the fit of the "off" times in (b) follows Eq. 4.7 with $\alpha_{\text{off}} = -1.71$ and $\Gamma_{\text{off}} = 21 \times 10^{-3}\ \text{s}^{-1}$. Reproduced from Krasselt, C., Schuster, J., and Borczyskowski, C. von (2011). Photoinduced hole trapping in single semiconductor quantum dots at specific sites at silicon oxide interfaces, *Phys. Chem. Chem. Phys.*, **13**, 17084, with permission from the PCCP Owner Societies.

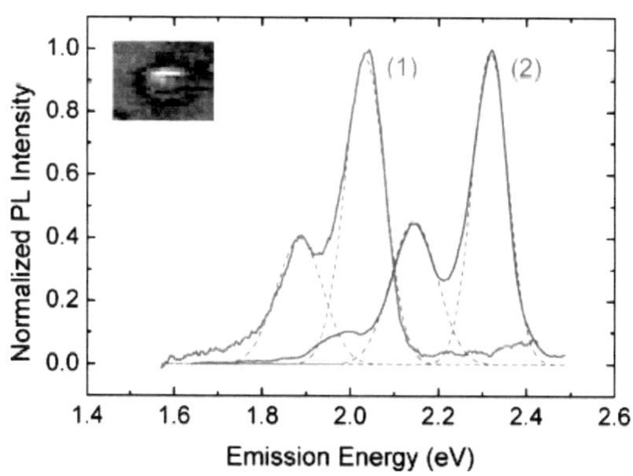

Figure 4.42 PL spectra of two single Si QDs. The broken lines show Gaussian fits to the experimental data. The insert shows a confocal scan over a single Si NC exhibiting PL blinking, as indicated by dark and bright stripes. Reprinted with permission from Martin, J., Cichos, F., Huisken, F., and Borczyskowski, C. von (2008). Electron-phonon coupling and localization of excitons in single silicon nanocrystals, *Nano Lett.*, **8**, 656–660. Copyright (2008) American Chemical Society.

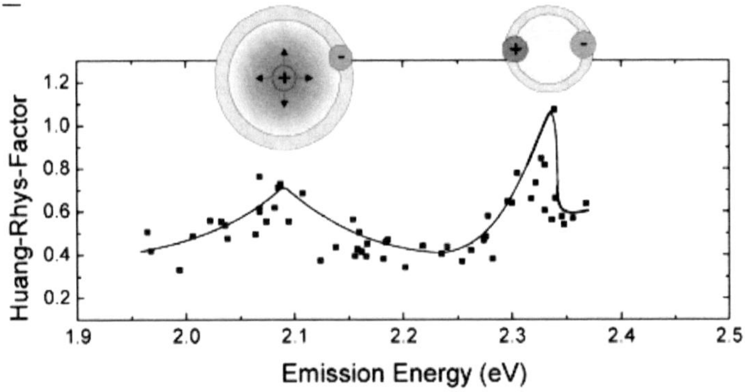

Figure 4.43 Dependence of the Huang–Rhys factor *S* on PL energy. The line serves only as a guideline. A scheme of the localization model separately for the electron (~2.1 eV) and the hole (~2.33 eV) is shown on top. Crystals are smaller at high PL energies (quantum confinement). Reprinted with permission from Martin, J., Cichos, F., Huisken, F., and Borczyskowski, C. von (2008). Electron-phonon coupling and localization of excitons in single silicon nanocrystals, *Nano Lett.*, **8**, 656–660. Copyright (2008) American Chemical Society.

4.5.2 Influence of Dielectric Properties

We have outlined in the previous sections that blinking is—at least in most of the cases—related to (photoinduced) charge separation, charge fluctuation, or charging. This naturally implicates that these charges "feel" the dielectric properties of the environment/interfaces since, as we have pointed out, the exciton wavefunction is leaking (or tunneling) out of the confined QD core. Consequently, blinking dynamics depend on the dielectric constant of the embedding environment (ligand shell and/or matrix), as has been shown in several publications [Issac et al., 2005, 2012; Schuster et al., 2005; El-Kork et al., 2011; Schmidt et al., 2012, 2014]. Very importantly, it should be emphasized that blinking dynamics (i.e., distributions of "on" and "off" times) depend very strongly on the dielectric features of the embedding environment, while PL energies and radiative rates are much less affected by external conditions [Schmidt et al.,2012, 2014]. This will be discussed in more detail in Chapter 6.

Figure 4.44 shows the influence of the embedding matrix on the "off" time statistics. Though the variation of the power-law exponent seems not to be dramatic we find a very systematic dependence upon the dielectric constant ε of the embedding matrix, as also shown in

Fig. 4.44. The power-law exponent (see Eq. 4.6) is decreasing with increasing ε. This can be modeled as already discussed in Section 4.4.1.7.3 in terms of self-localization of a (photo-) ejected charge [Issac et al., 2005]. A more dramatic influence of the matrix has been recently discovered comparing blinking dynamics of QDs in PS and poly(vinyl alcohol) (PVA) (showing additionally coupling to OH groups) [Schmidt et al., 2012, 2014]. Following the concept of self-localization [Issac et al., 2005, 2012], ejection of a charge into the matrix will be accompanied by a lowering of (optical) energy ΔE according to

$$\Delta E \approx f(\varepsilon) = (1 - 1/\varepsilon). \tag{4.16}$$

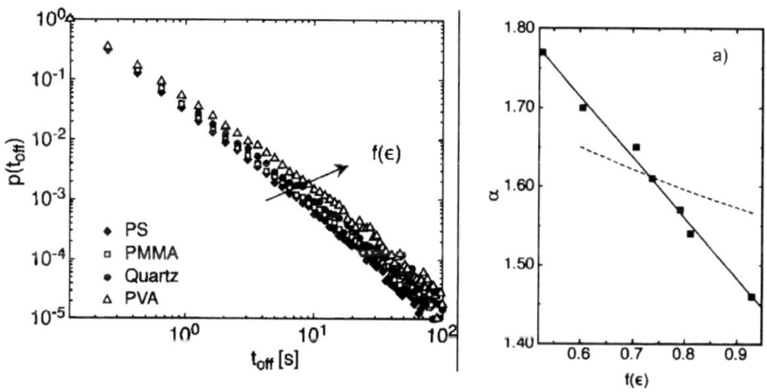

Figure 4.44 Left: "Off" time distributions for CdSe/ZnS QDs in PS, on PMMA, on PVA, and on quartz. The arrow indicates the direction of the increasing reaction field $f(\varepsilon)$. Right: Dependence of the power-law exponent α (see Eq. 4.6) of the "off" time statistics on the reaction field factor $f(\varepsilon)$ (squares), linear fit (solid line), and calculated dependence (dashed line) corresponding to the model of Verberk et al. (2012). Reprinted (figure) with permission from [Issac, A., Borczyskowski, C. von, and Cichos, F. (2005). Correlation between photoluminescence intermittency of CdSe quantum dots and self-trapped states in dielectric media, *Phys. Rev. B*, **71**, 161302.] Copyright (2005) by the American Physical Society.

Also "on" time distributions depend on the dielectric constant of the embedding matrix [Issac et al., 2012]. In case that signal/noise ratios are sufficiently high, the statistics of "on" times can be determined for a single QD instead of averaging "on" times of many single QDs, as is mostly reported in the literature. Figure 4.45 shows the distribution of the power-law exponent α_{on} and cutoff times Γ^{-1} determined from a truncated power according to Eq. 4.7 for individual CdSe/ZnS QDs in PS. Both parameters are broadly distributed, reflecting not only statistical errors but also differences among individual QDs and inhomogeneities of the embedding amorphous matrix. The distributions of cutoff times reflect to some extent asymmetric distributions of dynamics in non-Debye liquids described by Cole–Davidson or similar distribution functions. As can be followed in Section 4.4.1.7.8 and Chapter 6, there are ample examples that "on" times are not distributed according to (truncated) power laws. Nevertheless, CPA revealed that cutoff times are shortened by about a factor of 4 across the total intensity range during a blinking time trace when changing the matrix from PS (ε = 2.53) to PVA (ε = 14) [Schmidt et al., 2012, 2014], which is very similar to what is shown in Fig. 4.46.

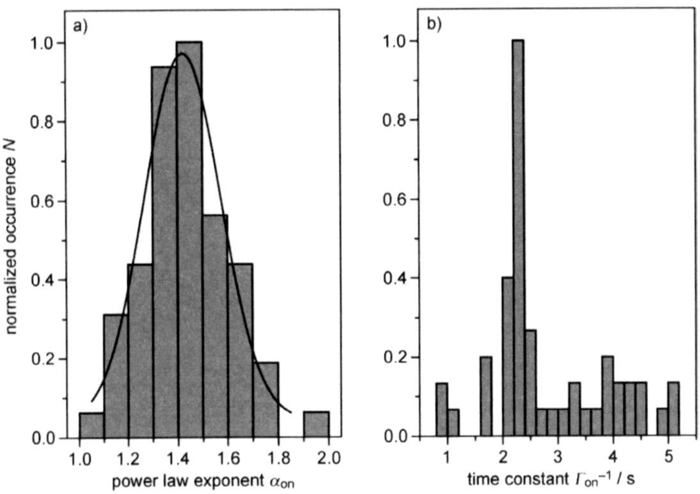

Figure 4.45 Power-law exponents α_{on} (left) and cutoff time Γ^{-1} (right) are shown for single CdSe/ZnS QDs in PS. The line corresponds to a Gaussian distribution. Cutoff times always show an asymmetric distribution depending on the matrix. Reproduced from Issac, A., Krasselt, C., Cichos, F., and Borczyskowski, C. von (2012). Influence of the dielectric environment on the photoluminescence intermittency of CdSe quantum dots, *ChemPhysChem*, **13**, 3223–3230, with permission from John Wiley & Sons, Inc.

Different from "off" times does the power-law exponent not depend on ε but the cutoff time Γ^{-1} decreases with increasing ε. The dependence is shown in Fig. 4.46. Cutoff times for "off" time distributions show the same dependence on ε, but since cutoff times are much longer than those of "on" time distributions, they are difficult to identify (see also Chapter 6). According to Fig. 4.46 (right), cutoff times decrease linearly with $1/\varepsilon$ in analogy to what is observed for the power-law coefficient.

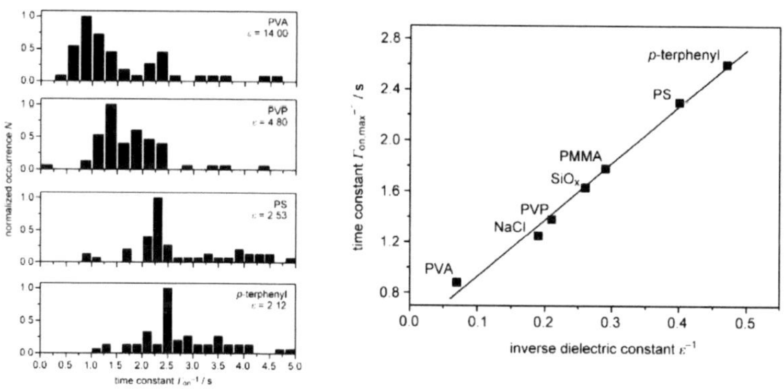

Figure 4.46 Left: Distribution of cutoff times for "on" times of single CdSe/ZnS QDs in the matrices *p*-terphenyl, PS, polyvinyl-propylene (PVP), and PVA. The respective dielectric constants are indicated. Right: Dependence of cutoff times $\Gamma_{on,max}{}^{-1}$ taken at the maximum of the distribution shown in the left graph on ε^{-1} for CdSe/ZnS (TOPO) QDs. Data for NaCl and PMMA have been added. The line is a linear fit to the matrix. Reproduced from Issac, A., Krasselt, C., Cichos, F., and Borczyskowski, C. von (2012). Influence of the dielectric environment on the photoluminescence intermittency of CdSe quantum dots, *ChemPhysChem*, **13**, 3223–3230, with permission from John Wiley & Sons, Inc.

Finally we would like to mention that also the electron–phonon coupling (expressed by the Huang–Rhys factor *S*) detected for single Si NCs in various matrices depends on the dielectric constant, as shown in Fig. 4.47 [El-Kork et al., 2011].

4.6 Influence of External Fields

Application of external electromagnetic fields is often very conclusive in materials science spectroscopy of any quantum system. At this point we concentrate on electromagnetic fields other than optical fields exciting the system in question, but we will rather discuss the influence of external electric and magnetic fields either of static or of alternating frequencies much lower than the ones characteristic for optical transitions. We will discuss separately the application of electric and magnetic fields but will omit the influence of an external field in combination with devices (which will be shortly touched in Zenkevich and Borczyskowski [2016]. The following sections will concentrate on spectroscopic aspects in single-QD detection. We will also discuss the influence of electromagnetic cavities, antenna, and near-field devices. The reader is referred to the introduction by Novotny and Hecht (2012).

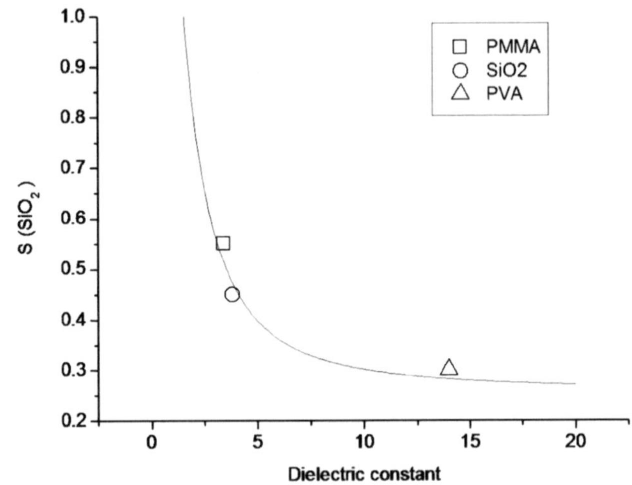

Figure 4.47 Variation of the Huang–Rhys factor S (SiO_2) with the dielectric constant ε of PMMA, SiO_2, and PVA. S reflects the coupling of a (localized) exciton at the Si/SiO_2 interface to Si-O phonons (see Fig. 4.42). The curve has been calculated according to El-Kork et al. (2011). Reprinted from El-Kork, N., Huisken, F., and Borczyskowski, C. von (2011). Dielectric effects on the optical properties of single silicon nanocrystals, *J. Appl. Phys.*, **110**, 074312, with the permission of AIP Publishing.

4.6.1 Electric Fields

The application of electric fields is related to two different approaches, namely the investigation of an AC or DC electric field (Stark effect) and the influence of charging in an electrochemical device on blinking dynamics, PL lifetimes, and spectra. The latter is motivated by the close relation of blinking with the presence or generation of charges.

4.6.1.1 Stark effect

Empedocles et al. (1997) were the first to apply the quantum confined Stark effect on single CdSe QDs (at low temperature) to investigate the influence on spectral diffusion and to probe the nature of the band edge states. For instance, excitonic states consisting out of delocalized hole and electron states are polarizable, while localized trap states are of dipolar character. Both can be distinguished from each other by application of external electric fields. Figure 4.48 shows the magnitude of the Stark shift of a single CdSe QD for electric fields between ± 350 kV/cm. In the given example all QDs showed a superposition of a quadratic (polarizability) and linear (dipole character) Stark effect. The linear part indicates that strong local electric fields are present. For the (at low temperature) high spectral resolution phonon side bands are observed. The electron–phonon coupling can be investigated, showing an increase with increasing Stark effect. Fluctuations of local electric fields result in spectral diffusion even at very low temperatures. The magnitude of local fields is in agreement with localized charges on the QD surface, which builds a bridge to blinking phenomena as being due to localized but effectively diffusing charges on the surface of a QD [Empedocles et al., 1999a; Krauss et al., 2001].

A further interesting question is how an external charge influences the radiative recombination rate of an electron–hole pair. An exemplary calculation has been reported by Wang (2001).

4.6.1.2 Charging of QDs and electric field–related blinking phenomena

While we have shown in Section 4.6.1.1 that there are several investigations on single QDs in terms of Stark effects, there are less than a dozen publications [Park et al., 2007; Jha et al., 2010; Galland et al., 2011; Galland et al., 2012; Qin et al., 2012; Qin and Guyot-Sionnest, 2012; Amecke, 2015; LeBlanc et al., 2014; Krasselt, 2015] which report on the influence of charging (application of electric fields) on blinking. The small number of investigations is somewhat surprising since it is generally assumed that charging of QDs (and molecules) is involved if not the origin of luminescence intermittency. Amecke (2015) investigated additionally the influence of electrode material on blinking controlled in an electrochemical cell. Besides the work of Park et al. (2007), LeBlanc et al. (2014), and Krasselt

(2015) in the Borczyskowski group all other publications used an electrochemical approach which allows for charging of QDs in an appropriate cell.

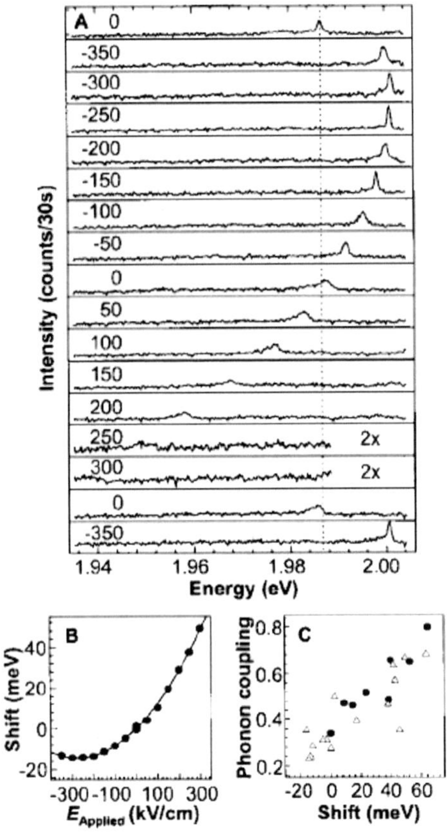

Figure 4.48 (A) PL spectra of a single CdSe/ZnS QD as a function of external electric field. Inset numbers indicate the applied field in kV/cm. (B) Plot of Stark shift versus electric field for the QD in (A) with fit to the sum of a linear and quadratic function of the applied field. (C) Phonon coupling versus Stark shift for a single QD in the presence of an applied electric field (closed circles) and versus spectral diffusion shift in the absence of an applied field (open triangles). Phonon coupling is measured as the ratio of the integrated intensity of the one LO phonon line to the ZPL. Stark and spectral diffusion data in (C) were taken from different QDs. Reprinted with permission from Empedocles, S. A., and Bawendi, M. G. (1999a). Spectroscopy of single CdSe nanocrystallites, *Acc. Chem. Res.*, **32**, 389–396. Copyright (1999) American Chemical Society.

We have outlined in Section 4.4.1.7 that one of the reasons for blinking is related to charging of QDs by one or more charges. An additional possibility is the trapping of charges (electron and holes) accompanied by trap relaxation processes. Trapping of, for example, an electron (hole) in a trap state can be described by reduction (oxidation) of the trap and oxidation (reduction) of the conduction (valence) band. Consequently, such redox processes can be investigated by electrochemical approaches.

One of the first investigations on the influence of an electric potential in an electrochemical cell upon single-QD blinking has been reported by Jha et al. [2010] for single CdSe/CdS and CdSe/ZnS QDs. Figure 4.49 summarizes the blinking statistics for a single CdSe/CdS QD as a function of the applied voltage. For voltages < –0.5 V blinking statistics can be approximated both for "on" and "off" times by a power law. At –0.6 V, "off" times are more close to truncated power laws with a considerably shortened cutoff times or a superposition of only a few exponential times, especially in the case of "on" times (see also Section 4.4.1.7.8). The authors argue that "on" and "off" times might be close to a one-step electron transfer process.

Qin et al. investigated CdSeS/ZnS [Qin et al., 2012] and CdSe/CdS [Qin and Guyot-Sionnest, 2012] QDS in electrochemical cells. Using electrochemistry with single-QD detection, the trion lifetime of CdSeS/ZnS QDs is 0.75 ns longer than the one for CdSe/ZnS (0.15 ns) [Qin et al., 2012]. In addition, the blinking behavior of QDs is in the trion state (negatively charged QD) to a large extent suppressed. Controllable charging at the single QD level allowed to monitor the (multiexponential) lifetime of a state with suppressed blinking, which has been assigned to the trion with a lifetime of 1.1 ns. The experiments show that the combination of synthesis of mixed core systems and controlled charging tunes the QY and blinking dynamics (see Fig. 4.50).

Similar effects have been found for single CdSe/CdS QDs with varying shell thickness [Qin Guyot-Sionnest, 2012]. Here, charging was possible with up to four charges. Additionally, electrochemically induced charging results in blue and red shifts of up to 40 meV; magnitude and direction are dependent on shell thickness.

Galland et al. (2011) found in (intermediate shell) CdSe/CdS QDs two types of blinking, namely A-type blinking due to charging and discharging of the QD core characterized by a correlation of PL intensities with PL lifetimes (see Eq. 4.9) and B-type blinking in

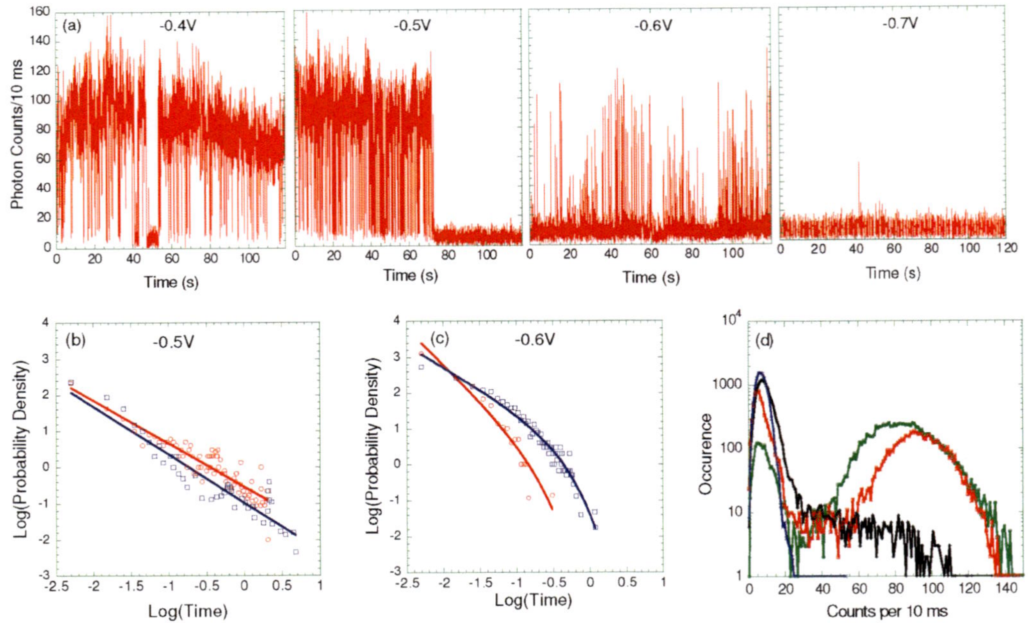

Figure 4.49 (a) PL counts for a single CdSe/CdS QD at −0.4, −0.5, −0.6, and −0.7 V. (b, c) Probability density of "on" (red circles) and "off" (blue squares) times at −0.5 and −0.6 V, respectively. The solid lines are fits discussed in the text. The "on"/"off" threshold is 12 counts/10 ms binning time. (d) The histograms of photon counts at −0.4 (green), −0.5 (red), −0.6 (black), and −0.7 (blue) are shown. The solid line connects the histogram data points. Reprinted with permission from Jha, P. P., and Guyot-Sionnest, P. (2010). Electrochemical switching of the photoluminescence of single quantum dots, *J. Phys. Chem. C*, **114**, 21138–21141. Copyright (2010) American Chemical Society.

which changes in the PL intensity are not accompanied by changes in the PL lifetime but probably by fluctuating charge distributions. The latter has been attributed to charge fluctuations in electron-accepting surface traps which might accept hot electrons (see also discussion in previous sections). Both blinking processes have been electrochemically controlled up to a completely suppression of blinking.

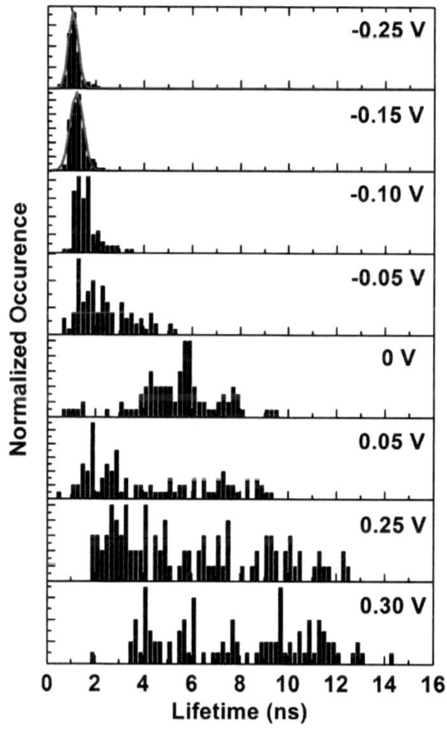

Figure 4.50 Histograms of PL lifetimes of a CdSeS/ZnS QD in an electrochemical cell at different voltages. Reprinted with permission from Qin, W., Shah, R. A., and Guyot-Sionnest, P. (2012). CdSeS/ZnS alloyed nanocrystal lifetime and blinking studies under electrochemical control, *ACS Nano*, **6**, 912–918. Copyright (2012) American Chemical Society.

In a later paper Galland et al. (2012) observed blinking of the PL lifetime by charging of (thick shell) CdSe/CdS QDs in an electrochemical cell without noticeable PL intensity blinking. They demonstrated that these lifetime variations are due to switching between the neutral and negatively charged QDs. (Multiple) charging can, as shown in Fig. 4.51, be induced by either an (electrochemical) electron injection or photoinduced charging. The reason for lifetime blinking at constant PL intensity is that negative charging results in faster radiative decay rates with only very small changes in PL intensity because of suppressed nonradiative Auger recombination. Both papers provide strong evidence that blinking is related to (photo- or electrochemically induced) processes.

Finally we would like to discuss the effect of static electric fields on blinking and PL decay times in isolating (dielectric capacitor type) devices where CdSe/ZnS QDs are embedded in thin PMMA films [Park et al., 2007; LeBlanc et al., 2014] or CdSe/CdS encapsulated in PMMA nanoparticles [Krasselt, 2015] applying external voltages of 50–840 kV/cm. LeBlanc et al. (2014) find with a confocal setup that PL lifetimes are shortened (decreased PL intensity) and that the intensity distribution between "on" and "off" intensities (gray states) is increased upon (static) electric fields up to 50 kV/cm. They attribute the changes in emission behavior to a redistribution of surface charges associated with the excited-state polarizability. This charge redistribution of photogenerated charges leads to an increased trapping rate and thus a redistribution of emitting (gray) states.

Park et al. (2007) swept periodically (with 0.1 Hz) an external electric field of 250 kV/cm. The PL intensity appears to be modulated both positively and negatively with respect to $E_{DC} = 0$, depending on the individual QD and differently during different observation times. The authors postulate charge-accepting trap states. The energies of the latter are modulated by the external field, thus controlling charge trapping and thereby modulating the PL QY.

Krasselt (2015) has investigated with a wide-field (low optical excitation rate) setup both alternating and static electric fields in a dielectric capacitor on CdSe/CdS QDs which have separately embedded into a PMMA nanoparticle to increase photostability. Here we will report some first results applying static fields. Electrodes are gold and ITO. PS is the isolator in the device. Electric fields up to $E_{DC} \approx 840$ kV/cm have been applied. Figure 4.52 shows changes in blinking dynamics together with the blinking activities and the "on" time fraction when switching the external field on and off. (For definition of the parameters, see Chapter 6). Basically, PL intensity is considerably increased upon application of an electric field. At the same time blinking frequency increases.

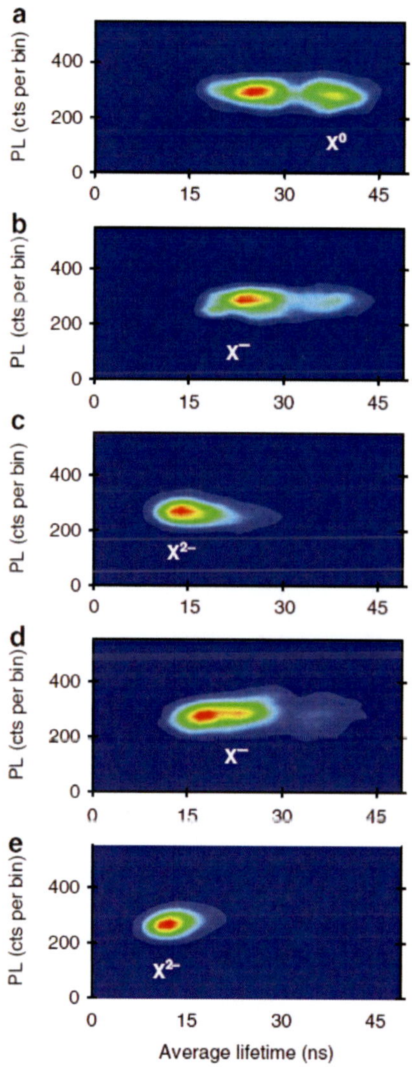

Figure 4.51 Electrochemical and photoinduced charging of a single NQD. (a–c) PL lifetime blinking for increasing negative potential (electron injection) and constant excitation fluence ($N_{ex} \approx 0.08$): (a) 0 V, (b) −0.5 V, and (c) −0.8 V. (d, e) The potential is kept at zero but the variation of fluence and zero voltage: (d) $N_{ex} \approx 0.22$ and (e) $N_{ex} \approx 0.7$. Reprinted by permission from Macmillan Publishers Ltd: [*Nature Communications*] (Galland, C., Ghosh, Y., Steinbrück, Hollingsworth, J. A., Htoon, H., and Klimov, V. I. (2012). Lifetime blinking in nonblinking nanocrystal quantum dots, *Nat. Commun.*, **3**, 908), copyright (2012).

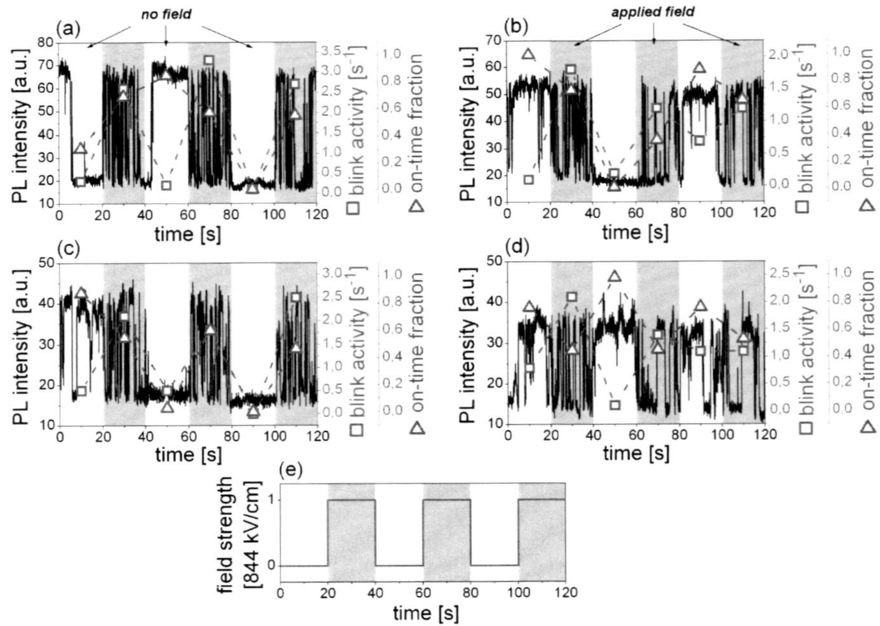

Figure 4.52 Blinking dynamics of four different CdSe/CdS QDs with a PMMA shell while switching periodically a DC electric filed between 0 and 844 kV/cm (see scheme at the bottom). Laser excitation power is 26 W/cm^2. Squares represent the average blinking activity during the cycling sequences, while triangles correspond to the averaged "on" time fraction. Broken lines are only guides for the eye. From Krasselt, C. (2015). *Dynamik der Photo-Limineszenze-Unterbrechung von Halbleiter-Nanokristallen in elektrischen Feldern,* PhD Thesis, TU Chemnitz, Germany, with kind permission from Cornelius Krasselt.

Blinking data have been evaluated according to Eq. 4.7 as a function of the amplitude of the electric field both for increasing and decreasing field cycles. As a general trend α_{on} (nearly linear from −1.6 to −1.4) and Γ^{-1}_{on} (from 200 s to 40 s) are independent of the "sweep" direction, decreasing with increasing E_{DC}. The α_{off} value is nearly constant (−1.45) in the upward direction but increases from −1.45 to −1.75 in the downward direction, while Γ^{-1}_{off} increases (decreases) from 18 s (25 s) to 25 s (100 s). This finding for the power-law exponents is qualitatively similar to the one of LeBlanc et al. (2014). Illustrative parameters (for a quantitative definition see Chapter 6) are the blinking activity A (events/s) and the "on" time fraction Φ (%) shown in Fig. 4.53. It is evident that the blinking activity is increasing and the "on" time fraction decreasing with increasing E_{DC}. However, the related processes are not reversible. After a field cycle blinking activity decreases and "on" time fraction increases compared to the initial situation which implements irreversibility which points toward charge acquisition close to a QD during cycling to high electric fields. The explanation of the reversible part is similar to the one which Park et al. (2007) and LeBlanc et al. (2014) have recently suggested. The observed strong increase of the blinking activity above 500 V is already in a range at which Stark effects become effective.

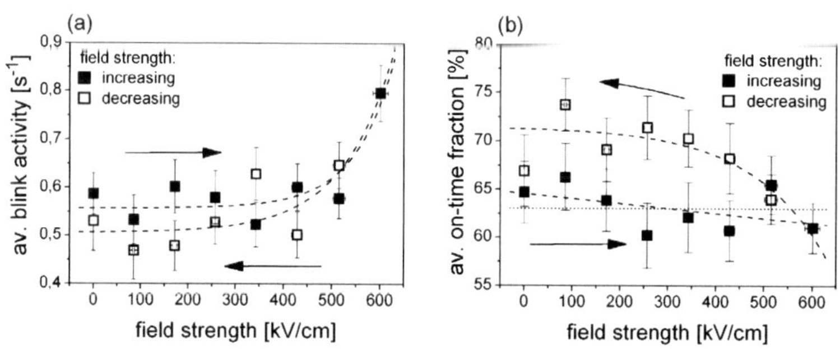

Figure 4.53 Average blinking activity (a) and "on" time fraction (b) as a function of applied external DC electric field for CdSe/CdS QDs in a PMMA shell. (For definitions see Chapter 6). Arrows show the direction of increasing (decreasing) field amplitude. Cycles have been started with increasing fields. Broken lines show fits according to a linear or exponential dependence. From Krasselt, C. (2015). *Dynamik der Photo-Limineszenze-Unterbrechung von Halbleiter-Nanokristallen in elektrischen Feldern*, PhD Thesis, TU Chemnitz, Germany, with kind permission from Cornelius Krasselt.

4.6.2 Magnetic Fields

In Chapter 1 we have outlined the general scheme of exciton states in CdSe QDs. According to this the lowest state is a dark state with a total spin projection of $J = \pm 2$ on the crystal c axis, which lies a few millielectron volts below the optically active $J = \pm 1^L$ state. These states can be mixed by (strong) external magnetic fields. Since the energy separation is very small compared to thermal energies, experiments have to be performed at temperatures below 4.2K. Only a few experiments have been performed on a single-QD level to unravel the band edge fine structure, coupling among the two lowest states, determining Landé factors and spin relaxation rates [Htoon et al., 2009; Biadala et al., 2010].

While the exciton states are, besides state mixing, due to the Zeeman effect not accessible to magnetic resonance spectroscopy, "on" trap states (at the QD surface) and "on" impurities in the core have been investigated by (optically detected) magnetic resonance spectroscopy, for example, by the groups of Lifshitz et al. (1998) and Weller [Fradkin et al., 2003]. Very recently a coherent magnetic resonance experiment has been reported on a few QDs, which helped to unravel single-QD blinking processes [Schooten et al., 2014].

4.7 Conclusion

In this chapter we described the basic principles of single-quantum-object detection, mainly of colloidal QDs. We touched upon most of the spectroscopic relevant topics from basic signatures of single QDs (bunching, antibunching, blinking, spectral diffusion) to charging of QDs in electrochemical cells and how these features can be investigated experimentally. We did not dwell on high-intensity excitations creating multiple excitons and also did not touch upon relaxation of hot charge carriers. Our special emphasis was on the background of tuning possibilities and influences outside the particle in a box. Especially, we focused on blinking phenomena and how they are influenced, characterized, controlled, and finally modeled. Still more concrete investigations of luminescence intermittency will be discussed in Chapter 6, including not yet published results.

To summarize the developments in blinking analysis, we have outlined that blinking, though characteristic for nearly all single quantum systems, does not have a unique origin which can be described by one unified model. The commonly used power-law characterization to describe blinking dynamics is a reasonable approach but not a universal law. Nevertheless, its application seems to be handsome since it is a simple approximation and qualitatively reliable with respect to the analysis on the influence of external variables. A breakthrough in a microscopic understanding of blinking beyond the search of universal models has merged in controlling blinking by charging and tuning the environment.

Whether blinking is friend or foe [Lee and Osborne, 2009] depends obviously on the point of view: it is both. In any case we should understand the underlying mechanisms and how to tune them, depending on the respective demands.

References

Amecke, N., and Cichos, F. (2011). Intermediate intensity levels duringthe emission intermittency of single CdSe/ZnS quantum dots, *J. Lumin.*, **131**, 375–378.

Amecke, N., Heber, A., and Cichos, F. (2014). Distortion of power law blinking with binning and thresholding, *J. Chem. Phys.*, **140**, 114306.

Amecke, N. (2015). *Characterization of Single Quantum Dot Blinking: Dwell Time Statistics and Electrochemical Control*, PhD Thesis, Leipzig University, Germany.

Antelman, J., Ebenstein, Y., Dertinger, T., Michalet, X., and Weiss, S. (2009). Suppression of quantum dot blinking in DTT-doped polymer films, *J. Phys. Chem. C*, **113**, 11541–11545.

Baker, T. A., Rouge, J. L., and Nesbitt, D. J. (2009). Single molecule studies of quantum dot fluorescence intermittency: evidence for both dark and light-assisted blinking dynamics, *Mol. Phys.*, **107**, 1867–1878.

Barkai, E., Jung, Y., and Silbey, R. (2004). Theory of single-molecule spectroscopy: beyond the ensemble average, *Annu. Rev. Phys. Chem.*, **55**, 457–507.

Betzig, E., Patterson, G. H., Sougrat, R., Lindwasser, O. W., Olenych, S., Bonifacino J. S., Davidson, M. W., Lippincott-Schwartz, J., and Hess, H. F. (2006). Imaging intracellular fluorescent proteins at nanometer resolution, *Science*, **313**, 1642–1645.

Biadala, L., Louyer, Y., Tamarat, P., and Lounis, B. (2010). Band-edge exciton fine structure of single CdSe/ZnS nanocrystals in external magnetic fields, *Phys. Rev. Lett.*, **105**, 157402.

Biebricher, A., Sauer, M., and Tinnefeld, P. (2006). Radiative and nonradiative rate fluctuations of single colloidal semiconductor nanocrystals, *J. Phys. Chem. B*, **110**, 5174–5178.

Blaise, G., and Sarjeant, W. J. (1998). Space charge in dielectrics: energy storage and transfer dynamics from atomistic to macroscopic scale, *IEEE Trans. Dielectr. Electr. Insul.*, **5**, 779–808.

Blanco, M., and Walter, N. G. (2010). Analysis of complex single-molecule FRET time trajectories, *Meth. Enzym.*, **472**, 153–178.

Blanton, S. A., Hines, M. A., and Guyot-Sionnest, P. (1996). Photoluminescence wandering in single CdSe nanocrystals, *Appl. Phys. Lett.*, **69**, 3905–3907.

Blaudeck, T. (2007). *Self-Assembly of Functionalized Porphyrin Molecules on Semiconductor Nanocrystal Surfaces*, PhD Thesis, TU Chemnitz, Germany.

Blaudeck, T., Zenkevich, E. I., Cichos, F., and Borczyskowski, C. von (2008). Probing wave functions at semiconductor quantum-dot surfaces by non-FRET photoluminescence quenching, *J. Phys. Chem. C*, **112**, 20251–20257.

Blaudeck, T., Zenkevich, E. I., Abdel-Mottaleb, Szwaykowska, M. K., Kowerko, D., Cichos, F., and Borczyskowski, C. von (2012). Formation principles and ligand dynamics of nanoassemblies of CdSe quantum dots and functionalised dye molecules, *ChemPhysChem*, **13**, 959–972.

Böhmer, M., and Enderlein, J. (2003). Orientation imaging of single molecules by wide-field epiuorescence microscopy, *J. Opt. Soc. Am. B*, **20**, 554–559.

Borczyskowski, C. von and Zenkevich, E. (2014). Formation principles and exciton relaxation in semiconductor quantum dot–dye nanoassemblies, in *Quantum Dot Molecules, Lecture Notes in Nanoscale Science and Technology* (Wu, J., and Wang, Z. M., eds.), Springer, Dordrecht, Heidelberg, New York, London, Vol. 14, pp. 77–148.

Bradac, C., Gaebel, T., Naidoo, N., Sellars, M. J., Twamley, J., Brown, L. J., Barnard, A. S., Plakhotnik, T., Zvyagin, A. V., and Rabeau, J. R. (2010). Observation and control of blinking nitrogen-vacancy centres in discrete nanodiamonds, *Nat. Nanotechnol.*, **5**, 345–349.

Brawand, N. P. Vörös, M., and Galli, G. (2015). Surface dangling bonds are a cause of B-type blinking in Si nanoparticles, *Nanoscale*, **7**, 3737–3744.

Brown,R.,Wrachtrup,J.,Orrit,M.,Bernard,J.,andBorczyskowski,C.von(1994). Kinetics of optically detected magnetic resonance of single molecules, *J. Chem. Phys.*, **100**, 7128–7191.

Bruhn, B., Valenta, Sangghaleh, J. F., and Linnros, J. (2011). Blinking statistics of silicon quantum dots, *Nano Lett.*, **11**, 5574–5580.

Chen, O., Jing Zhao, J., Chauhan, V. P., Cui, J., Wong, C., Harris, D. K., Wei, H., Han, H.-S., Fukumura, D., Jain, R. K., and Bawendi, M. G. (2013). Compact high-quality CdSe–CdS core–shell nanocrystals with narrow emission linewidths and suppressed blinking, *Nat. Mater.*, **12**, 445.

Chen, Y.,Vela, J., Htoon, H., Casson, J. L., Werder, D. J., Bussian, D. A., Klimov, V. I., and Hollingsworth, J. A. (2008). "Giant" multishell cdse nanocrystal quantum dots with suppressed blinking, *J. Am. Chem. Soc.*, **130**, 5026–5027.

Chen, O., Zhao, J. , Chauhan, V. P., Cui, I., Wong, C., Harris, D. K., Wei, H.,Hee-Sun Han, H.-S.,Fukumura, D., Jain, R. K., and Bawendi, M. G. (2013). Compact high-quality CdSe–CdS core–shell nanocrystals with narrow emission linewidths and suppressed blinking, *Nat. Mater.*, **12**, 445–451.

Chung, I., and Bawendi, M. G. (2004). Relationship between single quantum-dot intermittency and fluorescence intensity decays from collections of dots, *Phys. Rev. B*, **70**, 165304.

Cichos, F., Martin, J., and Borczyskowski, C. von (2004). Emission intermittency in silicon nanocrystals, *Phys. Rev. B*, **70**, 1153141.

Cichos, F., Borczyskowski, C. von, and Orrit, M. (2007). Power-law intermittency of single emitters, *Curr. Opin. Colloid Interface Sci.*, **12**, 272–284.

Cody, G. D. (1984). *Hydrogenated Amorphous Silicon, Semiconductors and Semimetals* (Pankove, J. I., ed), Academic, New York, Vol. 21, pp. 11–79.

Comeau, J. W. D., Costantino, S., and Wiseman, P. W. (2006). A guide to accurate fluorescence microscopy colocalization measurements, *Biophys. J.*, **91**, 4611–4622.

Cordonnes, A. A., Bixby, T. J., and Leon, S. R. (2011). Direct measurement of off-state trapping rate fluctuations in single quantum dot fluorescence, *Nano Lett.*, **11**, 3366–3369.

Cordones, A. A., and Leone, S. R. (2013). Mechanisms for charge trapping in single semiconductor nanocrystals probed by fluorescence blinking, *Chem. Soc. Rev.*, **42**, 3209–3221.

Crouch, C. H., Sauter, O., Wu, X., Purcell, R., Querner, C., Drudic, M., and Pelton, M. (2010). Facts and artifacts in the blinking statistics of semiconductor nanocrystals, *Nano Lett.*, **10**, 1692–1698.

Cui, J., Beyler, A. P., Bischof, T. S., Wilson, M. W. B., and Bawendi, M. G. (2014). Deconstructiing the photon stream from single nanocrystals: from binning to correlation, *Chem. Soc. Rev.*, **43**, 1287–1310.

Dabbousi, B. O., Rodriguez-Viejo, J., Mikulec, F. V., Heine, J. R., Mattoussi, H., Ober, R., Jensen, K. F., and Bawendi, M. G. (1997). (CdSe)ZnS core–shell quantum dots: synthesis and characterization of a size series of highly luminescent nanocrystallites, *J. Phys. Chem. B*, **101**, 9463–9475.

Debye, P., and Edwards, J. O. (1952). Long-lifetime phosphorescence and the diffusion process, *J. Chem. Phys.*, **20**, 236–239.

Dorfman, R. C., Tachiya, M., and Fayer, M. D. (1991). Forward and back photoinduced electron transfer in solid solutions: a comparison of theoretical methods, *Chem. Phys. Lett.*, **179**, 152–158.

Duke, C. B., and Fabish, T. J. (1976). Charge-induced relaxation in polymers, *Phys. Rev. Lett.*, **37**, 1075–1078.

Efros, A. L., and Rosen, M. (1997). Random telegraph signal in the photoluminescence intensity of a single quantum dot, *Phys. Rev. Lett.*, **78**, 1110–1113.

El-Kork, N., Huisken, F., and Borczyskowski, C. von (2011). Dielectric effects on the optical properties of single silicon nanocrystals, *J. Appl. Phys.*, **110**, 074312.

Frenzel, J., Joswig, J.-O., and Seifert, G. (2007). Optical excitations in cadmium sulfide nanoparticles, *J. Phys. Chem. C*, **111**, 10761–10770.

Empedocles, S. A., Norris, D. J., and Bawendi, M. G. (1996). Photoluminescence spectroscopy of single CdSe nanocrystallite quantum dots, *Phys. Rev. Lett.*, **77**, 3873.

Empedocles, S. A., and Bawendi, M. G. (1997). Quantum-confined Stark effect in single CdSe nanocrystallite quantum dots, *Science*, **278**, 2114–2117.

Empedocles, S. A., and Bawendi, M. G. (1999a). Spectroscopy of single CdSe nanocrystallites, *Acc. Chem. Res.*, **32**, 389–396.

Empedocles, S. A., and Bawendi, M. G. (1999b). Influence of spectral diffusion on the line shapes of singel CdSe nanocrystallite quantum dots, *J. Phys. Chem. B*, **278**, 2114–2117.

Enderlein, J., and Erdmann, R. (1997). Fast fitting of multi-exponential decay curves, *Opt. Commun.*, **134**, 371–378.

Fernee, M. J., Littleton, B. N., Plakhotnik, T., Rubinsztein-Dunlop, H., Gomez, D. E., and Mulvaney, P. (2012). Charge hopping revealed by jitter correlations in the photoluminescence spectra of single CdSe nanocrystals, *Phys. Rev. B*, **81**, 155307.

Fernee, M. J., Littleton, B. N., Cooper, S., Rubinsztein-Dunlop, H., Gomez, D. E., and Mulvaney, P. (2008). Acoustic phonon contributions to the emission spectrum of single CdSe nanocrystals, *J. Phys. Chem. C*, **112**, 1878–1884.

Fernee, M. J., Plakhotnik, T., Louyer, Y., Littleton, B. N., Potzner, C., Tamarat, P., Mulvaney, P., and Lounis, B. (2012). Spontaneous spectral diffusion in CdSe quantum dots, *J. Phys. Chem. Lett.*, **3**, 1717–1720.

Feynman, R. P., Vernon, F. L. Jr., and Hellwarth, R. W. (1957). Geometrical representation of the Schrödinger equation for solving maser problems, *J. Appl. Phys.*, **28**, 49–52.

Fischer, S. A., Crotty, A. M., Kilina, S. V., Ivanov, S. A., and Tretiak, S. (2012). Passivating ligand and solvent contributions to the electronic properties of semiconductor, *Nanoscale*, **4**, 904–914.

Fischer, U., Borczyskowski, C. von, Seiff, F., and Stehlik, D. (1983). Long-range energy transport in doped dibenzufurane crystals detected by intermolecular fluorescence fading, *Chem. Phys. Lett.*, **97**, 476–482.

Fisher, B. R., Eisler, H. J., Stott, N., and Bawendi, M. G. (2004) Emission intensity dependence and single-exponential behavior in single colloidal quantum dot fluorescence lifetimes, *J. Phys. Chem. B*, **108**, 143–148.

Fleury, L., Zumbusch, A., Orrit, M., Brown, R., and Bernard, J. (1993). Spectral diffusion and individual two-level systems probed by fluorescence of single terrylene molecules in a polyethylene matrix, *J. Lumin.*, **56**, 15–28.

Fomenko, V. and Nesbitt, D. J. (2008). Solution control of radiative and nonradiative lifetimes: a novel contribution to quantum dot blinking suppression, *Nano Lett.*, **8**, 287–293.

Fradkin, L., Langof, L., Lifshitz, E., Rogach, A. L., Gaponik, N. P., Weller, H., and Eychmüller, A. (2003). Magneto-optical studies of HgTe/HgxCd1-xTe(S) core–shell nanocrystals, *ChemPhysChem*, **4**, 1203–1210.

Frantsuzov, P. A., and Marcus, R. A. (2005). Explanation of quantum dot blinking withoutthe long-lived trap hypothesis, *Phys. Rev. B*, **72**, 155321.

Frantsuzov, P. A., Volkán-Kacsó, S., and Janko, B. (2009). Model of fluorescence intermittency of single colloidal semiconductor quantum dots using multiple recombination centers, *Phys. Rev. Lett.*, **103**, 207402.

Frantsuzov, P., Kuno, M., Janko, M., and Marcus, R. A. (2008). Universal emission intermittency in quantum dots, nanorods and nanowires, *Nat. Phys.*, **4**, 519–522.

Frantsuzov, P. A., Volkán-Kacscó, S., and Jankó, B. (2013). Universality of the fluorescence intermittency in nanoscale systems: experiment and theory, *Nano Lett.*, **13**, 402–408.

Fröhlich, H. (1954). Electrons in lattice fields, *Adv. Phys.*, **3**, 325–361.

Fu, Y., Zhang, J., and Lakowicz, J. R. (2007). Suppressed blinking in single quantum dots (QDs) immobilized near silver island films (SIFs), *Chem. Phys. Lett.*, **447**, 96–100.

Gady, B., Reifenberger, R., and Rimai, D. S. (1998). Contact electrification studies using atomic force microscope techniques, *J. Appl. Phys.*, **84**, 319–322.

Galland, C., Ghosh, Y., Steinbrück, A., Sykora, M., Hollingsworth, J. A., and Klimov, V. I. (2011). Two types of luminescence blinking revealed by spectroelectrochemistry of single quantum dots, *Nature*, **479**, 203–207.

Galland, C., Ghosh, Y., Steinbrück, Hollingsworth, J. A., Htoon, H., and Klimov, V. I. (2012). Lifetime blinking in nonblinking nanocrystal quantum dots, *Nat. Commun.*, **3**, 908.

Glennon, J. J., Tang, R., Buhro, W. E., and Loomis, R. A. (2007). Synchronous photoluminescence intermittency (blinking) along whole semiconductor quantum wires, *Nano Lett.*, **7**, 3290–3295.

Göhler, C. (2014). *Schnelle Spektralanalyse des Blinkens von Halbleiternanokristallen,* M.Sc. Thesis, TU Chemnitz, Germany.

Goméz, D. E., van Emden, J., and Mulvaney, P. (2006). Spectral diffusion of single semiconductor nanocrystals: the influence of the dielectric environment, *Appl. Phys. Lett.*, **88**, 154106.

Graaf, H., Vieluf, M., and Borczyskowski, C. von (2007). Selective binding of dye molecules and CdSe nanocrystals on nanostructures generated by AFM lithography of silicon surfaces, *Nanotechnology*, **18**, 265306.

Gruber, A., Dräbenstedt, A., Tietz, C., Fleury, L., Wrachtrup, J., and Borczyskowski, C. von (1997). Scanning confocal optical microscopy and magneticresonance on single defect centers, *Science*, **276**, 2012–2014.

Haase, M., Hübner, C. G., Reuther, E., Herrmann, A., Müllen, K., and Basché, Th. (2004). Exponential and power-law kinetics in single-molecule fluorescence intermittency, *J. Phys. Chem. B*, **108**, 10445–10450.

Hamada, M., Nakanishi, S., Itoh, T., Ishikawa, M., and Biju, V. (2010). Blinking suppressionin CdSe/ZnS single quantum dots by TiO_2 nanoparticles, *ACS Nano*, **4**, 4445–4454.

Hammer, N. I., Early, K. T., Sill, K., Odoi, M. Y., Emrick, T., and Barnes, M. D. (2006). Coverage-mediated suppression of blinking in solid state quantum dot conjugated organic composite nanostructures, *J. Phys. Chem. B*, **110**, 14167–14171.

Haran, G. (2004). Noise reduction in single-molecule fluorescence trajectories of folding proteins, *Chem. Phys.*, **307**, 137–145.

Heinzelmann, H., Hecht, B., Novotny, L., and Pohl, D. W. (2011). Forbidden light scanning near-field optical microscopy, *L. Microsc.*, **177**, 115–118.

Hellwege, H. (ed.) (1982). *Numerical Data and Functional Relationship in Science and Technology*, Landolt-Bornstein, New Series, Group III, Vol. 17, Pt. B Springer-Verlag, Berlin.

Hennessy, K., Badolato, A., Winger, A. M., Gerace, D., Atatüre, M., Gulde, S., Fält, S., Hu, E. L., and Imamoglu, A. (2007). Quantum nature of a strongly coupled single quantum dot–cavity system, *Nature*, **445**, 896–899.

Hollingsworth, J. A. (2013). Heterostructuring nanocrystal quantum dots toward intentional suppression of blinking and Auger recombination, *Chem. Mater.*, **25**, 1318–1331.

Holstein, T. (1959). Studies of polaron motion, *Ann. Phys.*, **8**, 325–389.

Hohng, S., and Ha, T. (2004). Near-complete suppression of quantum dot blinking ambient conditions, *J. Am. Chem. Soc.*, **126**, 1324–1325.

Hong, K. M., Noolandi, J., and Street, R. A. (1981). Theory of radiative recombination bydiffusion and tunneling in amorphous Si:H, *Phys. Rev. B*, **23**, 2967–2976.

Houel, J., Doan, Q. T., Cajgfinger, T., Ledoux, G., Amans, D., Aubret, A., Dominjon, A., Ferriol, S., Barbier, R., Nasilowski, M., Lhuillier, E., Dubertret, B., Dujardin, C., and Kulzer, F. (2015). Autocorrelation analysis for the unbiased determination of power-law exponents in single-quantum-dot blinking, *ACS Nano*, **9**, 886–893.

Htoon, H., Crooker, S. A., Furis, M., Jeong, S., Efros, A. L., and Klimov, V. I. (2009). Anomalous circular polarization of photoluminescence spectra of individual CdSe nanocrystals in an applied magnetic field, *Phys. Rev. Lett.*, **102**, 017402.

Issac, A., Borczyskowski, C. von, and Cichos, F. (2005). Correlation between photoluminescence intermittency of CdSe quantum dots and self-trapped states in dielectric media, *Phys. Rev. B*, **71**, 161302.

Issac, A. (2006). *Photoluminescence Intermittency of Semiconductor Quantum Dots in Dielectric Environments*, PhD Thesis, TU Chemnitz, Germany.

Issac, A., Krasselt, C., Cichos, F., and Borczyskowski, C. von (2012). Influence of the dielectric environment on the photoluminescence intermittency of CdSe quantum dots, *ChemPhysChem*, **13**, 3223–3230.

Jeong, S., Achermann, M., Nanda, J., Ivanov, S., Klimov, V. I., and Hollingsworth, J. A. (2005). Effect of the thiol-thiolate equilibrium on the photophysical properties of aqueous CdSe/ZnS nanocrystal quantum dots, *J. Am. Chem. Soc.*, **127**, 10126–10127.

Jha, P. P., and Guyot-Sionnest, P. (2009). Trion decay in colloidal quantum dots, *ACS Nano*, **3**, 1011–1015.

Jha, P. P., and Guyot-Sionnest, P. (2010). Electrochemical switching of the photoluminescence of single quantum dots, *J. Phys. Chem. C*, **114**, 21138–21141.

Kim, S., Fisher, B., Eisler, H. J., and Bawendi, M. (2003). Type-II quantum dots: CdTe/CdSe (core/shell) and CdSe/ZnTe (core/shell) heterostructures, *J. Am. Chem. Soc.*, **125**, 11466–11467.

Kilina, S., Ivanov, S., and Tretiak, S. (2009). Effect of surface ligands on optical and electronic spectra of semiconductor nanoclusters, *J. Am. Chem. Soc.*, **131**, 7717–7726.

Klimov, V. I. (2003). Charge carrier dynamics and optical gain in nanocrystal quantum dots: from fundamental photophysics to quantum dot lasing, in *Semiconductor and Metal Nanocrystals: Synthesis and Electronic and Optical Properties* (Klimov, V. I., ed.), Marcel Dekker, New York, pp. 159–214.

Ko, H. C., Yuan, C. T., Lin, S. H., and Tang, J. (2010). Blinking suppression of single quantum dots in agarose gel, *Appl. Phys. Lett.*, **96**, 012104.

Kowerko, D. (2010). *Dynamic Processes in Functionalised Perylene Bisimide Molecules, Semiconductor Nanocrystals and Assemblies*, PhD Thesis, TU Chemnitz, Germany.

Kowerko, D., Schuster, J., Amecke, N., Abdel-Mottaleb, M., Dobrawa, R., Würthner, F., and Borczyskowski, C. von (2010). FRET and ligand related non-FRET processes in single quantum dot-perylene bisimide assemblies, *Phys. Chem. Chem. Phys.*, **12**, 4112.

Krasselt, C. (2009). *Der Einfluss von Grenzflächen auf photoinduzierte Ladungslokalisierung,* Diploma Thesis, TU Chemnitz, Germany.

Krasselt, C., Schuster, J., and Borczyskowski, C. von (2011). Photoinduced hole trapping in single semiconductor quantum dots at specific sites at silicon oxide interfaces, *Phys. Chem. Chem. Phys.*, **13**, 17084.

Krasselt, C. (2015). *Dynamik der Photo-Limineszenze-Unterbrechung von Halbleiter-Nanokristallen in elektrischen Feldern*, PhD Thesis, TU Chemnitz, Germany.

Krause, S., Hartmann, M., Kahle, I., Neuman, M., Heidernätsch, M., Spange, S., and Borczyskowski, C. von (2013). Optical tracking of single Ag clusters in nanostructured water films, *J. Phys. Chem. C*, **117**, 24822–24829.

Krause, S., Heidernätsch, M., Hartmann, M., and Borczyskowski, C. von (2014). Optical investigation of diffusion of single Ag markers in confined water films, *J. Nanopart. Res.*, **16**, 2700.

Krauss, T. D., and Brus, L. E. (1999). Charge, polarizability, and photoionization of single semiconductor nanocrystals, *Phys. Rev. Lett.*, **83**, 4840.

Krauss, T. D., O'Brien, S. O., and Brus, L. E. (2001). Charge and photoionization properties of single semiconductor nanocrystals, *J. Phys. Chem. B*, **105**, 1725–1733.

Kuno, M., Fromm, D. P., Hamann, H. F., Gallagher, A., and Nesbitt, D. J. (2000). Nonexponentialblinking kinetics of single CdSe quantum dots: a universal power law behaviour, *J. Chem. Phys.*, **112**, 3117–3120.

Kuno, M., Fromm, D. P., Johnson, S. T., Gallagher, A., and Nesbitt, D. J. Modelling (2003). Distributed kinetics in isolated semiconductor quantum dots, *Phys. Rev. B*, **67**, 1253041.

Lakowicz, J. R. (2001). Radiative decay engineering: biophysical and biomedical applications, *Anal. Biochem.*, **298**, 1–24.

Leatherdale, C. A., and Bawendi, M. G. (2001). Observation of solvatochromism in CdSe colloidal quantum dots, *Phys. Rev. B*, **63**, 165315.

LeBlanc, S. J., McClanahan, M. R., Moyer, T., Jones, M., and Moyer, P. J. (2014). Fluorescence modulation in single CdSe quantum dots by moderate applied electric fields, *J. Appl. Phys.*, **115**, 034306.

Lee, S. F., and Osborne, M. A. (2009). Brightening, blinking, bluing and bleaching in the life of a quantum dot: friend or foe?, *ChemPhysChem*, **10**, 2174–2191.

Lewis, T. J. (1989). A simple general model for charge transfer in polymers, *Farad. Discuss Chem. Soc.*, **88**, 189–201.

Lewis, A., Taha, H., Strinkovski, A., Manevitch, A., Khatchatouriants, A., Dekhter, R., and Ammann, E. (2003). Near-field optics: from subwavelength illumination to nanometric shadowing, *Nat. Biotechnol.*, **21**, 1378–1386.

Li, X., Feng, D., Tong, H., Jia, T., Deng, L., Sun, Z., and Xu, Z. (2014). Hole Surface trapping dynamics directly monitored by electron spin manipulation in CdS nanocrystals, *J. Phys. Chem. Lett.*, **5**, 4310–4316.

Lifshitz, E., Dag, I., Litvitn, I. D., and Hodes, G. (1998). Optically detected magnetic resonance study of electron/hole traps on CdSe quantum dot surfaces, *J. Phys. Chem. B*, **102**, 9245–9250.

Lippitz, M., Kulzer, F., and Orrit, M. (2005). Statistical evaluation of single nano-object fluorescence, *ChemPhysChem*, **6**, 770–789.

Liptay, T. J., Marshall, L. F., Rao, P. S., Ram, R. J., and Bawendi, M. G. (2007). Anomalous Stokes shift in CdSe nanocrystals, *Phys. Rev. B*, **76**, 155314.

Lo, S. S., Khan Y., Jones, M., and Scholes, G. D. (2009). Temperature and solvent dependence of CdSe/CdTe heterostructure nanorod spectra, *J. Phys. Chem.*, **131**, 084714.

Lowell, J., and Rose-Innes A. C. (1980). Contact electrification, *Adv. Phys.*, **29**, 947–1023.

Lounis, B., Bechtel, H., Gerion, D., Alivisatos, P., and Moerner, W. E. (2000). Photon antibunching in single CdSe/ZnS quantum dot fluorescence, *Chem. Phys. Lett.*, **329**, 399–404.

Lounis, B., and Orrit, M. (2005). Single-photon sources, *Rep. Prog. Phys.*, **68**, 1129–1179.

Ma, X., Tan, H., Kipp, T., and Mews, A. (2010). Fluorescence enhancement, blinking suppression, and gray states of individual semiconductor nanocrystals closeto gold nanoparticles, *Nano Lett.*, **10**, 4166–4174.

Maestro, L. M., Rodríguez, E. M., Rodríguez, F. S., Iglesias-de la Cruz, M. C., Juarranz, A., Naccache, R., Vetrone, F., Jaque, D., Capobianco, J. A., and Solé, J. G. (2010). CdSe quantum dots for two-photon fluorescence thermal imaging, *Nano Lett.*, **10**(12), 5109–5115.

Mahler, B., Spinicelli, P., Buil, S., Quelin, X., Hermier, J. P., and Dubertret, B. (2008). Towards non-blinking colloidal quantum dots, *Nat. Mater.*, **7**, 659–664.

Makarov, D. E. (2015). *Single Molecule Science: Physical Principles and Models*, CRC Press.

Marcus, R.A. (1960). Exchange reaction and electron transfer reactions including isotopic exchange, *Discuss. Faraday Soc.*, **29**, 21–31.

Margolin, G., and Barkai, E. (2005). Nonergodicity of blinking nanocrystals and other Lévy-walk processes, *Phys. Rev. Lett.*, **94**, 080601.

Margolin, G., Protasenko, V., Kuno, M., and Barkai, E. (2006). Photon counting statistics for blinking CdSe–ZnS quantum dots: a Lévy walk process, *J. Phys. Chem. B*, **110**, 19053–19060.

Martin, J., Cichos, F., Chan, I. Y. F., Huisken, F., and Borczyskowski, C. von (2004). Photoinduced processes in silicon nanoparticles, *Israel J. Chem.*, **44**, 341–351.

Martin, J., Cichos, F., Huisken, F., and Borczyskowski, C. von (2008). Electron-phonon coupling and localization of excitons in single silicon nanocrystals, *Nano Lett.*, **8**, 656–660.

Martin, J., Cichos, F., and Borczyskowski, C. von (2012). Spectral diffusion of quasi localized excitons in single silicon nanocrystals, *J. Lumin.*, **132**, 2161–2165.

Messin, G., Hermier, J. P., Giacobino, E., Desbiolles, P., and Dahan, M. (2001). Bunching and antibunching in the fluorescence of semiconductor nanocrystals, *Opt. Lett.*, **26**, 1891–1893.

Moerner, W. E., and Kador, L. (1989). Optical detection and spectroscopy of single molecules in a solid, *Phys. Rev. Lett.*, **62**, 2535.

Moerner, W. E., Orrit, M., Wild, U. P., and Basché, T. (1998). *Single-Molecule Optical Detection, Imaging and Spectroscopy*, Verlag Chemie, Wiley.

Moerner, W. E., and Orrit, M. (1999). Illuminating single molecules in condensed matter, *Science*, **283**, 1670–1676.

Moerner, W. E. (2012). Microscopy beyond the diffraction limit using actively controlled single molecules, *J. Microsc.*, **246**, 213–220.

Montanari, G. C., and Morshuis, P. H. F. (2005). Space charge phenomenology in polymeric insulating materials, *IEEE Trans. Dielectr. Electr. Insul.*, **12**, 754–767.

Müller, J., Lupton, J. M., Rogach, A. L., Feldmann, J., Talapin, D. V., and Weller, H. (2004). Monitoring surface charge movement in single elongated semiconductor nanocrystals, *Phys. Rev. Lett.*, **93**, 167402.

Nadeau, J. L., Carlini, L., Suffern, D., Ivanova, O., and Bradforth, S. E. (2012). Effects of b-mercaptoethanol on quantum dot emission evaluated from photoluminescence decays, *J. Phys. Chem. C*, **116**, 2728–2739.

Neuhauser, R. G., Shimizu, K. T., Woo, W. K., Empedocles, S. A., and Bawendi, M. G. (2000). Correlation between fluorescence intermittency and spectral diffusion in single semiconductor quantum dots, *Phys. Rev. Lett.*, **85**, 3301–3304.

Nirmal, M., Dabbousi, B. O. Bawendi, M. G., Macklin, J. J., Trautman, J. K., Harris, T. D., and Brus, L. E. (1996). Fluorescence intermittency in single cadmium selenide nanocrystals, *Nature*, **383**, 802–804.

Novikov, D. S., Drndic, M., Levitov, L. S., Kastner, M. A., Jarosz, M. V., and Bawendi, M. G. (2005). Lévy statistics and anomalous transport in quantum-dot arrays, *Phys. Rev. B*, **72**, 075309.

Novotny, L., and Hecht, B. (2012). *Principles of Nanooptics*, Cambridge University Press, Cambridge.

Osad'ko, I. S., Eremchev, I. Y., and Naumov, A. V. (2015). Two mechanisms of fluorescence intermittency in single core/shell quantum dot, *J. Phys. Chem. C*, **119**, 22646–22652.

Orfield, N. J., McBride, J. R., Keene, J. D., Davis, L. M., and Rosenthal, S. J. (2015). Correlation of atomic structure and photoluminescence of the same quantum dot: pinpointing surface and internal defects that inhibit photoluminescence, *ACS Nano*, **9**, 831–839.

Orrit, M., and Bernard, J. (1990). Single pentacene molecules detected by fluorescence excitation in a p-terphenyl crystal, *Phys. Rev. Lett.*, **65**, 2716–2719.

Orrit, M. (2010). Chemical and physical aspects of charge transfer in the fluorescence intermittency of single molecules and quantum dots, *Photochem. Photobiol. Sci.*, **9**, 637–642.

Park, S. J., Link, S., Miller, W. L. Gesquiere, A., and Barbara, P. F. (2007). Effect of electric field on the photoluminescence intensity of single CdSe nanocrystals, *Chem. Phys.*, **341**, 169–174.

Peterson, J. J., and Krauss, T. D. (2006). Fluorescence spectroscopy of single lead sulphide quantum dots, *Nano Lett.*, **6**, 510–514.

Pfab, R. J., Zimmermann, J., Hettich, C., Gerhardt, I., Renn, A., and Sandoghdar, V. (2004). Aligned terrylene molecules in a spin-coated ultrathin crystalline film of p-terphenyl, *Chem. Phys. Lett.*, **387**, 490–495.

Pistol, M. E., Castrillo, P., Hessman, D., Prieto, J. A., and Samuelson, L. (1999). Random telegraph noise in photoluminescence from individual self-assembled quantum dots, *Phys. Rev. B*, **59**, 10725.

Plakhotnik, P., Fernée, M. J., Littleton, B., Rubinsztein-Dunlop, H., Potzner, C., and Mulvaney, P. (2010). Anomalous power laws of spectral diffusion in quantum dots: a connection to luminescence intermittency, *Phys. Rev. Lett.*, **105**, 167402.

Pohl, D. W., Denk, W., and Lanz, M. (1984). Optical stethoscopy: image recording with resolution $\lambda/20$, *Appl. Phys. Lett.*, **44**, 651.

Qin, W., Shah, R. A., and Guyot-Sionnest, P. (2012). CdSeS/ZnS alloyed nanocrystal lifetime and blinking studies under electrochemical control, *ACS Nano*, **6**, 912–918.

Qin, W., and Guyot-Sionnest, P. (2012). Evidence for the role of holes in blinking: negative and oxidized CdSe/CdS dots, *ACS Nano*, **6**, 9125–9132.

Haiyan Qin, H., Yuan Niu, Y., Meng, R., Lin, X., Lai, R., Fang, W., and Peng, X. (2013). Single-dot spectroscopy of zinc-blende CdSe/CdS core/shell nanocrystals: nonblinking and correlation with ensemble measurements, *J. Am. Chem. Soc.*, **136**, 179–187.

Reilly, P. D., and Skinner, J. A. (1995). Spectral diffusion of individual pentacene molecules in p-terphenyl crystal: stochastic theoretical model and analysis of experimental data, *J. Chem. Phys.*, **102**, 1540.

Rigler, R. Orrit, M., and Basche, T. (2012). *Single Molecule Spectroscopy: Nobel Conference Lectures*, Springer Series in Chemical Physics.

Riley, E. A. Hess, C. M., and Reid, P. J. (2012). Photoluminescence intermittency from single quantum dots to organic molecules: emerging themes, *Int. J. Mol. Sci.*, **13**, 12487–12518.

Rittweger, E., Han, K. Y., Irvine, S. E., Eggeling, C., and Hell, S. W. (2009). STED microscopy reveals crystal colour centres with nanometric resolution, *Nat. Photonics*, **3**, 144–147.

Rogez, B., Yang, H., Le Moal, E., Lévêque-Fort, S., Boer-Duchemin, E., Yao, F., Lee, Y.-H., Zhang, Y., Wegner, K. D., Hildebrand, N., Mayne, A., and Dujardin, G. (2014). Fluorescence lifetime and blinking of individual semiconductor nanocrystals on grapheme, *J. Phys. Chem. C*, **118**, 18445–18452.

Rosen, S., Schwartz, O., and Oron, D. (2010). Transient fluorescence of the off state in blinking CdSe/CdS/ZnS semiconductor nanocrystals is not governed by Auger recombination, *Phys. Rev. Lett.*, **104**, 157404.

Salvador, M. R., Graham, M.W., and Scholes, G. D. (2006). Exciton-phonon coupling and disorder in the excited states of CdSe colloidal quantum dots, *J. Chem. Phys.*, **125**, 184709.

Sauer, M., Hofkens, J., and Enderlein, J. (2011). *Handbook of Fluorescence Spectroscopy and Imaging: From Single Molecules to Ensembles: From Ensemble to Single Molecules*, Wiley-VCH, Weilheim.

Scher, H., and Montroll, E. W. (1975). Anomalous transit-time dispersion in amorphous solids, *Phys. Rev. B*, **12**, 2455–2477.

Schlegel, G., Bohnenberger, J., Potapova, I., and Mews, A. (2002). Fluorescence decay time of single semiconductor nanocrystals, *Phys. Rev. Lett.*, **88**, 1374011.

Schmidt, R. (2010). *Zeitaufgelöste Konfokalmikroskopie an einzelnen CdSe/ZnS-Halbleiternanokristallen,* Diploma Thesis, TU Chemnitz, Germany.

Schmidt, T., Chizhik, A. I., Chizhik, A. M., Potrick, K. Meixner, A. J., and Huisken, F. (2012). Radiative exciton recombination and defect luminescence observed in single silicon nanocrystals, *Phys. Rev. B*, **86**, 125302.

Schmidt, R., Krasselt, C., and Borczyskowski, C. von (2012). Change point analysis of matrix dependent photoluminescence intermittency of single CdSe/ZnS quantum dots with intermediate intensity levels, *Chem. Phys.*, **406**, 9–14.

Schmidt, R., Krasselt, C., Göhler, G., and Borczyskowski, C. von (2014). The fluorescence intermittency for quantum dot is not power-law distributed: a luminescence intensity resolved approach, *ACS Nano*, **8**, 3506–3521.

Schooten, K. J. van, Boehme, C., and Lupton, J. M. (2014). Coherent magnetic resonance of nanocrystal quantum-dot luminescence as a window to blinking mechanisms, *Chem. Phys. Chem.*, **15**, 1737–1746.

Schuster, R., Barth, M., Gruber, A., and Cichos, F. (2005). Defocused wide field fluorescence imaging of single CdSe/ZnS quantum dots, *Chem. Phys. Lett.*, **413**, 82–83.

Schuster, J., Brabandt, J., and Borczyskowski, C. von (2007). Discrimination of photoblinking and photobleaching on the single molecule level, *J. Lumin.*, **127**, 224.

Schuster, J., Cichos, F., and Borczyskowski, C. von (2005). Influence of self-trapped states on the fluorescence intermittency of single molecules, *Appl. Phys. Lett.*, **87**, 051915.

Shchukina, A. L., Eremchev, I. Y., and Naumov, A. V. (2015). Looking at a blinking quantum emitter through time slots: the effect of blind times, *Phys. Rev. E*, **92**, 032102.

Shimizu, K. T., Neuhauser, R. G., Leatherdale, C. A., Empedocles, S. A., Woo, W. K., and Bawendi, M. G. (2001). Blinking statistics in single semiconductor nanocrystal quantum dots, *Phys. Rev. B*, **63**, 205316.

Shimizu, K., Woo, W., Fisher, B., Eisler, H., and Bawendi, M. (2002). Surface-enhanced emission from single semiconductor nanocrystals, *Phys. Rev. Lett.*, **89**, 117401.

Shluger, A. L., and Stoneham A. M. (1993). Small polarons in real crystals: concepts and problems, *J. Phys. Condens. Matter*, **5**, 3049–3086.

Smith, J. M., Sher, P. H., Daniels, S. M, Pickett, N. L., O'Brien, P., Dalgarno, P. A., and Warburtone, R. J. (2006). Power-law decay of the fluorescence of semiconductor nanocrystals at nanosecond time-scales, arXiv:cond-mat/0604292v3.

Stefani, F. D., Knoll, W., Kreiter, M., Zhong, X., and Han, M. Y. (2005). Quantification of photoinduced and spontaneous quantum dot luminescence blinking, *Phys. Rev. B*, **72**, 125304.

Stefani, F. D., Hoogenboom, J. P., and Barka, E. (2009). Beyond quantum jumps: blinking nano-scale light emitters, *Phys. Today*, 34–39.

Sugisaki, M., Ren, H.-W., Nishi, K., and Masumoto, Y. (2001). Fluorescence intermittencyin self-assembled InP quantum dots, *Phys. Rev. Lett.*, **86**, 4883–4886.

Surya, C., and Hsiang, T. Y. (1986). Theory and experiment on the $1/f^{\gamma}$noise in p-channel

metal-oxide-semiconductor field-effect transistors at low drain bias, *Phys. Rev. B*, **33**, 4898–4895.

Synge, E. H. (1928). A suggested method for extending the microscopic resolution into the ultramicroscopic region, *Phil. Mag.*, **6**, 356.

Takagahara, T. (2002). *Semiconductor Quantum Dots: Physics, Spectroscopy and Applications* (Masumoto, Y., and Takagahara, T., eds.), Springer, New York.

Tang, J., and Marcus, R. (2005). Diffusion-controlled electron transferprocesses and power-law statistics of fluorescence intermittency of nanoparticles, *Phys. Rev. Lett.*, **95**, 107401.

Tachiya, M., and Mozumder, A. (1975). Kinetics of geminate-ion recombination by electron tunnelling, *Chem. Phys. Lett.*, **34**, 77–79.

Tittel, J., Göhde, W., Koberling, F., Mews, A., Kornowski, A., Weller, H., Eychmüller, A., and Basché, T. (1997). Investigations of the emission properties of single CdS-nanocrystallites, *Ber. Bunsenges. Phys. Chem.*, **101**, 1626–1630.

Trenkmann, I., Blaudeck, T., and Borczyskowski, C. von (2016). Static and dynamic quenching of quantum dot photoluminescence and dye molecules, in *Self-Assembled Organic-Inorganic Nanostructures: Optics and Dynamics* (Zenkevich, I., and Borczyskowski, C., von eds.), Pan Stanford, Singapore.

Vela, J., Htoon, H., Chen, Y., Park, Y.-S., Ghosh, Y., Goodwin, P. M., Werner, J. H., Wells, N. P., Casson, J. L., and Hollingsworth, J. A. (2010). Effect of shell thickness and composition on blinking suppression and the blinking mechanism in 'giant' CdSe/CdS nanocrystal quantum dots, *J. Biophotonics*, **3**, 706–717.

Verberk, R. Oijen, A. van, and Orrit, M. (2002). Simple model for the power-law blinking of single semiconductor nanocrystals, *Phys. Rev. B*, **66**, 233202.

Verberk, R., and Orrit, M. (2003). Photon statistics in the fluorescence of single molecules and nanocrystals: correlation functions versus distributions of on- and off-times, *J. Chem. Phys.*, **119**, 2214–2222.

Vogel, M., Gruber, A., Wrachtup, J., and von Borczyskowski, C. (1995). Determination of intersystem crossing parameters via observation of quantum jumps on single molecules, *J. Phys. Chem.*, **99**, 14915–14917.

Voznyy, O. (2011). Mobile surface traps in CdSe nanocrystals with carboxylic acid ligands, *J. Phys. Chem. C*, **115**, 15927–15932.

Voznyy, O., and Sargent, E. H. (2014). Atomistic model of fluorescence intermittency of colloidal quantum dots, *Phys. Rev. Lett.*, **112**, 157401.

Wang, L.-W. (2001). Calculating the influence of external charges on the photoluminescence of CdSe quantum dots, *J. Phys. Chem. B*, **105**, 2360–2364.

Wang, S., Querner, C., Emmons, T., Drndic, M., and Crouch, C. H. (2006). Fluorescence blinking statistics from CdSe core and core/shell nanorods, *J. Phys. Chem. B*, **110**, 23221–23228.

Wang, X., Ren, X., Kahen, K., Hahn, M. A., Rajeswaran, M., Maccagnano-Zacher, S., Silcox, J., Cragg, G. E., Efros, A. L., and Krauss, T. D. (2009). Non-blinking semiconductor nanocrystals, *Nature*, **459**, 686–689.

Wang, Y., Fruhwirth, G., Cai, E., Tony Ng, T., and Selvin, P. R. (2013). 3D super-resolution imaging with blinking quantum dots, *Nano Lett.*, **13**, 5233–5241.

Watkins, L. P., and Yang, H. (2005). Detection of intensity change points in time-resolved single molecule measurements, *J. Phys. Chem. B*, **109**, 617–628.

Watson, P. K., Schmidlin, F. W., and La Donna, R. V. (1992). The trapping of electrons in polystyrene, *IEEE Trans. Dielectr. Electr. Insul.*, **27**, 680–686.

Weisenburger, S., and Sandoghdar, V. (2015). Light microscopy: an ongoing contemporary revolution, *Contemp. Phys.*, **56,** 123–143.

Whitham, P. J. Knowles, K. E., Reid, P. R., and Gamelin, D. R. (2015). Photoluminescence blinking and reversible electron trapping in copper-doped CdSe nanocrystals, *Nano Lett.*, **15**, 4045–4051.

Willard, J. E. (1973). Trapped free radicals and electrons in organic glasses, *Science*, 553–561.

Wrachtrup, J., Borzyskowski, C. von, Orrit, M., Bernard, J., and Brown, R. (1993). Optical detection of magnetic resonance in a single molecule, *Nature*, **363**, 244–246.

Xia, X., and Silbey, R. J. (2006). Fluorescence intermittency of a single quantum systemand Anderson localization, arXiv:cond-mat/0601092v4.

Xu, C. S., Kim, H., Yang, H., and Hayden, C. C. (2007). Multiparameter fluorescence spectroscopy of single quantum-dot-dye FRET hybrids, *J. Am. Chem. Soc.*, **129**, 11008–11009.

Yaremko, A. M., Yakimchuk, V. O., Dzhagan, V. M., Valakh, M. Ya., Azhniuk, Yu. M., Baran, J., Ratajczak, H., and Drozd, M. (2007). Investigation of electron-phonon interaction in bulk and nanostructured semiconductors, *Semicond. Phys. Quantum Electron. Optoelectron.*, **10**, 1–5.

Yuan, C. T., Yu, P., and Tang, J. (2009a). Blinking suppression of colloidal CdSe/ZnS quantum dots by coupling to silver nanoprisms, *Appl. Phys. Lett.*, **94**, 243108.

Yuan, C. T., Yu, P., Ko, H. C., Huang, J., and Tang, J. (2009b). Antibunching single-photon emission and blinking suppression of CdSe/ZnS quantum dots, *ACS Nano*, **3**, 3051–3056.

Zenkevich, E. I., Stupak, A., Kowerko, D., and Borczyskowski, C. von (2012). Influence of single dye molecules on temperature and time dependent optical properties of CdSe/ZnS quantum dots: ensemble and single nanoassembly detection, *Chem. Phys.*, **406**, 21–29.

Zenkevich, E. I., Stupak, A., Goehler, C., Krasselt, C., and Borczyskowski, C. von (2015). Tuning electronic states of a CdSe/ZnS quantum dot by only one functional dye molecule, *ACS Nano*, **9**, 2886–2907.

Zenkevich, E. I. and Borczyskowski, C. von (2016). *Self-Assembled Organic-Inorganic Nanostructures: Optics and Dynamics*, Pan Stanford, Singapore.

Zhang, K., Chang, H., Fu, A., Alivisatos, A. P., and Yang, H. (2010). Continuous distribution of emission states from single CdSe/ZnS quantum dots, *Nano Lett.*, **6**, 843–847.

Zhao, J., Nair, G., Fisher, B. R., and Bawendi, M. G. (2010). Challenge to the charging model of semiconductor-nanocrystal fluorescence intermittency from off-state quantum yields and multiexciton blinking, *Phys. Rev. Lett.*, **104**, 157403.

Zbydniewska, E., Duzynska, A., Popoff, M., Hourlier, D., Lenfant, S., Judek, J., Zdrojek, M., and Melin, T. (2015). Charge blinking statistics of semiconductor nanocrystals revealed by carbon nanotube single charge sensors, *Nano Lett.*, **15**, 6349–6356.

Chapter 5

Optical Spectroscopy of Single Ag Nanoclusters

Stefan Krause
Institute of Physics, Center for Nanostructured Materials and Analytics, Technische Universität Chemnitz, Reichenhainer Str. 70, D-09107 Chemnitz, Germany
stefan.krause@physik.tu-chemnitz.de

Nanometer-sized, noble metal clusters are promising candidates for bright, stable, and nontoxic fluorescent labels in the field of bioimaging. Thereby, the size of the clusters as well as their surroundings will strongly influence the emission properties. A detailed understanding of these correlations is essential for further improvements and tailor-made fluorescent labels on the basis of noble metal clusters. Single-molecule techniques pave the way to study single nanoclusters and their photophysical properties, which often diverge strongly from the ensemble average. This review aims on an overview related to studies of Ag nanoclusters on the single-particle level.

Tuning Semiconducting and Metallic Quantum Dots: Spectroscopy and Dynamics
Edited by Christian von Borczyskowski and Eduard Zenkevich

ISBN 978-981-4745-24-6 (Hardcover), 978-981-4745-25-3 (eBook)
www.panstanford.com

5.1 Introduction

Fluorescence-based techniques have become standard methods in biological and medical applications for reasons of in vivo applicability and specific labeling [Johnson, 1998; Zhang et al., 2002; Resch-Genger et al., 2008]. As a consequence, the field of fluorescence applications is steadily increasing, which leads to a rising demand for suitable fluorescent probes and markers. Typically organic dyes are used as fluorescent labels as they are nontoxic and available for versatile applications. Unfortunately, they often lack photostability. Semiconducting quantum dots stipulate a more photostable and bright opportunity; however, they are strongly toxic and the comparably large size of a few nanometers prevents them from being sensitive for binding events [Resch-Genger et al., 2008]. Promising candidates to bridge the gap between photostability and nontoxicity are noble metal clusters, for example, Au and Ag clusters. At sizes above 1 nm these clusters show metallic characteristics and quantisized plasmon resonance effects [Genzel et al., 1975], making them ideal for photothermal detection [Zijlstra and Orrit, 2011]. In contrast, pronounced fluorescence can be observed when approaching cluster sizes beyond 1 nm [Ozin and Hugues, 1983; Treguer et al., 2005; Zheng et al., 2007; Zhang and Wang, 2014]. It is already known for decades that the absorption and emission characteristics of such clusters strongly depend on the cluster size—an effect similar to the quantum confinement in semiconducting nanocrystals. However, in contrast to semiconducting quantum crystals, which show fluorescence emission at sizes of thousands of atoms, Ag nanoclusters become emissive at atom numbers of tens of atoms and beyond [Choi et al., 2012]. This small size leads to a strong variation of the emission properties in the case of a small change in the atom number [Ozin and Hugues, 1983; Harb et al., 2008; Shang et al., 2011].

It is convenient that in the case of size dependence, uniform emission properties for all fluorescent labels within a specimen are favorable for most applications. This can in principle be achieved by mass spectroscopy, which yields a monodisperse size distribution. Unfortunately, it is not suitable for producing sufficient amounts for biolabeling. A more promising way is the synthesis within solution by means of laser ablation, colloidal growth, or organic scaffolds. These approaches typically result in a distribution of sizes and transition energies. In addition, Ag nanoclusters in solution must be prevented from agglomeration, which can be achieved by the scaffolds themselves, which serve as stabilizers, such as dendrimers [Zheng and Dickson, 2002], polymers or polyelectrolytes [Shang and Dong, 2008], proteins, peptides [Yu et al., 2007], DNAs [Ritchie et al., 2007; Vosch et al., 2007; Shang et al., 2011], or zeolites [De Cremer et al., 2008; Krause et al., 2013]. As a side effect, these different scaffolds influence the optical properties of the synthesized Ag nanoclusters themselves.

Several reviews and books summarize fundamental properties of noble metal clusters and aspects of their applications [Zheng et al., 2007; Shang et al., 2011; Choi et al., 2012; Zhang and Wang, 2014]. In the framework of this review we want to give an overview concerning studies of Ag nanoclusters at the single-molecule level. Single-molecule spectroscopy allows for a direct measurement of transition energies, fluorescence decays, intensities, and fluorescence intermittency (blinking) [Cichos et al., 2007] phenomena without any ensemble averaging and is therefore the ideal tool to study correlations between size, type of ligands, and the optical properties of Ag nanoclusters. This detailed understanding of the photophysics of noble metal nanoclusters is mandatory for the future synthesis of tailor-made, nontoxic, and stable fluorescent labels.

5.2 Single Ag Nanoclusters from Ag_2O Films

The first study of fluorescent Ag nanoclusters at the single-molecule level was published by Lynn et al. in 2001 [Peyser et al., 2001]. A particular feature was the photochemical activation of thin films of Ag_2O (~20 nm) at short wavelengths (450–480 nm), which resulted in a strong fluorescence emission of individual clusters. It was demonstrated that this photoactivation could be utilized for data storage as written images were stable in the dark for at least several days. The fluorescent clusters showed blinking phenomena as well as spectral diffusion processes over a wide range of wavelengths

(~100 nm). Emission patterns which were recorded with an integration time of 10 s confirmed the single-cluster emission as they showed typical shapes, as expected for a single radiating dipole (see Fig. 5.1C). Blinking could be reduced by utilizing green (510–550 nm) excitation light, which, in addition, increased the brightness of the observed clusters (see Fig. 5.1A,B). In this manner, spectral dependency of the blinking as well as of the spectral dynamics could be demonstrated. By measuring the saturation of absorption of the clusters, the absorption cross section could be determined to be at least 50 times higher in comparison to the best organic dye molecules. The fluorescence could be assigned to small Ag clusters produced by photoactivation of Ag_2O, which served as a building cage as well as a protective coating. Ag impurities resulted in an increased probability for photoreduction, which led to the formation of Ag^{2+}, Ag^2, and Ag^{3+}. Time-dependent measurements showed a nonlinear increase in fluorescence intensity, which was attributed to the creation and destruction of multiatom Ag_n nanoclusters [Peyser et al., 2002]. The complex time dependency of the fluorescence intensity favors multiphoton processes to be accountable for the formation of larger nanoclusters. According to the ratio of binding energy of the clusters and excitation energy, as well as the influence of unstable AgO, the clusters were likely to consist only of a few (2–8) atoms. Therefore, the observed strong spectral dynamics were addressed to addition/reduction of a single Ag atom. The dispersion of the fluorescence of a large number of emissive clusters at once showed broad spectra with at least five distinct features at 548, 594, 641, 673, and 725 nm (see Fig. 5.1, right). The relative contribution of each feature changed with proceeding excitation time. While the short-wavelength features dominated the spectra at the beginning of the excitation period, the major intensity contributions shifted toward longer wavelengths after tens of seconds of excitation. A final equilibrium was reached after ~60 s. The photochemical activation of the investigated Ag nanoclusters with UV and blue illumination and the possibility of read-out at green excitation without further activation makes these clusters ideal candidates for future data storage.

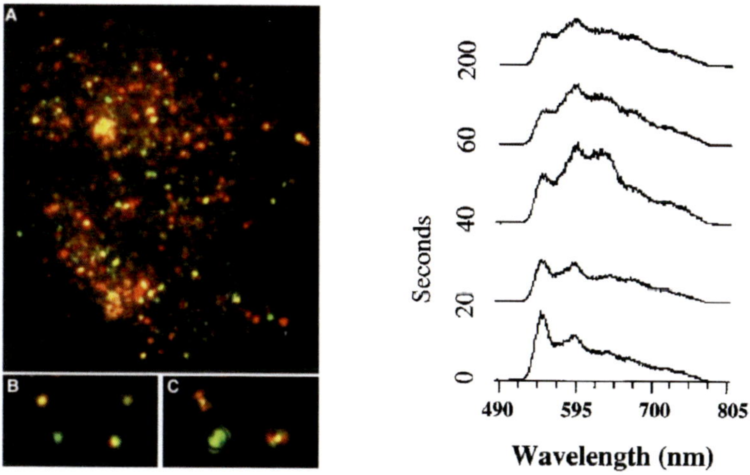

Figure 5.1 (A) Fluorescence from a 16.6 nm Ag/Ag_2O film excited at 514.5 nm. The image is 29.5 μm wide. (B) Four individual features of which (C) shows the respective emission patterns, confirming single-nanoparticle emission. The fluorescent nanoparticle in the upper-right corner of (B) bleached between photographs. These individual nanoparticles emit as single-quantum systems that vary widely in emission wavelength at Ar^+ laser excitation. Similar behavior was observed for 488 nm Ar^+ laser and blue (450 to 480 nm) mercury lamp excitation. In (B), the upper two molecules are separated by 4.0 μm; (C) is on the same scale as (B). (Right) Typical sequential emission spectra from a 600 nm diameter region of an AgO film excited at 514.5 nm (1 kW/cm^2) and collected through a long-pass filter. At least five distinct spectral features (548, 594, 641, 673, and 725 nm) can be seen, which also change in relative intensity as the Ag_n nanocluster size distribution is continuously modified through excitation. From Peyser, L. A., Vinson, A. E., Bartko, A. P., and Dickson, R. M. (2001). Photoactivated fluorescence from individual silver nanoclusters, *Science*, **291**, 103–106. Reprinted with permission from AAAS. Reprinted with permission from Peyser, L. A., Lee, T., and Dickson, R. M. (2002). Mechanism of Ag_n nanocluster photoproduction from silver oxide films, *J. Phys. Chem. B*, **106**, 7725–7728. Copyright (2002) American Chemical Society.

5.3 Colloidal Silver Particles

Another preparation approach of fluorescent silver clusters and colloids was introduced by Henglein et al. (1991) with radiolytic reduction of silver ions into charged silver clusters with a low number of atoms and colloidal silver particles. Thereby, a silver salt solution containing a polyanion for stabilization and an alcohol to convert the oxidizing OH radicals is exposed to γ-rays. The irradiation leads to a well-defined concentration of reducing or oxidizing radicals in the solution. These radicals transfer electrons to the silver ions, which form colloidal metal particles. The type of produced particles depends on the conditions, and either metallic silver clusters of several nanometers in diameter or very small fluorescent oligomeric clusters are stabilized. Treguer et al. were able to produce metallic particles without stabilizers by adding silver ions to a solution of $NaBH_4$ during vigorous stirring [Treguer et al., 2005]. The resulting solution was spin-casted onto a glass coverslip. The particles were excited at 476 nm and showed fluorescence at a manifold of colors. The left side of Fig. 5.2 shows a luminescence image of 10 μm × 10 μm of the spin-coated particles. Distinct broad emission bands were observed at about 550, 600, and 650 nm (Fig. 5.2a–d). The recorded spectra of individual particles were compared with the luminescence spectrum of silver oxide powder (Fig. 5.2e) and showed good agreement. Additionally, the luminescence increased with time, which gives rise to the assumption

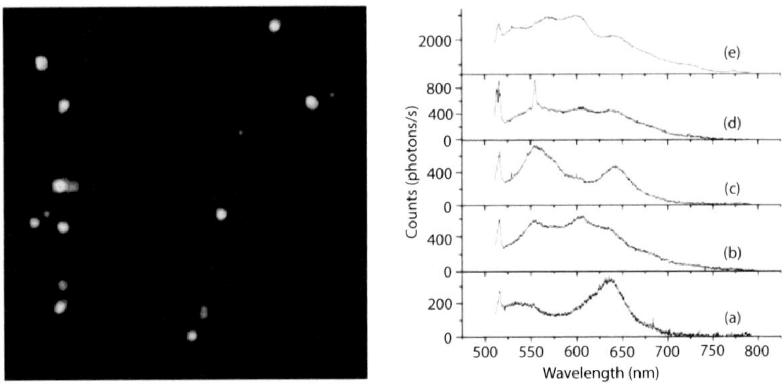

Figure 5.2 Left: 10 μm × 10 μm luminescence image of silver nanoparticles spin-coated onto glass cover slips. Right: (a–d) Luminescence spectra of four different single silver particles taken on the same sample. Excitation intensity = 10 kW/cm^2; integration time = 200 s. (e) Luminescence spectrum of silver oxide powder under laser illumination at 476 nm. Reproduced from Treguer, M., Rocco, F., Lelong, G., Nestour, A. L., Cardinal, T., Maali, A., and Lounis, B. (2005). Fluorescent silver oligomeric clusters and colloidal particles, *Solid State Sci.*, **7**, 812–818. Copyright © 2005, Elsevier Masson SAS. All rights reserved.

of continuous formation of emissive sites, as already observed for Ag_2O films [Peyser et al., 2001; Peyser et al., 2002]. Besides the strong variations in emission wavelength for different particles, time traces of individual particles showed strong intensity fluctuations and blinking. Surprisingly, lifetimes < 36 ps were observed for single emissive sites and attributed to the interaction of the small metallic clusters with the surface of bigger colloidal particles of the size of a few nanometers up to 50 nm. Despite the large amounts which can be produced by radiolytic production, the requirement of bigger colloidal particles for photoactivation complicates the application of these nanoparticles as fluorescence labels.

5.4 Dendrimer and DNA-Encapsulated Ag Nanoclusters

In the previous section we elucidated the properties of photoactivated Ag_n nanoclusters generated from thin films of Ag_2O as suitable candidates for data storage. However, these nanoclusters are surface bound and cannot be extracted for fluorescence labeling applications. In addition, nanoclusters of the size of a few nanometers tend to agglomerate in solution, which results in fluorescence quenching. Therefore, the clusters need to be prevented from agglomeration. Zheng et al. prepared dendrimer-encapsulated, water-soluble Ag nanoclusters dissolving poly(amidoamine) (PAMAM) G4-OH and G2-OH (fourth and second generations) together with $AgNO_3$ in water [Zheng and Dickson, 2002]. Without adding a reducing agent which typically forms larger nanoclusters (>3 nm) the dendrimers sequestered the metal ions from the solution, which resulted in small nanoclusters consisting of less than eight atoms. The solution had to be photoactivated by white light exposure to show absorption and emission of the photoreduced silver nanoclusters. Absorption occurred at 345 and 430 nm after photoactivation while a multicolor emission was observed and addressed to clusters of different sizes. Mass spectroscopy measurements showed that primary Ag_1–Ag_4 was formed within the G2 dendrimer. The microscopic investigation of a drop of the solution on a glass cover slide resulted in complete photoactivation within 6 s, followed by a bright and stable fluorescence. Individual clusters exhibited blinking as well as defined dipole emission patterns besides a stable fluorescence emission, which remained stable (~80%) for more than 30 min. The nanoclusters featured stable spectral characteristics without spectral diffusion, a property which was formerly not observed for the clusters produced from Ag_2O films. The measured fluorescence spectra could be addressed to five discrete species whose superposition resembles the bulk spectrum, as shown in Fig. 5.3. Peak intensities for the five species were found to occur at 533, 553, 589, 611, and 648 nm. The constant emission wavelengths were attributed to dendrimer-stabilized clusters, which were prevented from agglomeration as well as from additional photoreduction.

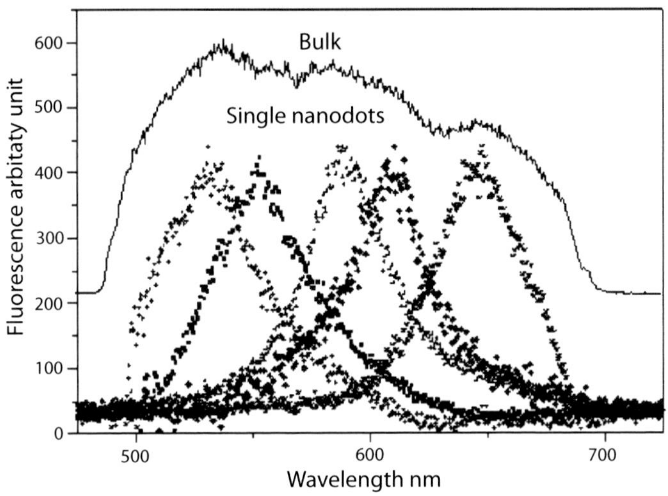

Figure 5.3 Room-temperature single nanocluster fluorescence spectra (476 nm Ar^+ laser excitation, 496 nm long-pass filter). Emission maxima for the five typical nanoclusters shown are 533, 553, 589, 611, and 648 nm. Indistinguishable from that on AgO surfaces, the ensemble fluorescence spectrum of bulk silver nanodot solutions (top) predominantly consists of these five spectra. Reprinted with permission from Zheng, J., and Dickson, R. M. (2002). Individual water-soluble dendrimerencapsulated silver nanodot fluorescence, *J. Am. Chem. Soc.*, **124**, 13982–13983. Copyright (2002) American Chemical Society.

Besides dendrimers another promising scaffold for the size-selective growth and stabilization of few-atom Ag nanoclusters has been demonstrated [Petty et al., 2004; Vosch et al., 2007]. Thereby, the affinity of cytosine bases of oligonucleotides for metal ions was harnessed to create Ag nanoclusters and encapsulate them within the single-stranded DNA (ssDNA), as shown in Fig. 5.4A. In the case of 12-cytosine base ssDNA the captured silver ions formed nanoclusters consisting of 1–2 atoms within the oligonucleotide. These synthesized complexes featured bright, highly polarizable, discrete transitions; good photostability; and small sizes within the biocompatible ssDNA scaffold. The emission occurred in the red and near-infrared (NIR) region and was attributed to two different species. The NIR-emitting species with an absorption maximum at ~650 nm and an emission maximum at 700 nm could be preferentially created by dissolution in phosphate-buffered saline (PBS) (see Fig. 5.4B). A fluorescence quantum yield of 17% and a fluorescence lifetime of 2.6 ns were determined. While mass spectroscopy supported the assumption that the fluorescence has to be addressed to Ag dimers, measurements in rare gas matrices [Schulze et al., 2004] suggest that Ag trimers are accountable for the emission. The cluster emission was compared with a typical organic fluorescent labeling molecule (Cy5.29) and appeared twice as bright. In addition the Ag clusters exhibited no blinking on experimentally relevant timescales (0.1 to >1000 ms), making them ideal for single-molecule (particle) reporter applications. The excellent photostability allowed for the detection of single clusters for minutes and hours at measured fluorescence intensities of 10 kcps. Applying fluorescence correlation spectroscopy the authors found fast, excitation-dependent blinking dynamics on the 10 μs timescale (see Fig. 5.4C), which is far beyond the relevant timescale of, for example, single-molecule tracking experiments. The observed intensity fluctuations resulted from excited-state transitions into a dark state via intersystem crossing (ISC). The decrease in the correlation time with increasing excitation intensity was found to be a consequence of shortened "on" times due to increased ISC rates and shortened "off" times due to dark-state absorption and back ISC, which transferred the clusters much faster back into the "on" state than observed for low excitation intensities.

In contradiction to organic dye molecules the blinking dynamics of the synthesized ssDNA-encapsulated Ag nanoclusters did not change upon oxygen exclusion, which is advantageous for fluorescence labeling applications and supports the assumption of ISC enhancement due to a heavy atom effect of silver. Figure 5.4D

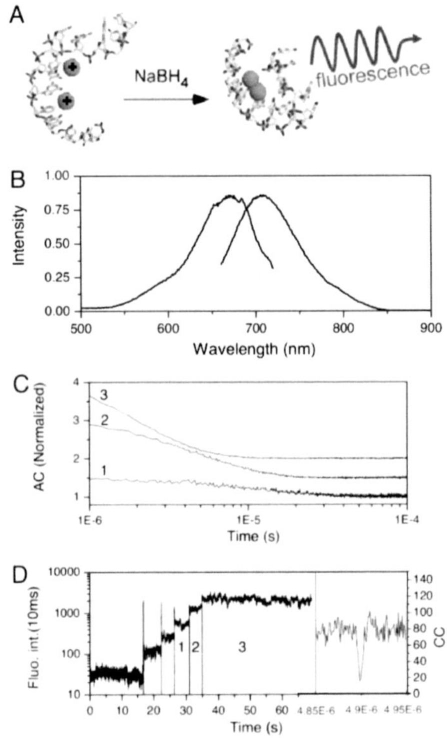

Figure 5.4 NIR-emitting Ag nanoclusters. (A) Schematic of the IR-emitting C_{12}–Ag_n formation. After complexation of C_{12} DNA with silver cations, the mixture is reduced with $NaBH_4$ and the NIR-emitting Ag nanocluster is studied. (B) Normalized excitation and emission spectra of the studied species. (C) Autocorrelation traces of the three brightest intensity levels from (D). The numbers from 1 to 3 correspond to the levels indicated in (D). The autocorrelation curves were constructed by using bin times of 100 ns. Curves 2 and 3 are for clarity vertically offset by 0.5 and 1.0, respectively. (D) Fluorescence intensity trajectory of a single Ag cluster excited under continuous-wave (cw) excitation at 647 nm as incident intensity is changed from 67 to 230, 670, 3000, 7200, and 23,000 W/cm^2, respectively. Vertical lines mark intensity changes. The right section shows the zero-delay portion (one channel is artificially delayed by ~4.9 μs with a DG-535 delay generator, Stanford Research, Sunnyvale, CA, to avoid detection dead times) of the photon arrival cross-correlation (CC) of a Ag cluster excited with 23 kW/cm^2 using a bin time of 1 ns. This region shows the antibunching feature of the fluorescence. From Vosch, T., Antoku, Y., Hsiang, Y. J., Richards, C. I., Gonzalez, J. I., and Dickson, R. M. (2007). Strongly emissive individual DNA-encapsulated Ag nanoclusters as single-molecule fluorophores, *Proc. Natl. Acad. Sci. USA*, **104**, 12616–12621. Copyright (2007) National Academy of Sciences, U.S.A. *Acad. Sci. USA*, **104**, 31 (2007).

shows how the detected fluorescence intensity increases with the excitation intensity. More than 200 kcps could be detected for a single cluster. At this high excitation intensity, long "on" and "off" time blinking occurred, which was addressed to a multiphoton origin. Nevertheless, at excitation intensities of 23 kW/cm^2 a single cluster passed through more than 10^{10} excitation cycles and the emission could be observed for several minutes before the cluster either transferred into a long-lived "off" state or was photochemically deactivated. Hooley et al. preceded with the optical investigation of Ag nanoclusters, encapsulate in ssDNA containing 24 cytosines [Hooley et al., 2015]. Emission maxima, fluorescence decay times, fluorescence intensities, blinking, and photon antibunching were measured simultaneously to probe the photophysical heterogeneities of these NIR-emitting silver nanoclusters. Thereby, the observation of photon antibunching guaranteed detection of single emitters, while the corresponding emission spectra revealed spectral diffusion processes without any obvious trends. Changes in the Ag nanoclusters geometry, local environment, or conformational changes of the DNA were discussed to be most likely responsible for the observed fluctuations.

Despite the efforts which have been made to synthesize stable and bright Ag clusters, the nature of their optical transitions is not fully understood and whether it is of molecule-like or collective (e.g., plasmon like) nature is still under debate. Oemrawsingh et al. addressed this question by means of low-temperature single-emitter spectroscopy. Thereby, ensemble line broadening as well as temperature line broadening can be effectively suppressed, which enables for a direct approach to the homogeneous line width. A detailed knowledge of the line widths and their origin is especially desirable for fluorescence applications as they limit the number of distinct color channels. In Oemrawsingh et al. (2012) two different Ag:DNAs excitable at 565–605 nm and with bright emission were analyzed. These Ag:DNAs were synthesized within 5′-CCG-CCA-CCC-CGC-GGT-3′ (Ag:DNA1) and 5′-CCG-CCC-CCC-TGC-GGT-3′ (Ag:DNA2), respectively. According to tandem high-performance liquid chromatography (HPLC)-mass spectrometry and in-line spectroscopy measurements [Schultz and Gwinn, 2012], the observed fluorescence was assigned to Ag:DNA complexes containing 14 to 16 Ag atoms. Considering the similarity of the synthesis in comparison to Zheng and Dickson (2002), which attributed the observed fluorescence to small Ag clusters of less than eight atoms, the big discrepancy in atom numbers seems questionable. However, while Ag:DNA1 showed pronounced fluorescence at 620 nm, Ag:DNA2 emitted light at 685 nm. Figure 5.5a shows emission as a function of excitation for Ag:DNA1 at 1.7 K. It is obvious that a change of the excitation wavelength is accompanied by a corresponding shift of the emission wavelength. This wavelength shift is a consequence of inhomogeneous (ensemble) line broadening. Figure 5.5b shows the emission spectra of Ag:DNA1 at three different temperatures. The change in temperature from 295 K to 1.7 K results in a fivefold increased fluorescence intensity but a negligible line narrowing from 50 nm to 41 nm. This is a result of the heterogeneous embedding matrix—glycerol or poly(vinyl alcohol) (PVA)—which influences the emission wavelength for every emitter according to its nanoscopic surrounding. The inhomogeneous line broadening could be circumvented by investigating single emitters. Oemrawsingh et al. (2012) observed single-emitter fluorescence from the Ag:DNAs at low temperatures by embedding the complexes at 20 nM concentrations in either glycerol or PVA, which forms a robust film. The two different host matrices had no significant influence on line shapes, line widths, and emission wavelengths. While at ambient conditions, blinking and bleaching within five minutes were observed, flushing the samples with He gas reduced both and favored oxygen to be accountable for these processes. Lifetime measurements on the same Ag:DNAs revealed an excited-state lifetime in the nanosecond range, which led Oemrawsingh et al. to the expectation of spectral line widths in the microelectron volt range if dephasing processes, accountable for line broadening, were absent. Figure 5.5c–e shows fluorescence spectra of three different individual emitters at different temperatures for 60 s integration time. The brightness of the emitters increases by a factor of 5 when changing temperature from 295 K to 1.7 K, which is in accordance with the ensemble data in Fig. 5.5b and was attributed to a decrease in nonradiative decay rates, as known for Au_{10} clusters [Yu et al., 2012]. Contrary to the ensemble line width, the single-emitter line width is reduced by a factor of 0.5 upon temperature lowering, which results in about a 26 nm spectral line width. This width is still much broader than expected from the fluorescence lifetime, which leads to the conclusion that dephasing mechanisms must be involved in the spectral broadening. By measuring emission spectra at different excitation wavelengths,

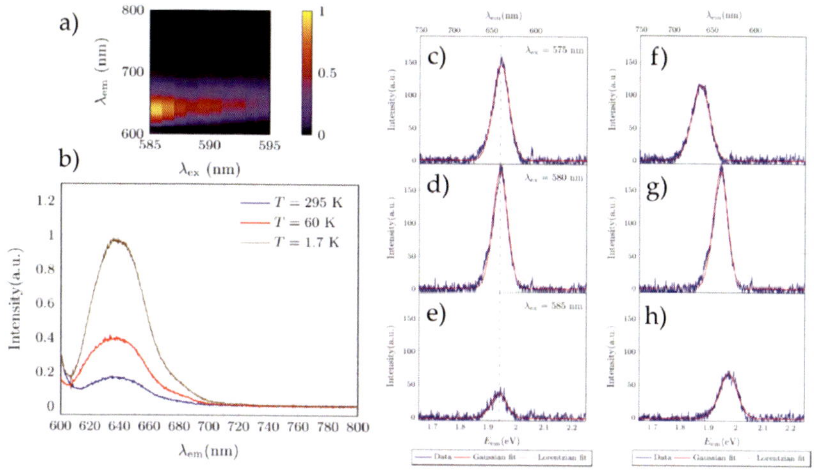

Figure 5.5 (a) Emission intensity of an Ag:DNA1 ensemble at 1.7 K as a function of the excitation and emission wavelengths. Results for Ag:DNA2 were similar. (b) Emission spectra of an Ag:DNA1 ensemble at different temperatures, excited at 585 nm. The width and position of the emission peak are only very weakly dependent on temperature. Nevertheless, the efficiency of emission is increased by a factor of 5. (c) Emission spectra plotted versus photon energy of three different Ag:DNA1 emitters, with both Gaussian and Lorentzian fits. At room temperature, the single emitters are generally very dim, with an average fwhm of 50 nm or 0.15 eV. (d) At 223 K the brightness of a single emitter increases. The energy of the luminescence band does not correlate to temperature but is rather changed due to differences of the individual emitters and their environment, as discussed in relation with Fig. 5.7. (e) The spectrum of another individual Ag:DNA1 emitter, now at 1.7 K. As compared to the spectrum in (c), the brightness has increased by a factor of 5 and the line width has decreased to 23 nm or 0.070 eV for the specific emitter shown in (e). Averaging over 17 emitters, we find a full-width at half-maximum (FWHM) of 26 nm or 0.079 eV at 1.7 K. The line shape is closer to Gaussian than Lorentzian, which is most apparent in the wings. Emission spectra as a function of photon energy of the same Ag:DNA1 emitter at 1.7 K, for the excitation wavelengths (f) 575 nm, (g) 580 nm, and (h) 585 nm. The emission peak remains at 1.94 eV, or 639.1 nm, as indicated by the dashed line. The spectrum in (h) appears less bright, due to the fact that the emitter bleached during data acquisition. Reprinted with permission from Oemrawsingh, S. S. R., Markešević, N., Gwinn, E. G., Eliel, E. R., and Bouwmeester, D. (2012). Spectral properties of individual DNA-hosted silver nanoclusters at low temperatures, *J. Phys. Chem. C*, **116**, 25568–25575. Copyright (2012) American Chemical Society.

Oemrawsingh et al. could exclude fluctuations of the surrounding medium as the responsible process for the spectral diffusion. Figure 5.5f–h shows emission spectra of a single emitter which was excited at three different wavelengths. It is obvious that the emission is independent of the excitation wavelength, which would not occur in the case of spectral diffusion processes. In addition, the authors found the line widths to be independent of the embedding medium (glycerol, PVA). This eliminates different conformers to be accountable for the broad emission line widths as glycerol would favor compact conformers in contrast to a polymer, which should allow for a multitude of conformers due to the free volume between neighbored polymer chains. Furthermore, the featureless Gaussian line shapes presented in Fig. 5.5c–h are striking arguments that the observed optical transitions do not have a typical molecular character which would lead to asymmetric line shapes, either because of electron–phonon coupling or as a result of vibronic states. Therefore, the authors conclude that the investigated Ag:DNAs have a collective, multielectron excitation comparable to plasmon-like excitation. This is further supported by the observation that Ag:DNA2 emitters exhibit twice the line width of Ag:DNA1 emitters. The two different DNAs influence the detailed cluster structure, which has a strong influence on the dephasing and line widths, as known from the varying line widths in different rare gas matrices [Harbich et al., 1990; Harbich et al., 1992; Fedrigo et al., 1993; Félix et al., 1999; Sieber et al., 2004].

5.5 Ag Nanoclusters from Zeolite Y Cages

A further possibility to produce fluorescent noble metal nanoclusters is given by zeolite cages, which serve as a size director and an encapsulating material at once [Wöhrle and Schulz-Ekloff, 1994; Warnken et al., 2001; Yumura et al., 2011; Miyanaga et al., 2013]. Thereby, zeolite Y can be loaded with a Ag(I) precursor ($[AgO_2C(CH_2OCH_2)_3H]$) by incipient wetness impregnation to create fluorescent Ag nanoclusters [Kahle et al., 2014]. As shown by Krause et al. the embedded clusters can be extracted by bringing larger zeolite Y particles into aqueous solution [Krause et al., 2013]. Due to the channel size of the zeolite Y cage, clusters of 0.78 nm in diameter and less should be able to escape the zeolite cages. The dispersion was spin-casted onto mica substrates and the size distribution of the clusters was measured by atomic force microscopy (AFM), as can be seen in the height image and the distribution of heights in Fig. 5.6a,b. The distribution led to an average cluster diameter of (0.90 ± 0.17) nm, which is above the zeolite channel size and could be explained by the contribution of cluster agglomerates whose existence was proofed by the AFM phase images (not shown here). Taking a face-centered cubic Ag lattice into account, the authors estimated an upper limit of 22 Ag atoms which contribute to the cluster formation.

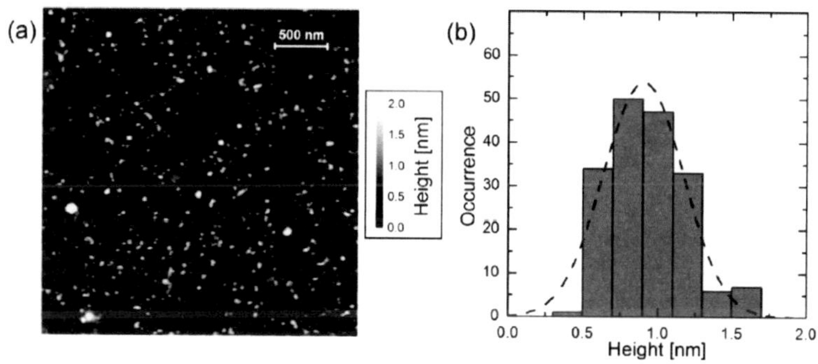

Figure 5.6 (a) AFM height image of immobile Ag nanoclusters on mica. (b) Histogram of cluster diameters and a Gaussian fit with a peak value of (0.90 ± 0.17) nm. Reprinted with permission from Krause, S., Hartmann, M., Kahle, I., Neumann, M., Heidernätsch, M., Spange, S., and Borczyskowski, C. von (2013). Optical tracking of single Ag clusters in nanostructured water films, *J. Phys. Chem.* C, **117**, 24822–24829. Copyright (2013) American Chemical Society.

As can be seen in Fig. 5.6a, the dispersion on mica which was used for AFM investigations led to a high cluster density, which was not suitable for single-emitter investigations. Instead the solution was diluted and spin-casted onto a SiO_2 substrate.

In the present contribution we would like to discuss some unpublished results on the photostability of Ag clusters. It has been recently shown that blinking of a single quantum object is related to (reversible) photobleaching at the ensemble level [Cichos et al., 2004; Schuster et al., 2007]. We investigated the fluorescence intensity for a large number of particles averaged as a function of time. Figure 5.7a shows the overall nanoparticles integrated fluorescence intensity across an area of 25 × 25 μm^2 as a function of time. At t = 2 s the excitation light was switched on, followed by an instant increase of the fluorescence intensity up to I_0. At t = 14 s the excitation light was switched off for a recovering period of t_R = 1811 s. At t = 1825 s the excitation light was switched on again and the fluorescence intensity recovered entirely to $I_R = I_0$. From the two fluorescence intensity values follows the ratio I_R/I_0 = 1. It has to be mentioned that time t = 1825 s for a complete recovery might be overestimated as a total recovery could occur already at shorter recovering periods which were not verified in the experiments. Nevertheless, photochemical changes of Ag nanoclusters which are irreversible can be safely excluded to be responsible for the observed decay in fluorescence intensity. The same experiment was performed for two other recovering periods, namely 100 s and 300 s. The ratio I_R/I_0 is shown in Fig. 5.7b as a function of the length of the recovery period. With the above-mentioned overestimation of the time for a complete fluorescence recovery, the behavior is likely to show a linear trend in the log-log plot. A similar dependency has been observed for silica nanoparticles [Cichos et al., 2004]. This similarity leads to the conclusion that blinking phenomena with high probabilities for the "off" states are accountable for the decreasing fluorescence intensity at constant excitation. A still further similarity indicates that the initial intensity can be completely recovered. This implies that no permanent photoinduced chemistry takes place, contrary to what is observed for many (single) molecules [Cichos et al., 2007]. Probably the observed reversible bleaching is related to charging processes. We would like to mention that we have ruled out that the luminescence recovery is due to the already mentioned diffusion of the particles. If such recovery via diffusion would apply, a recovery of intensity should start from the edge of the bleached spot, which has, however, not been observed to a noticeable extent.

Spectral investigations of the zeolite Y–synthesized Ag nanoclusters were performed at a single-particle level. To estimate the number of Ag atoms the aqueous suspension was diluted by a factor of 0.01 and spin-casted onto silicon dioxide. Single emitters were excited at 488 nm. The Ag nanoclusters showed a bright emission without the necessity of photoactivation. In agreement with the previous work on Ag nanoclusters, blinking dynamics at long timescales (1 to 10 s) and photobleaching for several 100 s could be observed. Figure 5.7c shows a representative Ag nanocluster spectrum (red) as well as the sum spectrum (dark gray) of 40 single Ag nanocluster spectra. It is obvious that the sum spectrum is barely broadened in comparison to the single-cluster spectrum. As the emission properties are preferentially determined by the number of atoms within the cluster, the negligible broadening corroborates the assumption that the investigated clusters are of very similar size. All emission spectra showed a maximum at (545 ± 3) nm and a pronounced side band, which was attributed to vibronic surface interaction with probably OH groups [Zhuravlev, 2000].

Comparison with the spectrum obtained from a micrometer-sized zeolite Y particle (light gray) reveals a much broader emission spectrum of the latter, probably caused by a bigger variety of cluster sizes. Zheng et al. demonstrated for gold clusters that the emission energy E_{em} is directly connected to the atom number n of the cluster via the jellium model through $n = \sqrt[3]{\frac{E_F}{E_{em}}}$, with E_F being the Fermi energy of the bulk material [Zheng et al., 2004]. Krause et al. applied this model to the zeolite Y–synthesized Ag nanoclusters and received an atom number of n = 14 atoms, which corresponds well to the atom number estimated from a cluster size of 0.78 nm (size of the pore channel of zeolite Y). Further comparison with ensemble and single-particle data from the literature yielded an atom number of n = 8–15.

In addition to the investigation of the emission properties which allow for an estimation of the contributing atom number, the authors surprisingly observed diffusion phenomena of the Ag nanoclusters on pure SiO_2. For the first time the fluorescence of single Ag nanoclusters was utilized for material sciences and single-particle tracking revealed diffusion coefficients of up to 0.4 $\mu m^2/s$. Humidity-dependent measurements of the diffusion indicated that it takes place within a water film of the size of a few nanometers at the SiO_2 surface. With increasing humidity, the amount of clusters which contributed to fast and slow diffusion processes increased. It was found that this increase appears in a step-wise manner at about 60%–65% relative humidity, which is in good agreement for the humidity at which an establishment of bulk-like water on top of the surface near ice-like water begins [Asay and Kim, 2005; Verdaguer et

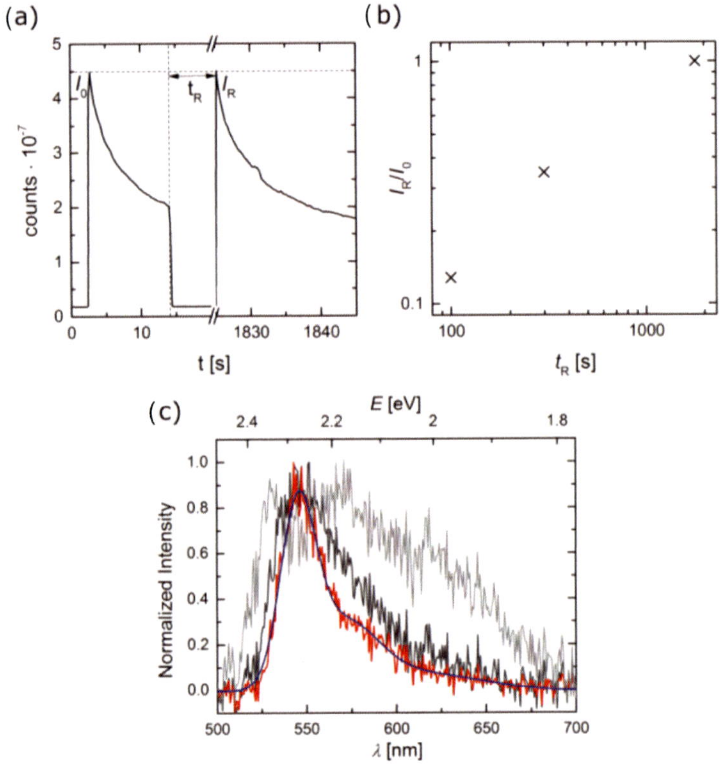

Figure 5.7 (a) Sum of the fluorescence intensity of a small ensemble of Ag nanoclusters collected from a SiO_2 substrate area of 25 × 25 μm^2 as a function of time. The time axis is interrupted between 20 s and 1825 s. I_0 indicates the maximum intensity at the beginning of the experiment, and I_R the maximum recovered fluorescence intensity, while t_R indicates the time of recovery in the dark. (b) Ratio of I_R and I_0 as a function of recovery time t_R in a log-log presentation. The data point at 1800 s was calculated from the data of (a). The fluorescence curves for the other two data points are not shown. The corresponding recovery times are 100 s and 300 s. (c) Normalized spectrum of a single Ag nanocluster (red line) with a three-component Gaussian fit to the data (blue line) and the sum spectrum of 40 different single clusters (black line). The broad light-gray line represents the spectrum taken from a zeolite particle, including a variety of Ag clusters with different sizes. This spectrum is very similar to the one reported by Zheng et al. [Zheng and Dickson, 2002]. Reprinted with permission from Krause, S., Hartmann, M., Kahle, I., Neumann, M., Heidernätsch, M., Spange, S., and Borczyskowski, C. von (2013). Optical tracking of single Ag clusters in nanostructured water films, *J. Phys. Chem. C*, **117**, 24822–24829. Copyright (2013) American Chemical Society.

al., 2007]. Additionally, a reduction of the humidity to 0% resulted in a complete suppression of diffusion. A further influence was shown for the surface treatment, which either increased or decreased the amount of surface silanol groups and thus the wettability of the substrate.

5.6 Conclusion

Noble metal nanoclusters, for example, silver nanoclusters, constitute promising alternatives as fluorescent labels instead of toxic semiconductor nanodots and nontoxic but often unstable organic dyes. During the past two decades a variety of synthesis methods with respective advantages and disadvantages emerged. In this context single-molecule spectroscopy techniques allowed for detailed characterization of the synthesized species and their size and spectral distributions. While the photoactivation of Ag_2O films [Peyser et al., 2001; Peyser et al., 2002] and nanoclusters [Treguer et al., 2005] offers great potential for data storage (switchable, stable), the produced silver nanocluster exhibits a lack of monodispersity and their application as fluorescent labels is questionable due to the size (above several nanometers) and insufficient amount which are produced, especially at Ag_2O surfaces. In contrast, the usage of organic scaffolds provides sufficient amounts of monodisperse, water-soluble, stable, and nonaggregating nanoclusters. Dendrimers allow for the synthesis of encapsulated nanoclusters with a diameter of a few nanometers which show discrete emission species of whose sum spectrum resembles the bulk spectrum of Ag_2O [Zheng and Dickson, 2002]. Even more promising is the synthesis through the cytosine bases of an ssDNA [Petty et al., 2004; Vosch et al., 2007; Oemrawsingh et al., 2012]. Besides the bright and stable emission these nanoclusters are biocompatible, show no blinking on relevant timescales, and emit in the red to NIR region, which is favorable for biological applications. Spectroscopic investigations of single emitters revealed spectral fluctuations [Vosch et al., 2007] and a collective multielectronic nature of excitation, which is known from plasmon resonance effects in novel metal particles [Genzel et al., 1975]. A further possibility to produce Ag nanoclusters of a defined size is introduced by the synthesis within a zeolite cage, which serves as size director. It was shown that individual nanoclusters with a diameter of about 1 nm could be extracted from the zeolite [Krause et al., 2013]. These clusters showed a bright emission at 545 nm and could be utilized to study dynamics within ultrathin water films by single-particle tracking. Additional fluorescence recovery after photobleaching experiments revealed a complete recovery after bleaching and a subsequent, sufficient waiting time. These results imply that blinking and corresponding long "off" times are responsible for the apparent decrease in fluorescence intensity—a process which is known from silica nanoparticles [Cichos et al., 2004].

References

Asay, D. B., and Kim, S. H. (2005). Evolution of the adsorbed water layer structure on silicon oxide at room temperature, *J. Phys. Chem. B*, **109**, 16760–16763.

Choi, S., Dickson, R. M., and Yu, J. (2012). Developing luminescent silver nanodots for biological applications, *Chem. Soc. Rev.*, **41**, 1867–1891.

Cichos, F., Martin, J., and Borczyskowski, C. von (2004). Emission intermittency in silicon nanocrystals, *Phys. Rev. B*, **70**, 115314.

Cichos, F., Borczyskowski, C. von, and Orrit, M. (2007). Power-law intermittency of single emitters, *Curr. Opin. Colloid Interface Sci.*, **12**, 272–284.

De Cremer, G., Antoku, Y., Roeffaers, M. B. J., Sliwa, M., Van Noyen, J., Smout, S., Hofkens, J., De Vos, D. E., Sels, B. F., and Vosch, T. (2008). Photoactivation of silver-exchanged zeolite A, *Angew. Chem.*, **120**, 2855–2858.

Fedrigo, S., Harbich, W., and Buttet, J. (1993). Optical response of Ag_2, Ag_3, Au_2, and Au_3 in argon matrices, *J. Chem. Phys.*, **99**, 5712.

Félix, C., Sieber, C., Harbich, W., Buttet, J., Rabin, I., Schulze, W., and Ertl, G. (1999). Fluorescence and excitation spectra of Ag_4 in an argon matrix, *Chem. Phys. Lett.*, **313**, 105–109.

Genzel, L., Martin, T. P., and Kreibig, U. (1975). Dielectric function and plasma resonances of small metal particles, *Z. Physik B*, **21**, 339–346.

Harb, M., Rabilloud, F., Simon, D., Rydlo, A., Lecoultre, S., Conus, F., Rodrigues, V., and Félix, C. (2008). Optical absorption of small silver clusters: Ag_n, (n = 4–22), *J. Chem. Phys.*, **129**, 194108.

Harbich, W., Fedrigo, S., and Buttet, J. (1992). The optical absorption spectra of small silver clusters (n = 5–11) embedded in argon matrices, *Chem. Phys. Lett.*, **195**, 613–617.

Harbich, W., Fedrigo, S., Meyer, F., Lindsay, D. M., Lignieres, V., Rivoal, J. C., and Kreisle, D. (1990). Deposition of mass selected silver clusters in rare gas matrices, *J. Chem. Phys.*, **93**, 8535.

Henglein, A., Mulvaney, P., and Linnert, T. (1991). Chemistry of Ag_n aggregates in aqueous solution: non-metallic oligomeric clusters and metallic particles, *Faraday Discuss.*, **92**, 31–44.

Hooley, E., Paolucci, V., Liao, Z., Carro Temboury, M. R., and Vosch, T. (2015). Single-molecule characterization of near-infrared-emitting silver nanoclusters, *Adv. Opt. Mater.*, **3**, 1109–1115.

Johnson, I. (1998). Fluorescent probes for living cells, *Histochem. J.*, **30**, 123–140.

Kahle, I., Krause, S., Krasselt, C., Jakob, A., Oehlke, A., Georgi, C., Schulze, S., Lang, H., Hietschold, M., Spange, S., and Borczyskowski, C. von (2014). Assemblies from metallic and semiconducting nanocrystals, *Appl. Phys. A*, **115**, 617–625.

Krause, S., Hartmann, M., Kahle, I., Neumann, M., Heidernätsch, M., Spange, S., and Borczyskowski, C. von (2013). Optical tracking of single Ag clusters in nanostructured water films, *J. Phys. Chem. C*, **117**, 24822–24829.

Miyanaga, T., Suzuki, Y., Matsumoto, N., Narita, S., Ainai, T., and Hoshino, H. (2013). Formation of Ag clusters in zeolite X studied by in situ EXAFS and infrared spectroscopy, *Micropor. Mesopor. Mater.*, **168**, 213–220.

Oemrawsingh, S. S. R., Markešević, N., Gwinn, E. G., Eliel, E. R., and Bouwmeester, D. (2012). Spectral properties of individual DNA-hosted silver nanoclusters at low temperatures, *J. Phys. Chem. C*, **116**, 25568–25575.

Ozin, G. A., and Hugues, F. (1983). Silver atoms and small silver clusters stabilized in zeolite Y: optical spectroscopy, *J. Phys. Chem.*, **87**, 94–97.

Petty, J. T., Zheng, J., Hud, N. V., and Dickson, R. M. (2004). DNA-templated Ag nanocluster formation, *J. Am. Chem. Soc.*, **126**, 5207–5212.

Peyser, L. A., Lee, T., and Dickson, R. M. (2002). Mechanism of Ag_n nanocluster photoproduction from silver oxide films, *J. Phys. Chem. B*, **106**, 7725–7728.

Peyser, L. A., Vinson, A. E., Bartko, A. P., and Dickson, R. M. (2001). Photoactivated fluorescence from individual silver nanoclusters, *Science*, **291**, 103–106.

Resch-Genger, U., Grabolle, M., Cavaliere-Jaricot, S., Nitschke, R., and Nann, T. (2008). Quantum dots versus organic dyes as fluorescent labels, *Nat. Methods*, **5**, 763–775.

Ritchie, C. M., Johnsen, K. R., Kiser, J. R., Antoku, Y., Dickson, R. M., and Petty, J. T. (2007). Ag nanocluster formation using a cytosine oligonucleotide template, *J. Phys. Chem. C*, **111**, 175–181.

Schultz, D., and Gwinn, E. G. (2012). Silver atom and strand numbers in fluorescent and dark Ag:DNAs, *Chem. Commun.*, **48**, 5748–5750.

Schulze, W., Rabin, I., and Ertl, G. (2004). Formation of light-emitting Ag_2 and Ag_3 species in the course of condensation of Ag atoms with Ar, *Chem. Phys. Chem.*, **5**, 403–407.

Schuster, J., Brabandt, J., and Borczyskowski, C. von (2007). Discrimination of photoblinking and photobleaching on the single molecule level, *J. Lumin.*, **127**, 224–229.

Shang, L., and Dong, S. (2008). Facile preparation of water-soluble fluorescent silver nanoclusters using a polyelectrolyte template, *Chem. Commun.*, 1088–1090.

Shang, L., Dong, S., and Nienhaus, G. U. (2011). Ultra-small fluorescent metal nanoclusters: synthesis and biological applications, *Nano Today*, **6**, 401–418.

Sieber, C., Buttet, J., Harbich, W., Félix, C., Mitrić, R., and Bonačić-Koutecký, V. (2004). Isomer specific spectroscopy of metal clusters trapped in a matrix: Ag_9, *Phys. Rev. A*, **70**, 041201.

Treguer, M., Rocco, F., Lelong, G., Nestour, A. L., Cardinal, T., Maali, A., and Lounis, B. (2005). Fluorescent silver oligomeric clusters and colloidal particles, *Solid State Sci.*, **7**, 812–818.

Verdaguer, A., Weis, C., Oncins, G., Ketteler, G., Bluhm, H., and Salmeron, M. (2007). Growth and structure of water on SiO_2 films on Si investigated by Kelvin probe microscopy and in situ X-ray spectroscopies, *Langmuir*, **23**, 9699–9703.

Vosch, T., Antoku, Y., Hsiang, Y. J., Richards, C. I., Gonzalez, J. I., and Dickson, R. M. (2007). Strongly emissive individual DNA-encapsulated Ag nanoclusters as single-molecule fluorophores, *Proc. Natl. Acad. Sci. USA*, **104**, 12616–12621.

Warnken, M., Lázá, K., and Wark, M. (2001). Redox behaviour of SnO_2 nanoparticles encapsulated in the pores of zeolites towards reductive gas atmospheres studied by in situ diffuse reflectance UV/Vis and Mössbauer spectroscopy, *Phys. Chem. Chem. Phys.*, **3**, 1870–1860.

Wöhrle, D., and Schulz-Ekloff, G. (1994). Molecular sieve encapsulated organic dyes and metal chelates, *Adv. Mater.*, **6**, 875–880.

Yu, J., Patel, S. A., and Dickson, R. M. (2007). In vitro and intracellular production of peptide-encapsulated fluorescent silver nanoclusters, *Angew. Chem.*, **119**, 2074–2076.

Yu, P., Wen, X., Toh, Y., and Tang, J. (2012). Temperature-dependent fluorescence in Au_{10} nanoclusters, *J. Phys. Chem. C*, **116**, 6567–6571.

Yumura, T., Nanba, T., Torigoe, H., Kuroda, Y., and Kobayashi, H. (2011). Behavior of Ag_3 clusters inside a nanometer-sized space of ZSM-5 zeolite, *Inorg. Chem.*, **50**, 6533–6542.

Zhang, J., Campbell, R. E., Ting, A. Y., and Tsien, R. Y. (2002). Creating new fluorescent probes for cell biology, *Nat. Rev. Mol. Cell Bio.*, **3**, 906–918.

Zhang, L., and Wang, E. (2014). Metal nanoclusters: new fluorescent probes for sensors and bioimaging, *Nano Today*, **9**, 132–157.

Zheng, J., and Dickson, R. M. (2002). Individual water-soluble dendrimer-encapsulated silver nanodot fluorescence, *J. Am. Chem. Soc.*, **124**, 13982–13983.

Zheng, J., Nicovich, P. R., and Dickson, R. M. (2007). Highly fluorescent noble metal quantum dots, *Annu. Rev. Phys. Chem.*, **58**, 409–431.

Zheng, J., Zhang, C., and Dickson, R. M. (2004). Highly fluorescent, water-soluble, size-tunable gold quantum dots, *Phys. Rev. Lett.*, **93**, 077402.

Zhuravlev, L. T. (2000). The surface chemistry of amorphous silica. Zhuravlev model, *Colloids Surf A Physicochem Eng Asp.*, **173**, 1–38.

Zijlstra, P., and Orrit, M. (2011). Single metal nanoparticles: optical detection, spectroscopy and applications, *Rep. Prog. Phys.*, **74**, 106401.

Chapter 6

Tuning Luminescence Intermittency

Ines Trenkmann, Clemens Göhler, Cornelius Krasselt, and Christian von Borczyskowski
Institute of Physics, Centre for Nanostructured Materials and Analytics, Technische Universität Chemnitz, D-09107 Chemnitz, Germany
borczyskowski@physik.tu-chemnitz.de

6.1 Introduction

It is well documented in the previous chapters of this book that semiconductor quantum dots (QDs) exhibit luminescence intermittency (blinking). Blinking has been qualitatively explained by various models, as outlined in Chapter 4. Independent of the model, the presence or (photoinduced) creation of charges plays a crucial role. It has been shown in many former publications [Cichos et al., 2007; Riley et al., 2012; Cui et al., 2014] that blinking depends on various factors such as the concrete composition of the QD, the ligand (surfactant) shell, the embedding matrix or supporting substrate, the excitation power, and the excitation energy. Also external (electric) fields have a considerable influence on the blinking processes [Jha and Guyot-Sionnest, 2010; Galland et al., 2011; Krasselt, 2015].

Tuning Semiconducting and Metallic Quantum Dots: Spectroscopy and Dynamics
Edited by Christian von Borczyskowski and Eduard Zenkevich

ISBN 978-981-4745-24-6 (Hardcover), 978-981-4745-25-3 (eBook)
www.panstanford.com

Several attempts have been reported to suppress blinking in order to achieve an increased quantum yield (QY) to improve sensitivity in labeling experiments [Hohng and Ha, 2004; Spinicelli et al., 2009; Vela et al., 2010]. To proceed successfully, a systematic understanding of the underlying mechanism causing blinking is mandatory. Moreover, if blinking is truly understood it might on the other hand serve as a sensoric tool to investigate subtle changes in the vicinity of single QDs. In any case, it is necessary to emphasize that blinking can be systematically tuned. This becomes even more important as blinking plays a dominant role in super-resolution microscopy for which the Nobel Prize in Chemistry was awarded in 2014 [Hell and Wichmann, 1994; Betzig et al., 2006; Dickson et al., 1997].

In this chapter, we will concentrate on those aspects which are not related to intrinsic QD properties, such as composing materials, size, and semiconducting capping layers, but on external factors. Special emphasis will be on a generalized model of blinking in combination with not yet published experimental data (Sections 6.2 and 6.4). We will also discuss the recently investigated influence of excitation power and wavelength (Section 6.3). We will report new results on the effects of substrate surfaces and capping ligands on blinking dynamics. A prerequisite to understand the manifold of blinking processes is the assumption that colloidal QDs constitute, together with the embedding environment, a strongly disordered (heterogeneous) system [Cichos et al., 2007]. We will mainly concentrate on CdSe/ZnS QDs.

6.2 Identification of Different Types of Blinking

6.2.1 Blinking Analysis

6.2.1.1 Truncated power-law blinking of CdSe/ZnS

Figure 6.1a shows a typical threshold-determined blinking time trace with 1 ms binning time of CdSe/ZnS in a polystyrene (PS) matrix. At the bottom we show the reconstructed blinking time trace according to change point analysis (CPA; see also Chapter 4) [Watkins and Yang, 2005; Schmidt et al., 2012; Schmidt et al., 2014]. It is immediately evident that at these short binning times very short blinking events with a very high photoluminescence (PL) intensity show up besides the previously often identified blinking events in the range of 20 ms and longer [Cichos et al., 2017]. In Fig. 6.1b we show the conventional threshold-based analysis of the "on" time distribution in the form of a truncated power-law analysis. Additionally, we present the distribution of clustered PL intensities observed during a blinking time trace in Fig. 6.1c [Schmidt et al., 2014], which we have grouped into four qualitatively different ranges *L*, *I*, *N*, and *S*. In close analysis to blinking time traces higher PL intensities (up to a factor of 2 compared to *N* states) occur in the form of spikes rarely spaced in the range *S*. Except in ranges *L* and *S*, intensities are more or less equally separated, as demonstrated in the intensity density curve also shown in the graph.

Since blinking is identical with intensity fluctuations, CPA can immediately identify intensity jumps ΔI also at intermediate PL intensities (dim or gray intensities). As discussed (see Chapter 4) the identification of intensity jumps is only limited by the respective signal-to-noise ratio. A typical jump, as shown for an exemplary single QD (QD9) in PS in Fig. 6.2h, exhibits that predominantly small jumps occur at ±25% of the maximum intensity at the approximate half-width of the jump distribution. Jumps over the total range of PL intensities are rare. The spike in the center of the distribution is related to noise.

Figure 6.1d collects the PL decay times τ_{micro} in the nanosecond range for every clustered PL intensity. As an approach, fits have been obtained via a monoexponential decay function, which is, however, only a rough approximation below $I \approx 15$ kcps.

Via CPA it is possible not only to determine intensities and intensity jumps but also to identify relations among the various intensities. To facilitate such a procedure we have introduced a PL intensity index *n*, which numbers consecutively each identified intensity from $n = 1$ to $n = n_{\max}$ for the highest intensity. Such a scale is also indicated in Fig. 6.1c. For the example of QD9 in PS the relations between the intensities n_i and n_k, if n_k follows (or precedes) n_i in time (in other words n_i and n_k follow each other directly in time) are collected in Fig. 6.2a–d.

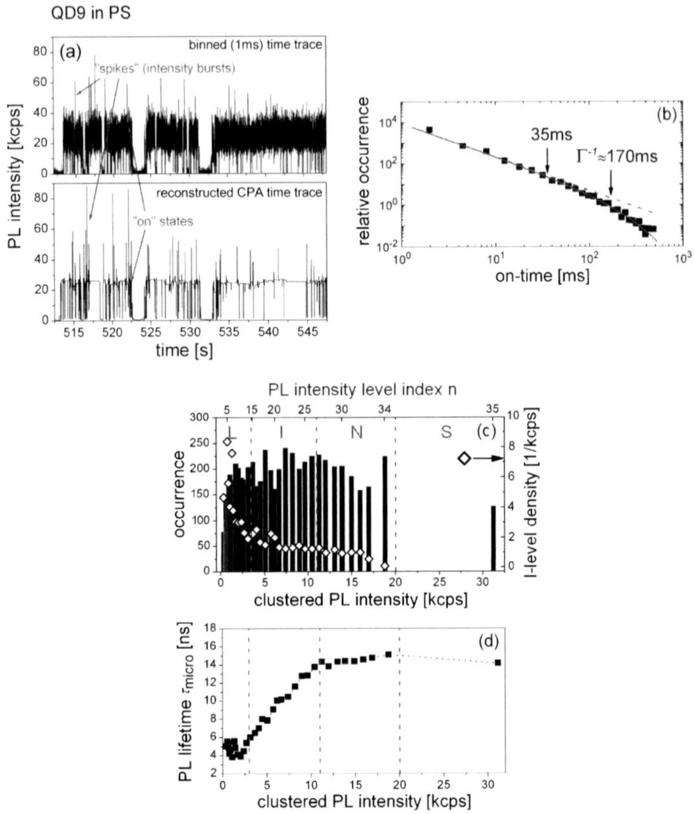

Figure 6.1 PL dynamics of a single CdSe/ZnS QD (e.g., QD9 in polystyrene [PS]). (a) Long PL time trace of (conventional threshold-based) 1 ms binned (top) and CPA-reconstructed (bottom) data. Strong intensity spikes are visible in both data sets. (b) Conventional (threshold-separated) "on" time histogram and truncated power-law analysis following Eq. 6.2 (Chapter 4) (solid line). Deviations from a power law (dashed line) at 35 ms and the (exponential) truncation time $\Gamma^{-1} \approx 170$ ms are indicated. (c) CPA histogram of constant intensities after PL intensity clustering. Besides the intensity given in kcps, we also include an intensity-level index *n* (top). Typical intensity ranges *L, I, N,* and *S* are indicated. Symbols mark the density of intensities. (d) PL decay times τ_{micro} as a function of clustered intensities. The times stem from fits corresponding to a monoexponential data treatment, which is only an approximation for intensities below $I \approx 15$ kcps (see Fig. 6.3 for details). Reprinted with permission from Schmidt, R., Krasselt, C., Göhler, G., and Borczyskowski, C. von (2014). The fluorescence intermittency for quantum dot is not power-law distributed: a luminescence intensity resolved approach, *ACS Nano,* **8**, 3506–3521. Copyright (2014) American Chemical Society.

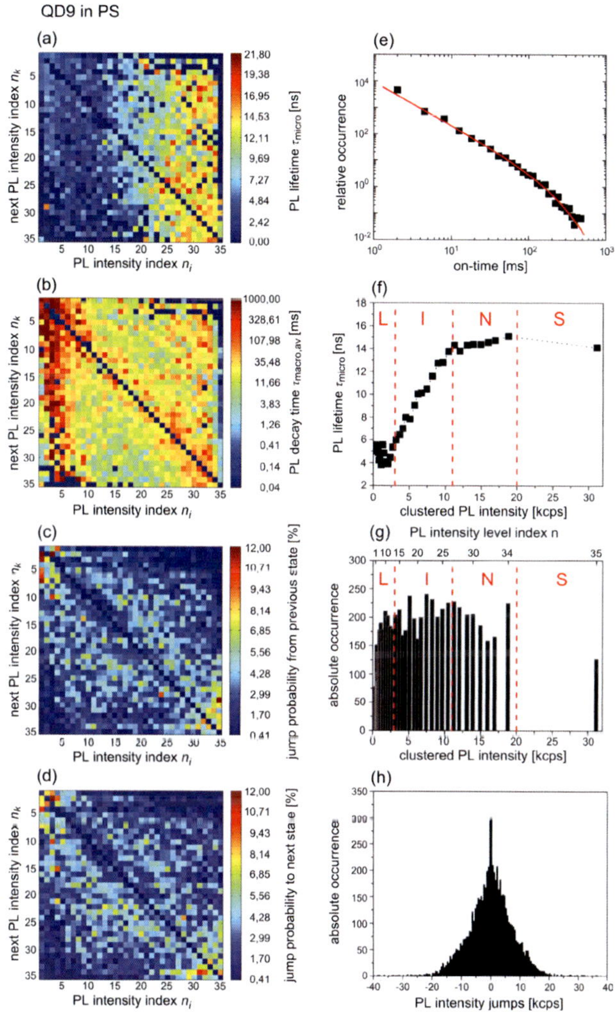

Figure 6.2 (a) Monoexponentially fitted PL decay time (lifetime) in the nanosecond range as a function of the PL intensity n_i (compare (f)). (b) Average "on" time (PL decay time) τ_{macro} before jumping in a single step to the next intensity n_k. (c) Probability to leave intensity n_j to the next intensity n_k. (d) Reverse process of (c). (c, d) Two different types of intensity correlations (blinking) among high intensities (*S* type) and low intensities (*L* type) can be identified (see color code). (e) Truncated power-law behavior of "on" times obtained by threshold analysis fitted by a truncated power law. (f) Fit of the PL intensity–dependent decay by a monoexponential function. (g) Clustered intensity distribution determined by CPA. (h) Distribution of intensity jumps among the intensities shown in (g).

Four different features have been investigated, namely (a) the PL decay time τ_{micro} (determined by a monoexponential fit of the PL decay in the nanosecond range for every intensity), (b) the arithmetically averaged macrolifetime $\tau_{\text{macro,av}}$ in n_i before the next intensity jump takes place, (c) the probability of jumps from a selected PL intensity to the directly proceeding one, and (d) the probability for a reverse jump. Data in (c) and (d) have been inspected for the influence of changing the direction of jumps (relation n_{i+1} or n_{i-1}), which turned out to be in most cases nearly symmetric within experimental accuracy. Data are represented in a false color code. Since time resolution is experimentally limited, $\tau_{\text{macro,av}}$ values are only reliable for times longer than about 500 μs. The right side in Fig. 6.2 shows (e) the "on" time distribution obtained by the threshold-based approach, (f) the (monoexponentially fitted) decay time τ_{micro} of the respective PL intensity, (g) the clustered PL intensity distribution, and (h) the intensity jump distribution.

What can we learn from such an analysis of interintensity correlations? We will discuss the related features exemplary for a specific single QD. Although the examined observations vary quantitatively among individual QDs they are qualitatively quite similar. We identify certain intensity ranges, as defined in Fig. 6.1. Firstly, jump probabilities from n_i to n_k are not equally distributed. From Fig. 6.2c three features immediately emerge: (i) Jumps of intensities in the *S* range occur predominantly to the highest intensities of the *N* range and vice versa. There are almost no jumps from the highest to the lowest intensities (blue lower-left and upper-right corners with probabilities < 0.5%). We name the related blinking events *S* blinking. (ii) Jumps in the *I* range (and partly in the *N* range) take predominantly place via relatively small intensity jumps, as can be concluded from the higher probabilities appearing in close proximity (parallel) to the diagonal. This type of blinking we name *I* blinking. (iii) Finally we observe an enhanced intensity jump probability in the *L* range (and lower part of the *I* range), which we name *L* blinking. Some care has to be taken in this range since intensities are very small and might therefore be influenced by noise.

We would like to point out that discrimination between only two basically different types of blinking has been recently reported [Galland et al., 2011; Schmidt et al., 2014]. Now we introduce a third blinking type at very low PL intensities (also supported by the increased density of intensities in this range). Correlations with additional observables presented later will provide confidence that identification of three different types of blinking is reliable.

The average time period $\tau_{\text{macro,av}}$ of remaining in an intensity level n_i before changing to the next intensity level (Fig. 6.2b) shows basically three time regimes, which are nearly independent of the jump width, that is, the consecutive intensity levels, contrary to the probabilities. The following interrelations of intensities are identified the following: (i) times between 1 and 30 ms (transitions from S to upper N levels); (ii) long times close to 100 ms and 30 ms at the intersection of the L–I range and the N–S range, respectively; and (iii) smoothly varying times between 10 and 20 ms across the remaining I–N range with an indication of a minimum close to the I–L intersection.

The analysis of τ_{macro} reveals that "on" times are not power-law-distributed [Schmidt et al., 2014]. A similar finding has been reported previously for the different QD system CdSe/CdS [Gómez et al., 2009] for which they found (by a threshold-based analysis) "on" times in the range from 15 to 24 ms. Additionally they assigned an exciton decay time of 44 ns compared to 5.2 ns for an intermediate intensity level, which they assigned to a (after electron trapping) positively charged trion. They could assign two blinking classes ("on" to intermediate and intermediate to "off" intensities, respectively).

A further investigated parameter is the PL decay time τ_{micro} shown in Fig. 6.2a as a function of the PL intensities. Here we observe that τ_{micro} is constant (~14 ns) in the S and upper N ranges (not immediately seen for the S range in this graph but in Fig. 6.2f). Then τ_{micro} decreases continuously down to ~5 ns with decreasing intensity across all ranges. However, close to intensity $n_i \approx 8$ we find a shorter decay time of ~2 ns.

Remarkably, we find that both $\tau_{\text{macro,av}}$ and τ_{micro} do not depend on interrelated intensities, but the jump probability does (see Fig. 6.2c,d). On the other hand, both times depend drastically on the PL intensity itself. The intensity dependence of the PL decay time is shown in Fig. 6.2f in comparison to the PL intensity distribution in Fig. 6.2g. The latter is shown on an absolute intensity scale together with classified intensity ranges.

It is generally accepted that PL intensity is closely related to the PL decay time according to

$$I \approx k^{\text{r}}\tau_{\text{micro}} = k^{\text{r}}/(k^{\text{r}} + \Sigma_i k_i^{\text{nr}}) \tag{6.1}$$

with the radiative rate k^r and the nonradiative rates k_i^{nr}. One of the nonradiative types of rates is identified with broadly distributed quenching rates controlling the PL intensity during a blinking time trace. According to Eq. 6.1 and assuming a constant radiative rate we expect a linear relationship between τ_{micro} and I, which obviously does not hold in the present case. Moreover, a detailed analysis reveals that the PL decay is clearly nonexponential below about $I \approx 1/3 \cdot I_{max}$ in the N range. However, for simplicity we present monoexponential fits in Figs. 6.1 and 6.2. Deviations parameterized in the form of a stretched exponential function are additionally shown in Fig. 6.3.

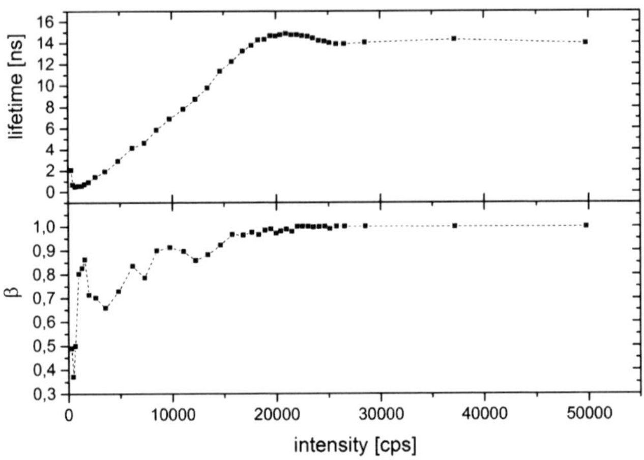

Figure 6.3 Fit of the PL decay at every PL intensity for QD10 (in PS) by a stretched exponential function, according to $I \approx \exp[-(t/\tau)^\beta]$ with the parameters τ (top) and β (bottom). Note that this QD shows a slight maximum for τ = 15 ns at about I = 20 kcps. Reproduced with kind permission from Robert Schmidt.

Most remarkably, opposite to Eq. 6.1 τ does not depend linearly on the intensity in the S–N range. In the N–I range the dependence is nearly linear over a limited intensity range, but at least two different τ/I increments can be assigned, which has been observed and explained by a change of k^r [Zhang et al., 2010]. Since the decay function is multiexponential (see Fig. 6.3), this implies that even when selecting a well-defined PL intensity at least a biexponential approach has to be considered in the I range, as we have reported recently [Schmidt et al., 2014]. τ reaches a minimum (with $\beta \approx 0.9$!)

close to I = 2500 cps and increases again in the L range. We have assigned this minimum to the presence of a third decay component (related to a state with very low PL intensity) with a decay time in the range of 1 ns [Schmidt et al., 2014]. Similar low-level intensities accompanied by fast decay times have been reported in the literature [Galland et al., 2011; Amecke and Cichos, 2011; Cordones et al., 2011b].

Assembling all observations which we find for the interrelation of PL intensities, the intensity jumps, and PL decay times during a typical blinking time trace, a surprisingly consistent picture emerges, which is sketched in Fig. 6.4. Especially, we find that (i) despite large intensity variations a constant PL decay time in the range S to N goes in line with S-type blinking with (large) intensity jumps only in the N–S range, (ii) the minimum of τ_{micro} is close to a maximum in $\tau_{\text{macro,av}}$ near the L–I intersection (see Fig. 6.2f), while (iii) intensity jumps in the I range are relatively small. A still more diversified analysis will be discussed later in this chapter.

According to these observations we assign blinking phenomena to three different (overlapping) blinking classes, as indicated in Fig. 6.4 at the bottom. The observations at the L–I intersection indicate the presence of a rather well-defined state with low emission intensity (and very short PL decay time) somewhat above the zero background intensity. Similar observations have been already reported for CdSe/CdS and large CdSe/ZnS QDs [Cordones et al., 2011a; Galland et al., 2012; Amecke and Cichos, 2011]. Blinking in the L range predominantly occurs between the zero intensity level and L intensities with related long "on" times. The "on" times become shorter and moderately intensity dependent in the I–N range. At most five "on" times are necessary to explain blinking according to a threshold-based analysis [Schmidt et al., 2014].

S-type blinking (in the N–S range) is probably due to hot-electron excitation, which results (for part of the observation time) in an alternative radiationless recombination path. Sometimes this path is blocked and all excitation populates the radiative state. Since this state is the one which is monitored a constant PL decay time will be observed in the N and S ranges. There are strong arguments for this explanation according to recent experiments [Galland et al., 2011] and the PL excitation spectra shown in Fig. 6.35 which demonstrate that the overall QY is reduced upon excitation into energetically high electronic states. We have related S-type blinking with the occurrence of a PL intensity–independent PL decay times with intermittency in the population pathways. This observation is probably not a singular finding since indications of intensity-independent PL decay times have been reported but not interpreted in the literature [Biebricher et al., 2006].

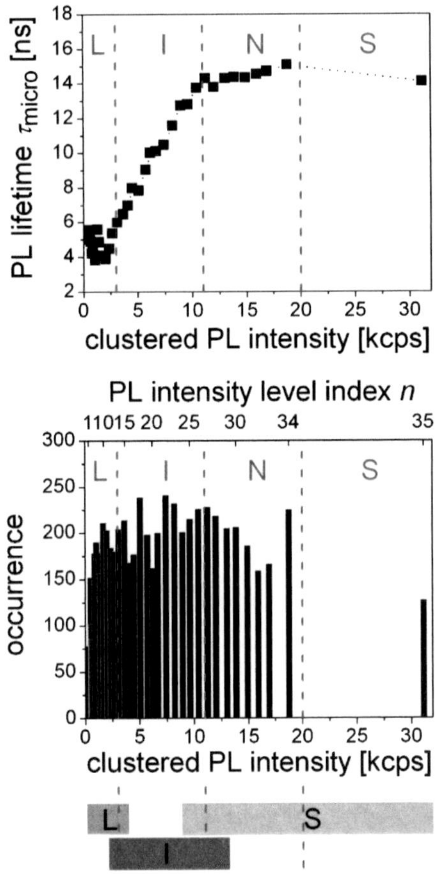

Figure 6.4 Definition and relation of different types of blinking (bottom) according to the PL decay times (top) and in comparison to the clustered intensities (middle) for the example of QD9 in PS. At the bottom, the various blinking ranges *L*, *I*, and *S* are shown. The overlap of ranges shown in the graph indicates that a precise identification of the respective range is experimentally not possible.

At this point we would like to mention that the longest "on" times for QD9are close to 30 ms in the *N* range and to 100 ms in the L range. If we compare these findings with those of a threshold-based "on" time analysis (neglecting extremely short "on" times, which cannot be observed at binning times typically longer than 10 ms), we suggest that the longest "on" times in the S–N range go in line with the cutoff times in the "on" time power-law analysis. It is intriguing to observe that the longest "on" times in the L range correspond closely to the cutoff times typically observed for an "off" time power-law distribution [Issac et al., 2012]. This might point to a systematically different identification of "on" times comparing the two analytical approaches.

6.2.1.2 Anomalous power-law blinking of CdSe/ZnS

While many CdSe/ZnS QDs in PS (see Section 6.2.1.1) show for the "on" times a truncated power-law behavior some of them exhibit a strong deviation, as shown in Fig. 6.5, which can be described by a superposition of a truncated power law and an exponential function according to [Krasselt et al., 2011]:

$$A_1 t_{on}^{-\alpha} \cdot \exp(-\Gamma_{on} t_{on}) + A_2 \exp(-\gamma_{on} t_{on}). \quad (6.2)$$

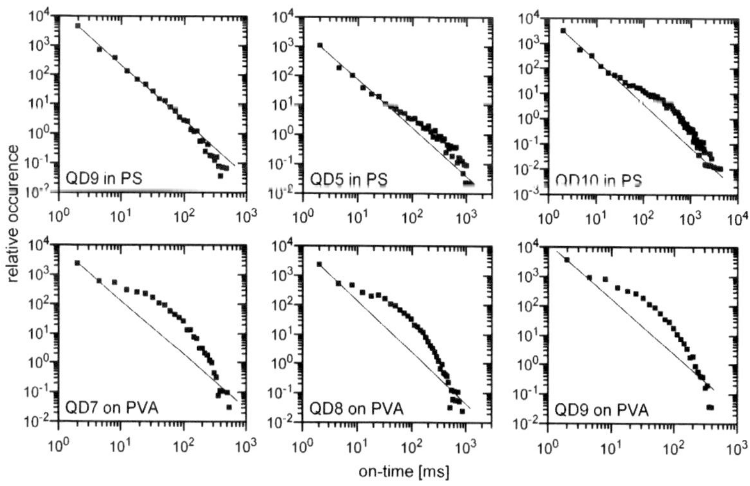

Figure 6.5 Truncated power-law blinking (analyzed by the threshold-based approach; binning time 1 ms) for three different QDs in PS (top row) and on poly(vinyl alcohol) (PVA) (bottom row). Straight lines show power-law behavior (without truncation). Reprinted from Schmidt, R., Krasselt, C., and Borczyskowski, C. von (2012). Change point analysis of matrix dependent photoluminescence intermittency of single CdSe/ZnS quantum dots with intermediate intensity levels, *Chem. Phys.*, **406**, 9–14, Copyright (2012), with permission from Elsevier.

This deviation is dramatically increased for CdSe/ZnS QDs on a thin poly(vinyl alcohol) (PVA) film [Schmidt et al., 2012]. Similar features have been reported by us recently [Krasselt et al., 2011] for tri-(n-octyl)phosphinoxid (TOPO)-capped CdSe/ZnS QDs on various SiO_x substrates and have been attributed to the interaction of QDs with silanol (OH) groups on the SiO_x surface as will be discussed in more detail in Section 6.4.

Additionally, this deviation results also in several typical new features derived from CPA. These features are collected in Fig. 6.6:

- The clustered intensity distribution (Fig. 6.6g) shows an increased density in the *N* range.
- The jump distribution (Fig. 6.6h) reveals side bands which correspond to enhanced probabilities for large intensity jumps.
- The (monoexponential) PL decay time (Fig. 6.6f) in the *N* range is at its maximum by about 1 ns longer compared to the one of QDs with "on" times described by a truncated power law. The minimal decay time at very low PL intensities becomes shorter by up to a factor of 2 down from about 2 ns to 1 ns.
- The threshold determined truncated power law (Fig. 6.6e) shows an anomalous behavior corresponding to the above-described deviation (Eq. 6.2).

Naturally, these features have also influence on the interintensity correlations shown in Fig. 6.6 (left). While for the truncated power-law behavior, most of the jumps in the *I–N* range occur nearly equally distributed along the diagonal (since they are small), we now observe in Fig. 6.6c,d increased jump probabilities at the top-left range of the indicated diagonal (e.g., $n_i = 7$ to $n_k = 15$) and large jumps (e.g., $n_i = 30$ to $n_k = 7$). Simultaneously, the monoexponentially fitted τ_{macro} becomes shorter, as compared in Fig. 6.7. An exponential fit to every intensity reveals that the fit is to some extent biexponential in the high intensity range *N/S* at which we observe an increased density of intensities. Typically the biexponential contribution is only in the range of the *N* intensities and shows decay times of 20 ms and 150 ms. The latter can be assigned to the OH-related states.

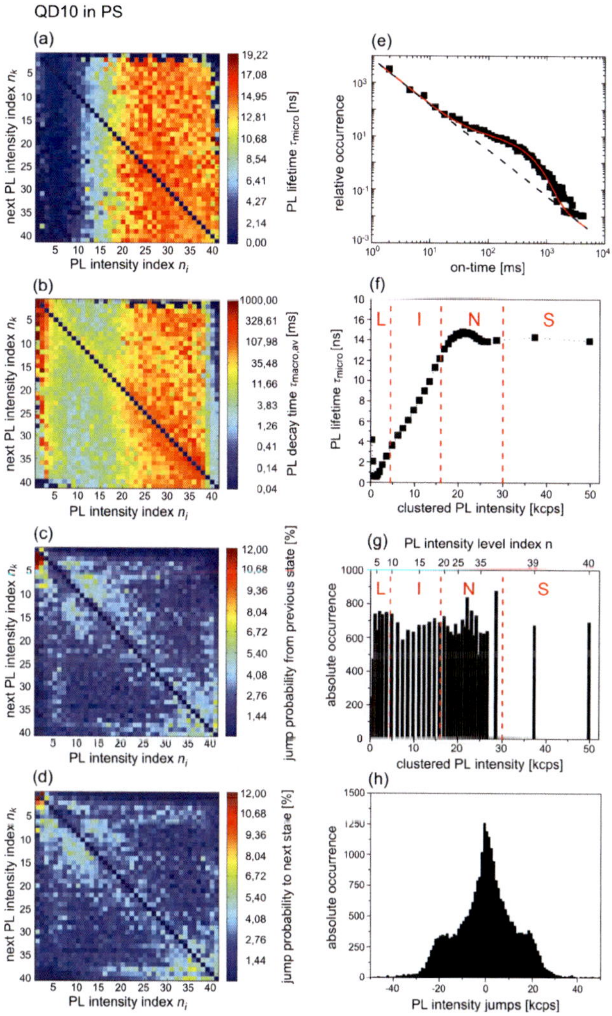

Figure 6.6 Dynamics during a blinking time trace of QD10 in PS. Left (in false color code): (a) Monoexponentially fitted PL decay time in the ns range as a function of the PL intensity n_i (compare (f)). (b) Average "on" time τ_{macro} in intensity level n_i before jumping in a single step to the next intensity n_k. (c) Probability to leave intensity n_i to the next intensity n_k. (d) Reverse process of (c). Right: (e) Truncated power-law behavior (black dashed line) of blinking "on" times with deviation corresponding to Eq. 6.2 (red line) obtained by threshold analysis. (f) Fit of the PL intensity-dependent decay by a monoexponential function. (g) Clustered intensity distribution determined by CPA. (h) Distribution of intensity jumps among the intensities shown in (g).

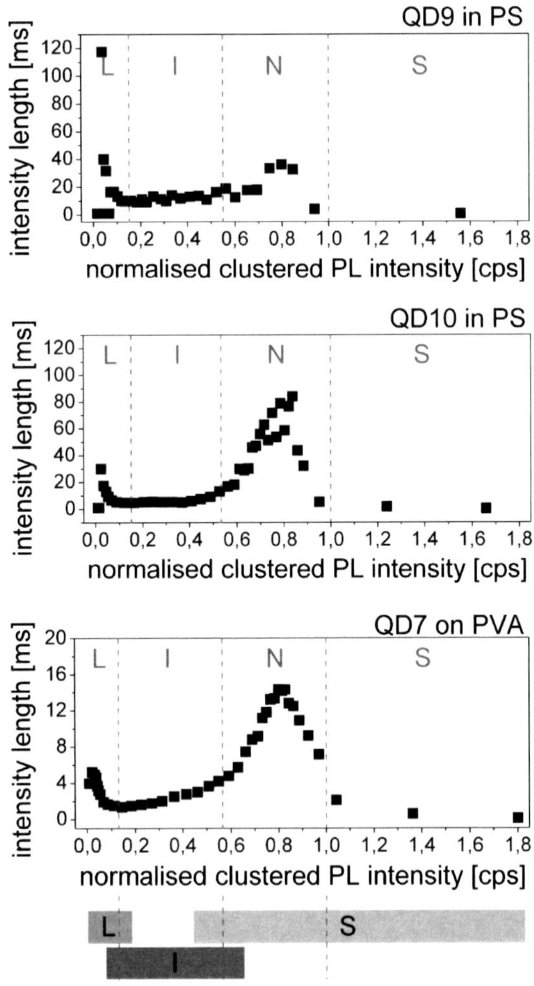

Figure 6.7 PL intensity dependence of "on" times (intensity decay time τ_{macro}) as determined by monoexponential fits from CPA. The types of blinking processes are indicated at the bottom. Intensities of the three QDs have been normalized to 1 at the *N*–*S* intersection. Top: QD9 in PS (truncated power law). Middle: QD10 in PS (anomalous power law). Bottom: QD7 on PVA (anomalous power law). Note that the "on" times for QDs in PS become only shorter (longer) in the case of anomalous behavior in the *L*/*S* (*N*/*S*) range but remain similar in the *I* range. For QDs on PVA all "on" times are shorter across the total range compared to PS.

As already mentioned, the manifestation of an anomalous (and truncated) power law becomes much more dramatic (Fig. 6.5) when changing the embedding environment from PS to the much more polar PVA, which contains OH groups. A sketch of a possible interaction of QDs with OH groups is shown in Fig. 6.8. This interaction might be very similar to attachment of H_2O, which has shown to increase the PL QY [Gómez et al. 2009]. The difference to a strong ligation with (the small) H_2O molecule probably is that in the present case the coupling is much weaker and thus subject to fluctuations.

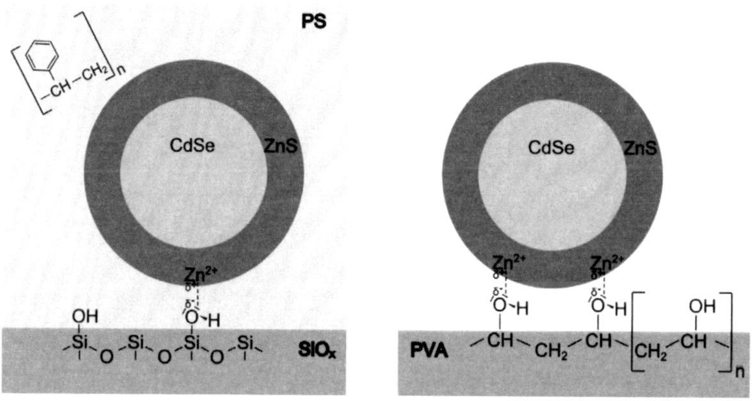

Figure 6.8 Scheme of interactions of QDs with OH groups on a SiO_x (left) and OH groups on a PVA polymer film.

In the case of interactions with OH groups large-intensity jumps become more prominent both in the intensity jump distribution (see Fig. 6.9) and in the intensity correlation diagram in which strong correlations exist, for example, from n_i = 21 to 5 or n_i = 6 to 16 (Fig. 6.10). It becomes evident that the jump probabilities (see circles and squares) in Fig. 6.10 are now no longer symmetric to each other, as will be explained in Section 6.2.3. The "on" times of the respective (CPA) intensities are now much shorter compared to the ones in the case of power-law behavior (see Fig. 6.7). Additionally, deviations from a monoexponential "on" time distribution are observed especially in the *N*–*S* intersection (but also in the *L*–*I* intersection).

It is interesting to note that though the magnitudes of intensity jumps seem to differ among various QDs (see Fig. 6.9) they are nearly identical if normalizing intensity jumps to the absolute PL intensities of the respective QDs which differ up to a factor of 2.

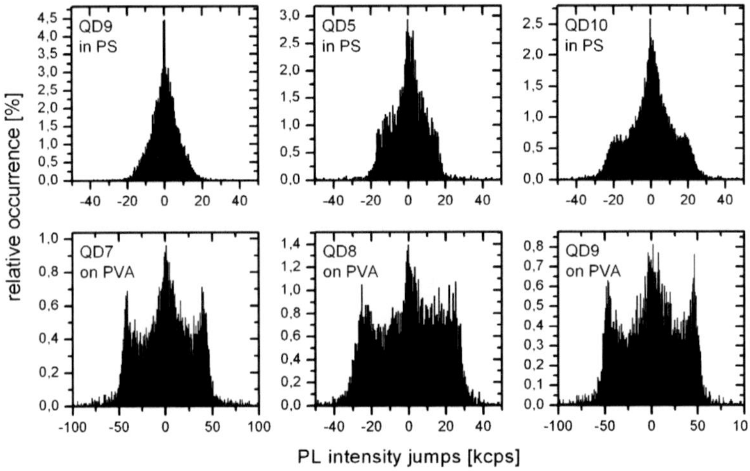

Figure 6.9 PL intensity jump distribution for three different QDs in PS (top) and on PVA (bottom). Note that for normalized PL intensities the relative magnitudes of the respective intensity jumps do not depend on the kind of matrix or the kind of power law. Reprinted from Schmidt, R., Krasselt, C., and Borczyskowski, C. von (2012). Change point analysis of matrix dependent photoluminescence intermittency of single CdSe/ZnS quantum dots with intermediate intensity levels, *Chem. Phys.*, **406**, 9–14, Copyright (2012), with permission from Elsevier.

Interestingly, we observe a clear built-up time of the PL intensity in the *L*–*I* intersection ($n_i \approx 6$) with time < 2 ms. A closer inspection of the data from QDs in PS, showing an anomalous power law, also indicates a small built-up component with a somewhat longer built-up time (<3 ms). The correlation diagrams indicate that each intensity can be populated from a large distribution of intensities. Since the overall decay of intensities is related to a large distribution of "on" times, a selected intensity increase will be broadly distributed in time and will be experimentally missed. However, if one state (intensity) predominantly populates a well-defined other state a clear built-up component might be identified. Additionally, the built-up components are observed in the same intensity range in which we find the contribution of a very short PL decay time (see Section 6.2.1.3) close to the minimum of τ_{micro} in Figs. 6.2f and 6.6f. Presently we do not have an explanation for this finding.

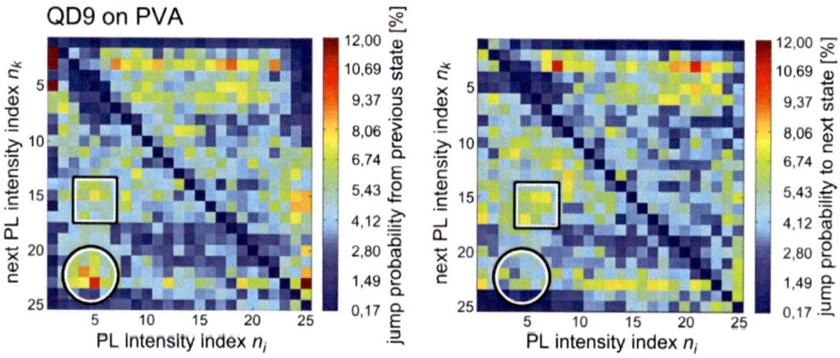

Figure 6.10 Probabilities of intensity jumps from previous (left) to next (right) intensities for QD 9 in PVA. Circles and squares indicate changes in jump probabilities compared to QDs with merely truncated power-law behavior. We note that in this presentation the numbering *n* of intensities is different compared to some previous presentations.

Though τ_{macro} changes considerably upon variation of the matrix, τ_{micro} remains nearly the same, as convincingly shown in Fig. 6.11. In this figure we have superimposed the exponentially deconvoluted PL decay for CdSe/ZnS QDs for various matrix conditions, as discussed before. To achieve such a comparison absolute PL intensities of the respective QDs have to be normalized (e.g., to the *N*–*S* intersection, as shown here). As described in detail recently [Schmidt et al., 2014], three-exponential PL decay functions are needed to describe the PL decay uniquely across the total (CPA-resolved) intensity range. They have to be assigned to at least three different electronic QD states (see Section 6.2.3). Two of these states (with a localized hole) show a broad distribution of intensities which are intrinsically related to the decay times $\tau_{\text{micro},i}$ according to Eq. 6.1. The third state is a short-lived one (with low PL intensity), which we have assigned recently to the quenched exciton state. A mechanistic model has been recently presented in Schmidt et al. (2014).

6.2.1.3 Interrelation of PL decay time blinking and intensity blinking

Up to here we have mainly discussed the distribution of PL intensities as determined by CPA and the related "on" times τ_{macro} combined with the interintensity relations separately for each intensity. We found that the "on" times are not power-law-distributed but have to be described by a set of typically less than five (exponential)

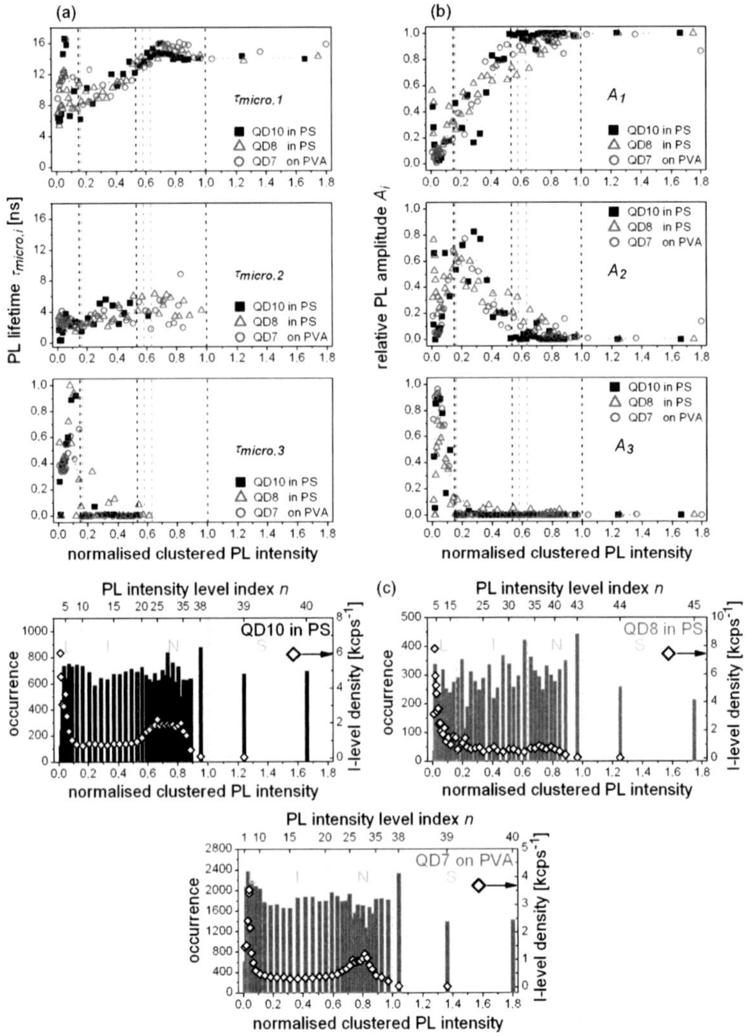

Figure 6.11 Three-exponential fits of the PL decay of single CdSe/ZnS QDs (QD8 and QD10) in PS and (QD7) on PVA as a function at the *N*–*S* intersection normalized clustered intensity with (a) decay times $\tau_{\text{micro},i}$ and (b) corresponding (relative) amplitudes A_i. Dashed lines separate intensity ranges *L*, *I*, *N*, and *S*. (c) Corresponding distributions of normalized clustered PL intensities, including marked intensity ranges, PL intensity-level densities (diamond symbols), and intensity-level indices n_i.

decay times in the range of 0.5–100 ms [Schmidt et al., 2014]. This implies that the (truncated) power-law behavior, which is usually assigned to the distribution of "on" times [Cichos et al., 2007] cannot be considered as a general "law"—at least not for CdSe/ZnS QDs. We suggest that this finding applies also for many other kinds of QDs since deviations from a (truncated) power law are often observed though not explicitly discussed. Our statement supports the recently suggested model of multiple relaxation centers of Frantzusov et al. (2009).

Equally important is the finding that blinking processes can be characterized by at least three classes which have different origins since different electronic states or relaxation pathways of the excited states are involved (see also the discussion in Section 6.2.3). The first indications for such a finding have been reported by Galland et al. (2011). At the present stage whether blinking assigned to these classes is governed by one and the same microscopic process remains open.

We like to emphasize that under certain conditions dynamics of well-defined relaxation centers (e.g., silanol or OH groups) can be identified to cause an anomalous power law of the "on" times [Krasselt et al., 2011]. Typical examples of modifications of a power law (by a well-defined relaxation process) and the related manifestation on the intrinsic PL intensity distribution and PL decay times are shown in Fig. 6.12 for CdSe/ZnS QDs in the same (PS) sample. This comparison shows the heterogeneity of the interaction of an individual QD with its disordered environment.

To get more insight into the mechanism governing blinking processes the responsible relaxation centers should be modified systematically, as we will discuss next. The nature, number, and dynamics of active relaxation centers [Frantzusov et al., 2009] can, for example, be tuned by modifications of the surrounding, such as the polarity of the matrix [Issac et al., 2005; El-Kork et al., 2011; Issac et al., 2012; Schuster et al., 2005] or interactions with specific sites (e.g., silanol groups) at the QD solvent interface [Krasselt et al., 2011; Schmidt et al., 2012; Schmidt et al., 2014]. With respect to the above-identified different blinking classes previous reports can now be discussed more rigorously, for example, the independence of the

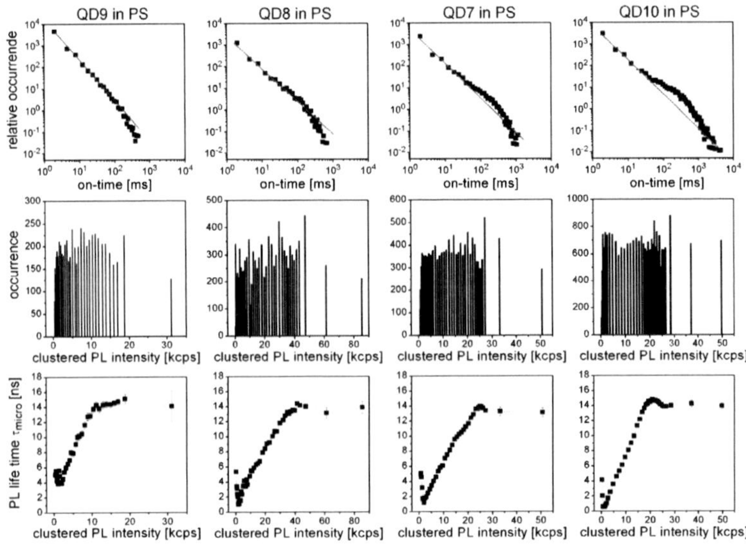

Figure 6.12 "On" time distributions (top), (clustered) PL intensity distributions (middle), and intensity dependent PL decay times τ_{micro} (bottom) for four different CdSe/ZnS QDs in PS. The anomalous power-law behavior increases from left to right and results both in higher PL intensity densities at high intensities as well as in (about 1 ns) longer PL decay times in the *N* range. PL decay has been fitted by a monoexponential function.

"on" time power-law exponent and the dependence of the "on" time cutoff time on the solvent polarity [Issac et al., 2012]. This applies also to the dependence of the "off" time power-law exponent and the respective cutoff time on the matrix polarity [Issac et al., 2005; Issac et al., 2012]. By changing the embedding polymers from PS (ε = 2.53) to PVA (ε = 14), we observe (see Fig. 6.7) that upon increase of polarity "on" times become shorter by a factor of about 5, which shifts the truncation of the power law to shorter times but leaves their distribution with respect to PL intensity very similar (constant power-law exponent). Merely in the *N* range contributions from OH groups add up. The "on" times related to the OH groups are obviously longer (~70 ms) compared to the ones without silanol groups (~40 ms) in the same solvent PS (see Fig. 6.7). But also in this range "on" times are shortened by a factor of about 5 if the surrounding matrix of the QDs is changed to PVA. These findings have to be compared to an "on" time shortening of a factor of 3 (as determined by a threshold-based approach) when comparing PS with PVA [Issac et al., 2012]. Differences of the absolute "on" time shortening upon increase in solvent polarity, as reported in the literature, might be explained by the fact that the experimentally determined cutoff times depend critically on the threshold setting in power-law analysis. Moreover, the influence of OH groups depends critically on the type and (thermal) pretreatment of the substrate, the kind of QD ligands, the ligand coverage, and the water content of the sample [Krasselt et al.,2012], as will be discussed in Section 6.4 in more detail. Concluding, besides the shortening of "on" times upon increase of solvent polarity the PL intensity–related distribution of "on" times remains basically the same, which might explain the reported invariance of the "on" time power-law exponent upon polarity changes [Issac et al., 2012]. However, this shows that both polarity and coordinative features of the environment influence blinking phenomena.

Besides the elaborate investigation of "on" times as a function of PL intensity and embedding surrounding, we have also investigated the PL decay in the nanosecond regime as a function of PL intensity (see Figs. 6.1, 6.2f, 6.3, 6.4, 6.6f, and 6.11). It is evident that the PL decay of a single QD is multiexponential, even at a well-defined PL intensity, as has already been previously explained in several publications [Schlegel et al. 2002; Fisher et al., 2004; Zhang et al., 2010] and more recently in an elaborate study [Schmidt et al., 2014]. Galland et al. (2012) report blinking of PL decay times for a nonblinking PL intensity. Consequently, the total PL from a single QD (and naturally also that of a QD ensemble) is multiexponential. Surprisingly, even after CPA with extremely well-defined PL intensities the (multiexponential) PL decay remains basically the same. These PL times do almost not depend on the embedding matrix, besides a small increase of τ_{micro} in the N range in the case of binding to silanol (OH) groups (see Fig. 6.12).

Recently, we have analyzed this intensity–time relationship carefully and found that the intensity-dependent PL decay of a single QD can be described sufficiently by three different decay times [Schmidt et al., 2014]. Figure 6.11 collects the findings for intensity-independent PL decays for several QDs under different environmental conditions (PS or PVA). A clear behavior emerges, namely the presence of three characteristic times $\tau_{micro,i}$. $\tau_{micro,3} \approx 1$ ns is present only at very low intensities (L range) within a narrow distribution of intensities. $\tau_{micro,2} \approx 5$ ns (and shorter) is found in the I–L range (we assign this dependency to a state 2 [A_2]), while $\tau_{micro,1} \approx 14$ ns (and shorter) is distributed across the total intensity range from S to L (which we assign to state 1 [A_1]). Obviously, $\tau_{micro,1}$ and $\tau_{micro,2}$ decrease linearly with decreasing PL intensity in the N–I range. Due to sensitivity reasons it is difficult to follow the exact behavior in the L range in which the additional short component $\tau_{micro,3}$ shows up. Two interpretations are possible: (i) The times $\tau_{micro,1}$ and $\tau_{micro,2}$ continue to become shorter toward the lowest intensity or (ii) they level off and remain constant at about 2 and 9 ns, respectively. We tentatively prefer explanation (ii), which would imply that the nonradiative rate k^{nr} (vide infra and Eq. 6.2) remains constant both at high and low PL intensities. As a consequence the PL intensity is not completely "off" in the L range (finite nonradiative rate) but masked by noise.

Most remarkably, $\tau_{micro,1}$ remains constant in the N/S range above $I \approx 0.8\, I_N$ which needs an independent and detailed explanation given in Section 6.2.2. Moreover, also $\tau_{micro,2}$ remains probably constant—though the data are very noisy—above the relative intensity $I \approx 0.5\, I_{max}$. However, the contribution to the overall intensity is drastically smaller for this component. Roughly speaking, a constant PL decay time with increasing intensity can only be realized if the population rates of the emitting states are varying in time (S blinking). The presence of S-type blinking events becomes very clear at very high intensities (appearance of "spikes"). However, analysis of PL decay times strongly indicates that S-type blinking extends across the total intensity range for which τ_{micro} remains constant. On the other hand, from comparison with the correlation diagrams we conclude that the "on" times in the S-type blinking regime extend at least from the maximum of the "on" time distribution in the N range (see Fig. 6.7) down to less than 1 ms in the S range.

The linear decrease of τ_{micro} with decreasing intensity can be described by an increase of nonradiative rates k^{nr} which determines the PL intensity according to Eq. 6.1. The nonradiative rate is defined as $k^{nr} = [k^{nr}(I_{max}) + k^{nr}(I_n)]$. In this description $k^{nr}(I_{max})$ is defined at the maximum PL intensity at a constant population rate, while $k^{nr}(I_n)$ describes the dependence on the PL intensity I_n. Since we find across a considerable intensity range at least two PL decay times for each PL intensity, this can be explained only by assuming two different radiative and nonradiative rates. Concluding, the finding of two different intensity-dependent decays features can be understood by assuming different ratios of $k^{nr}(I)/k^r$ for each of the components. Describing the intensity-dependent PL decay by a single decay time (or a stretched exponential function), the intensity-dependent overlap of two separate decay channels will result in a deviation and a change of the increment of a linear I/τ relationship. Such a deviation is in fact observed in several examples in Fig. 6.12. Moreover, Zhang et al. (2010) have reported a similar behavior and explained it by a change of k^r, which is in principle in line with our arguments.

We will now discuss the relation of the two identified (blinking) states 1 (A_1) and 2 (A_2) in the I range (see Fig. 6.11), for which we find a nearly linear behavior of the PL intensities between 14 ns and 6 ns for state 1 and 5 ns and 2 ns for state 2. The longest decay time of 14 ns (at high PL intensities) is somewhat too short compared to what is expected for a QD of this kind (~20 ns) [Zenkevich et al., 2014]. Following Eq. 6.1, we therefore conclude that the effective nonradiative decay rate $k^{nr}(I_N)$ at the maximum of the PL intensity of state 1 in the N range is close to $k^{nr}(I_N) = (1 - \text{QY})/\tau = 1 \times 10^7\ \text{s}^{-1}$ (assuming QY ≈ 80%), which results in a radiative decay rate $k^r = \text{QY}/\tau = 4 \times 10^7\ \text{s}^{-1}$. The nonradiative rate found at this intensity is assumed to be independent of those processes which cause blinking (B). In the course of blinking, $k^{nr}(B)$ becomes successively larger (assuming a constant k^r throughout the total intensity range) and finally approaching $k^{nr}(B) \approx 1.3 \times 10^8\ \text{s}^{-1}$ in the L range.

Comparing the increments of I/τ in Fig. 6.11 we obtain for state 2 a radiative rate $k^r = 7.6 \times 10^7\ \text{s}^{-1}$, which is about a factor of 2 higher than the one of state 1. A similar finding has been reported recently [Galland et al., 2012]. The authors attribute PL decay times of the two identified states as being due to the neutral exciton 1 and a negatively charged QD with negligible Auger-type PL quenching, which results in doubling of the radiative decay rate and a doubled exciton population, finally giving rise to PL lifetime blinking without changes in the PL intensity. In this consideration nonradiative decay channels are neglected. Since we observe at least two PL decay times at any given PL intensity, this corresponds to lifetime blinking, as has also been reported previously [Zhang et al., 2006]. However, we have to consider simultaneously nonradiative contributions in the present experiment and obtain $k^{nr}(I_N) = 1.2 \times 10^8\ s^{-1}$ at the highest intensity of state 2 (resulting in QY ≈ 38%). In the L range, $k^{nr}(B)$ becomes $3.0 \times 10^8\ s^{-1}$, which is a factor of 2.3 larger compared to $k^{nr}(B)$ of state 1. An increase in the radiative decay rate might be explained alternatively by an increased overlap of the electron and hole wavefunctions. Presently we cannot decide which of the two explanations applies. However, since similar PL decay time distributions are observed across a wide range of excitation intensities [Zenkevich et al., 2015], we expect the formation of multiple excitons to be less important in the present case. This is also supported by findings in Section 6.3.1 where it is demonstrated that blinking properties depend only moderately on excitation rates. However, one has to consider that also moderate excitation powers may result in multiple exciton formation [Peterson and Nesbitt, 2009].

It is interesting to follow what has been published in this respect. Several single QD data show that during blinking time traces, one and the same (low) PL intensity might be related to two PL decay times [Cordones et al., 2011b; Galland et al., 2011; Galland et al., 2012; Amecke and Cichos, 2012], which they interpret as a breakdown of the charging hypothesis as an explanation for blinking. However, such detailed investigations as the present one [Schmidt et al., 2014] have not yet been reported. Moreover, ensemble experiments might also elucidate some of the above-discussed aspects. Petrov et al. (2006) have shown that the PL decay of a (somewhat larger, emitting at 588 nm in liquid toluene solution at 300 K) CdSe/ZnS QD can be described by three prominent broadly distributed decay times centered at 1 ns, 10 ns, and 22 ns, respectively (see Fig. 6.13). Note that the distribution of PL decay times is shown on a logarithmic timescale. Thus the 1 ns distribution is much narrower than the other two.

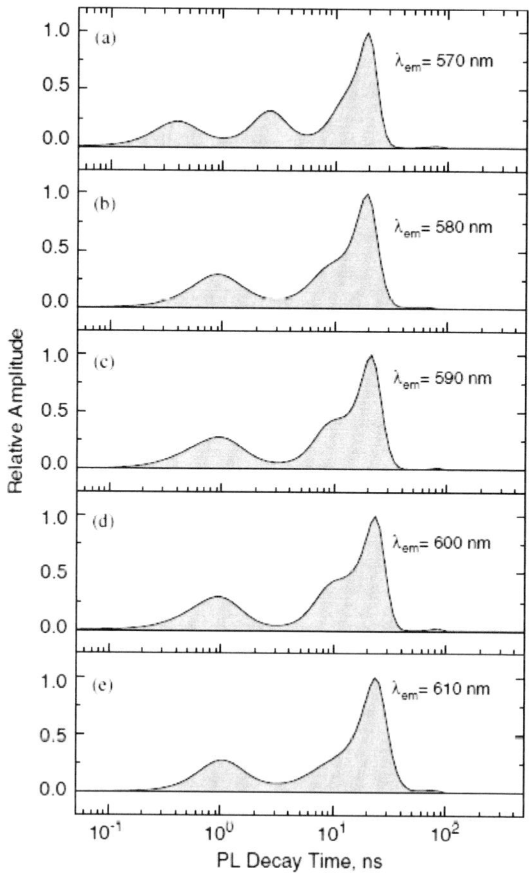

Figure 6.13 Multicomponent analysis of PL decay times of TOPO-capped CdSe/ZnS in liquid toluene at different detection wavelengths λ_{em} reveals a complex distribution of decay times. Note the logarithmic timescale and slight wavelength dependence of the relative contributions. Decay distributions have been normalized to the maximum close to 22 ns. The 10 ns component is slightly blue-shifted relative to the 22 ns component. A marginal shift to longer decay times is observed for the 22 ns component with an increasing detection wavelength, which we attribute to the size distribution within the QD ensemble. Additional very long-lived (>100 ns) decay components are observed, which we assign to the PL of deep traps (see also Chapter 2). Excitation is at λ_{exc} = 575 nm, which is energetically below the lowest absorption band for the smallest QDs. (a) Influence of anti-Stokes transitions (not discussed here). Reprinted from Petrov, E. P., Cichos, F., and Borczyskowski, C. von (2006). Intrinsic photophysics of semiconductor nanocrystals in dielectric media: formation of surface states, *J. Lumin.*, **119–120**, 412– 417, Copyright (2006), with permission from Elsevier.

Recently we have reported the tuning of PL properties caused by attachment of a single dye molecule studied via temperature-dependent ensemble experiments [Zenkevich et al., 2015]. One finding is that the PL decay is described by at least three decay times close to 1 ns, 5 ns, and 15 ns. These times are very close to the findings for single CdSe/ZnS QDs via CPA and reported values (see references in Zenkevich et al. [2015]). In that paper, the influence of dye attachment to the surface has been investigated in detail. Interestingly, though the dye attachment strongly quenches the overall PL intensity of the QD this is basically caused by reshuffling of the relative amplitudes of the three PL decay components (see also Trenkmann et al. [2016]).

How can all these findings be described in a unified model? Though a comprehensive model will be discussed in Section 6.2.3, we already suggest here that PL blinking of CdSe/ZnS QDs results (for single QDs or ensembles) in most environments in shortening of the average PL decay time, which is inherently related to fluctuating PL intensities. Additionally, at least three distinct and, in principle by quantum mechanical approaches, discernible electronic states (with different intrinsic PL dynamics and radiative rates) are explored during a blinking time trace. Two of these states are related to a broad distribution of possible PL intensities which themselves are linearly related to the respective PL decay times. The overall blinking features ("on" times, "off" times, and their distributions) are often but not necessarily characterized by power laws and are controlled via exploring statistically a few abundant intrinsic QD states with respect to energies and dynamics. These intrinsic QD features (PL decay times and energies) are only marginally influenced by the embedding environment. What is strongly influenced (or even controlled) is the timescale of jumps among the manifold of all intrinsic QD states. The principal relationship of the density of intensities and related states (1–3) is shown in Fig. 6.14, both in relation to blinking classes *L*, *I*, and *S* and intensity-dependent "on" times, which has to be interpreted in combination with Figs. 6.7 and 6.11.

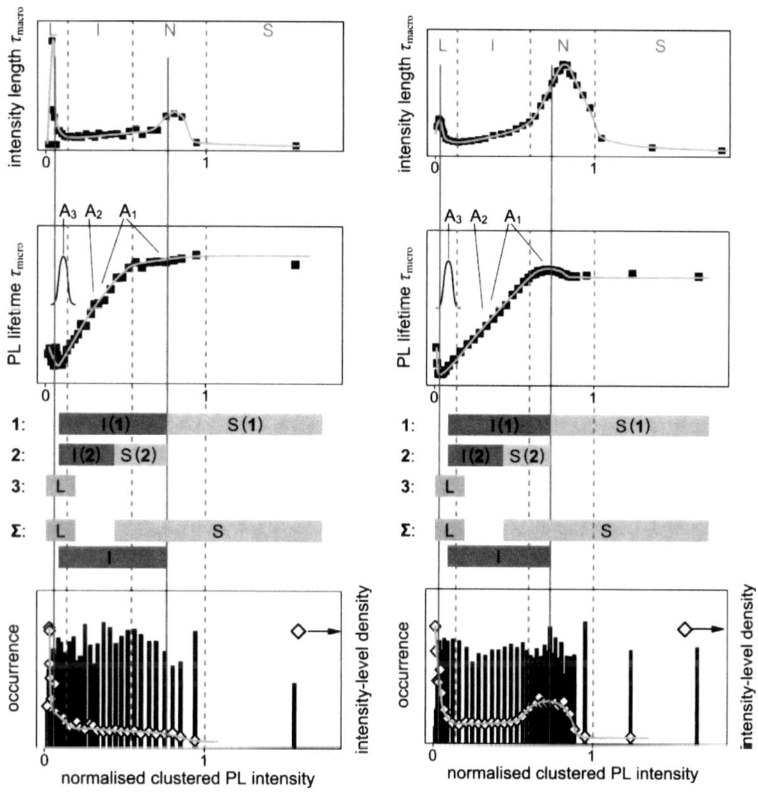

Figure 6.14 Left: QD9 in PS. Right: QD10 in PS. Detailed assignment of blinking processes to the PL intensity–related properties: (top) τ_{macro} ("on" times), (middle) τ_{micro} (three-exponential PL decay times, but presented according to a monoexponential fit), and (bottom) PL intensities (clustered and density of intensities). We make use of the assignment of three emissive QD states 1, 2, and 3 and assign to each of them a blinking range on the intensity scale. States 1 and 2 follow the same "on" time dynamic [Schmidt et al., 2014], while state 3 and the OH group–related state show different dynamics since τ_{macro} becomes biexponential in the respective intensity range.

The question immediately arises whether "on" times differ for states 1 and 2. In Fig. 6.14 we have indicated the intensity (10000, 15000, 25000) at which the emission of state 2 sets in. We do not observe a marked change of τ_{macro} (see Fig. 6.7) at this intensity. The intensity dependence of the "on" times remains rather smooth down to the minimum of "on" times in the case of PS and somewhat steeper for PVA. From this we conclude that the jump time ("on" time) does not depend significantly on the type of state 2 or 1. The intensity range N of constant intensities goes in line with constant PL decay times, even up to the very clear spike intensities, which does not change dynamic parameters on the microtimescale. All this intensity range (N and S) is therefore related to S-type blinking. Probably, this applies to both states 1 and 2, though it is less obvious for state 2, since it is buried under the dominant PL intensities of state 1 (see Fig. 6.14).

Interestingly, the lengths of "on" times in the S-type blinking range depend drastically on PL intensity in that the intensity separation ΔI_S of clustered intensities is approximately proportional to $1/\tau_{macro}$. At the highest S intensity, the "on" times τ_{macro} are $\sim$1 ms and shorter, while they are $\sim$30 ms at the lowest S intensity, depending on the individual QD and the embedding matrix.

6.2.1.4 Assignment of three types of intensity blinking

We now continue our discussion of different classes (types) of blinking. We relate the findings of the distribution of τ_{macro} with the one of τ_{micro} comparing Figs. 6.7, 6.11, and 6.14. We conclude that the distribution of "on" times in the S range extends (in PS) from some tens of milliseconds (N range) to 1 ms (spikes). I–N-type blinking extends from 10 ms to at most 20 ms with increasing intensity which is only a small variation across that respective intensity range. The third blinking regime in the L range is accompanied by long "on" times from 20 ms to 100 ms, depending on the specific QD. The timescale of S-type blinking decreases rapidly with increasing intensity from a few tens of milliseconds down to <1 ms. Moreover, blinking in the I range relates to a rather constant density of intensities, whereas the density of intensities in the S range is very inhomogeneous. The density of states is compared for QD9 and QD10 in PS in Fig. 6.14, in which we have also included the various blinking regimes. QD10 shows an anomalous power-law blinking (coupling to OH groups, see Fig. 6.12).

All "on" times depend strongly on solvent polarity (see Fig. 6.7). Superimposed is the OH group (silanol)-related blinking occurring predominantly more or less from "on" to "off" intensities and vice versa [Krasselt et al., 2011]. Its quantitative contribution to an anomalous power-law-type distribution (see Eq. 6.2) is across a narrow time range at the highest *N* intensities (and is therefore expressed by a superimposed exponential term). We have recently suggested [Schmidt et al., 2014] that the three apparent states are related to the strongly quenched exciton (state 3) and two different trapped hole states (1 and 2) which are both highly emissive. The resulting model is shown in Chapter 4.

Figure 6.15 shows schematically how various PL intensities are interrelated via intensity blinking processes (intensity jumps). As discussed we find besides the regular (*I*) blinking two additional pronounced intensity regions which we have assigned to (i) *S* blinking (caused via intermittence in the population pathways or in elevated electronic states) and (ii) *L* blinking among intensity zero and intensity *L*, which we tentatively assign to the (quenched) exciton emission. The majority of jumps (*I*-type blinking) take place along the diagonal in Fig. 6.15 by jumps among states 1 or 2 with intermediate intensity. The interrelation of intensities is shown schematically in Fig. 6.15.

The presence of the specific (OH group–related) PL intensities for some of the QDs in PS and all QDs on PVA characterized by a deviation from a truncated power law (according to Eq. 6.2, see Fig. 6.5) is accompanied by an about 1 ns increased PL decay time in the high-intensity *N*–*S* range and an about 10 meV blue-shifted PL emission in the respective intensity range (see Section 6.2.2). This is further accompanied by an increase of intensity jump probabilities (of large jumps) among high intensities (*N* range) and low intensities (*L* range), as is demonstrated in Fig. 6.9.

Figure 6.15 shows (marked by circles) that in the presence of OH group–related states the probability of large jumps between high (n = 20–23) and low (n = 5–8) PL intensities is increased. The low intensities are in the intensity range of the state 3. This correlation indicates that these intensities are directly related by a single intensity jump [Krasselt et al., 2011]. In Fig. 6.15 (bottom) we have indicated by broken lines a further intensity correlation which shows up in the presence of OH groups. It closely corresponds with the intensity range expected for state 2 and the probably related *S*-type blinking. This implies that both states 1 and 2 are subject to the influence of OH groups. This will be discussed in more detail in Section 6.2.3.

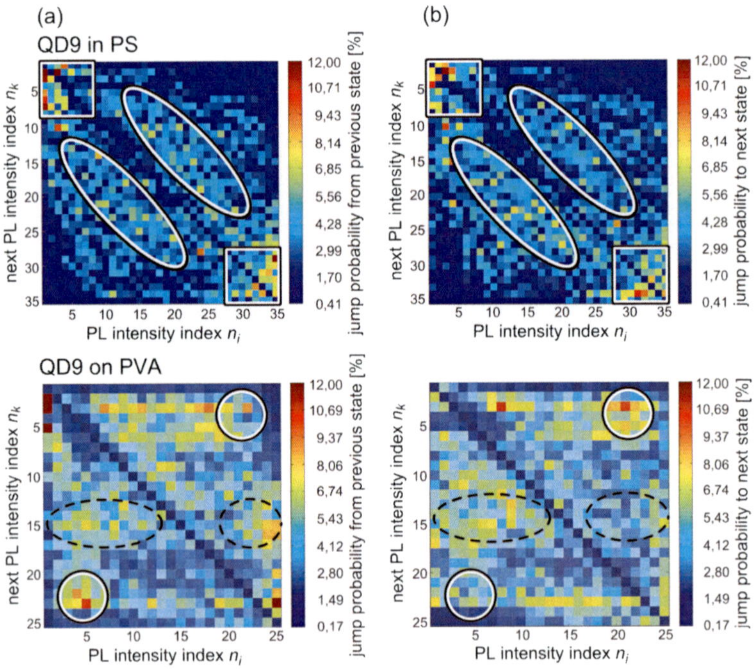

Figure 6.15 Comparison of jump probabilities (a) from and (b) to next intensities for QD9 in PS (top) and for QD9 in PVA with OH group coupling (bottom). The OH group–related jumps are marked by circles close to the off-diagonal corners. Prominent *S*- and *L*- type jump probabilities are marked by squares in the probability diagrams for QD9 in PS. *I*-type blinking is marked by ovals parallel to the diagonal. Broken lines indicate jumps among OH group–related intensities of state 2, both of *I*-type (left) and *S*-type (right). Note that the numbering of intensities of QD9 (PVA) is different from the one of QD9 (PS).

Moreover, we speculate that the OH group–related intensity in the *N* range corresponds to the (nonquenched) excitonic state which we have otherwise identified with state 3 as the strongly quenched excitonic state. This is not necessarily a contradiction, since a single QD undergoes jumps among many electronic state configurations. One of them is the OH group–related one. If during this realization hole trapping takes place only with low probability we expect strong emission from the true excitonic state (with a, as observed, slightly increased PL decay time and a 10 meV blue-shifted PL vide infra). If hole trapping occurs during the excitation cycle, we will observe an intensity jump to the strongly quenched state 3 (or close to it), which is also slightly blue-shifted with respect to states 1 and 2, as we show in Section 6.2.2 and as we have shown recently [Schmidt et al., 2014]. In the following section we will discuss whether there is evidence to validate this speculation.

6.2.2 Correlation of Intensity Blinking and Energy Jumps of Emitting States

In the previous section, we have shown that blinking is related to a distribution of PL intensities and vice versa with a set of PL decay times. We argued that these relationships can be unraveled. As a result, we identified on the base of the PL decay at least three basically different emitting QD states. Now the question arises whether the postulated states can be also related to well-defined PL energies.

To proceed, we applied CPA and separated the QD PL into two different channels (red and blue, respectively) via a dichroit positioned as close as possible to the center of the averaged spectral PL band of a single QD. The basic scheme of the setup is shown in Fig. 6.16. Details of the analysis are described in Chapter 4. In case the PL wavelength is fluctuating in time this fluctuation will be detected as an alternating intensity in the respective spectrally separated detection channels. Such effects have been repeatedly observed [Empedocles and Bawendi, 1999; Biebricher et al., 2006; Gómez et al., 2006; Plakhotnik et al., 2010; Fernée et al., 2012; Martin et al., 2012; Cui et al., 2014] and have been termed "spectral diffusion" or "jitter" of energies. The advantage of the CPA-related approach is the possibility to relate spectral jumps (in energy) during a blinking time trace to jumps in intensity and PL decay times simultaneously. Such a comprehensive investigation has not yet been reported in the case of blinking QDs and results in a quasi-3D data analysis [Biebricher et al., 2006].

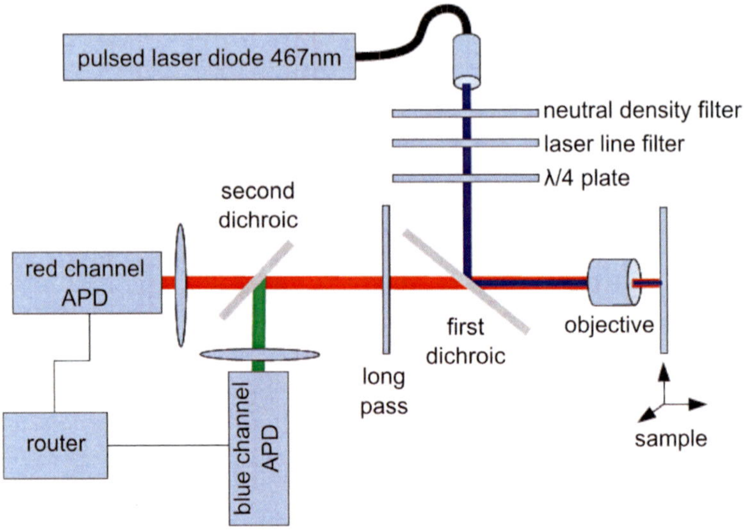

Figure 6.16 Schematic representation of a confocal microscope with spectral selectivity. After excitation of the QD sample via a pulsed 467 nm laser diode the PL is spectrally separated from the excitation by a first dichroic beam splitter. A second dichroic beam splitter splits the QD emission into a red and a blue channel, which are separately monitored.

We have analyzed two different types of QDs, namely TOPO-capped CdSe/ZnS QDs and CdSe/CdS QDs. The latter are incorporated into poly(methyl methacrylate) (PMMA) nanoparticles of about 50 nm diameter for better stability [Negele, 2013]. QDs have been diluted in toluene and then spin coated onto SiO_x substrates. The distribution of "on" and "off" times determined via CPA (using an artificial threshold) is shown in Fig. 6.17. For blinking times longer than 1 ms (the time resolution of the experiment), the distributions of blinking times are not correlated with different PL wavelengths.

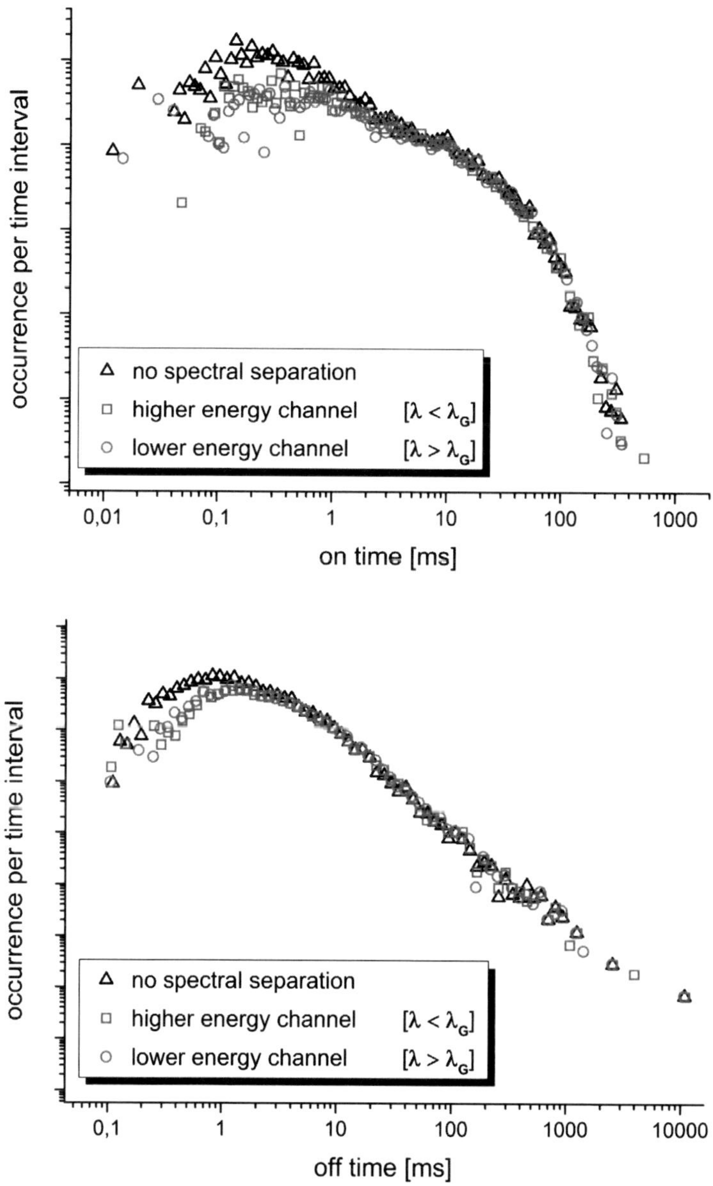

Figure 6.17 Distributions of "on" (top) and "off" times (bottom) have been determined by CPA (and an artificial threshold) in 3 different modes of spectral detection: total, "square," and "open circle" PL, respectively. Data represent a single CdSe/ZnS QD.

A typical PL intensity distribution of TOPO-capped CdSe/ZnS during a blinking time trace and analyzed by CPA is shown in Fig. 6.18 before (solid black line) and after clustering (gray bars).

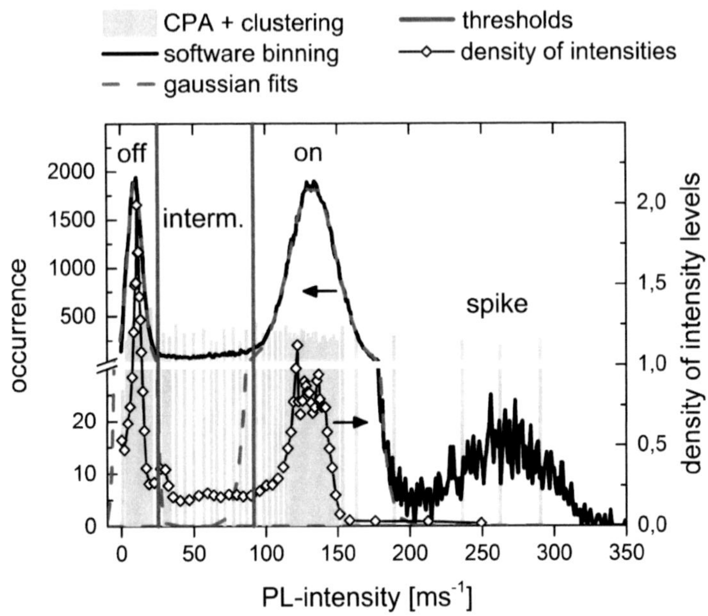

Figure 6.18 Intensity related probability of photon detection without spectral separation as detected conventionally (binned, black line) and following CPA with intensity clustering (gray bars). The density of states is shown by open diamonds. The two black lines mark the intermediate intensity range I (assigned as "threshold").

Details of spectral assignments of spectral shifts from intensity fluctuations are given in Chapter 4. Figure 6.19 shows a typical example of spectral shifts ΔE related to the PL intensity following CPA. Spectral shifts are given with respect to the spectrally separating dichroic beam splitter. The figure contains also the density of intensities at the bottom. Referring to the blinking assignment in the previous section we assign the spectral shifts as follows. The range of almost constant ΔE above the PL intensity of about 250 ms^{-1} is assigned to state 1 and the state related to OH groups. According to Fig. 6.11, state 2 should start to contribute noticeably only below the right blue line that is close to the small density peak at $I \approx 160$ ms^{-1}.

From this we estimate that the energy difference $\Delta E_1 - \Delta E_2$ between states 1 and 2 is in the range of –10 meV.

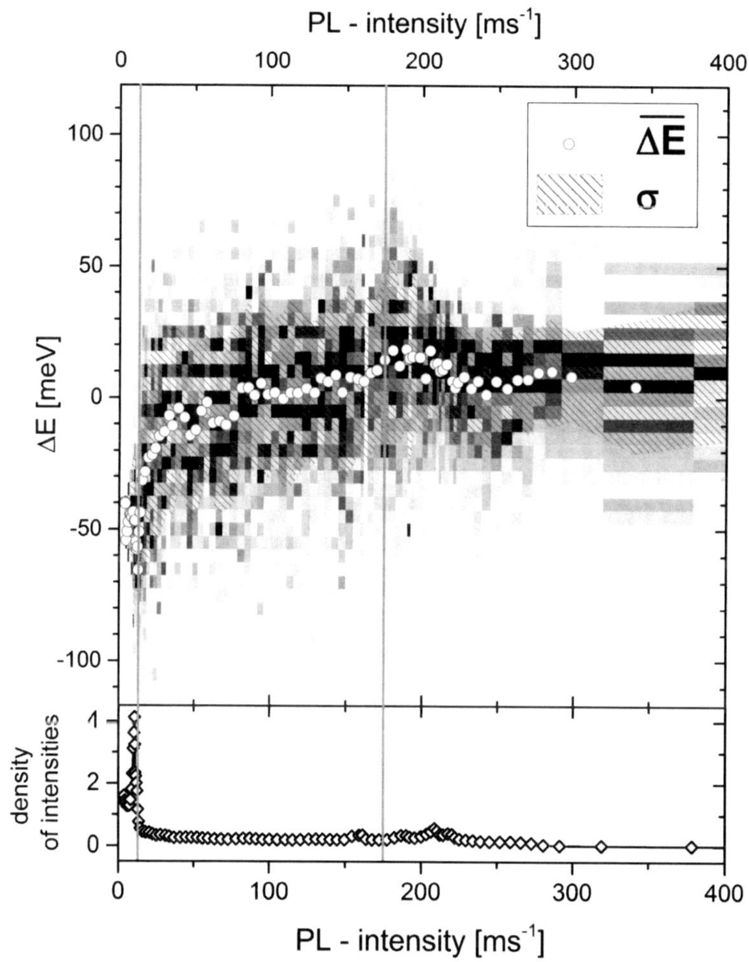

Figure 6.19 Top: Spectral shifts ΔE of a single TOPO-capped CdSe/ZnS QD on SiO_x with respect to the position of the dichroic beam splitter as a function of PL intensity. Intensity identifications during a total blinking time trace of 20 min result in a distribution σ of energies (hatched area). The arithmetic average $\Delta E(I)$ of $\sigma(I)$ is indicated by the white dots. Bottom: Density of PL intensities. The gray bars limit the intermediate intensity range I. The PL decay is monoexponential at intensities above the gray bars on the right.

Figure 6.20 shows (in a different presentation) a comparison of the spectral shifts of a single CdSe/ZnS QD and a single CdSe/CdS/PMMA QD. Still, the qualitative behavior is quite similar, though the related spectral shifts are somewhat smaller compared to CdSe/ZnS. Note, that we have chosen the numbered intensities (intensity-level index n) instead of experimental PL intensity values in this presentation. The density of intensities is also shown. The spectral distribution (width and maximum) at each intensity has been fitted by a Gaussian function and the results are also shown in Fig. 6.20 (top) as a function of the PL intensity. Furthermore, Fig. 6.20 (bottom) shows three examples of spectral fluctuations at three different PL intensity values.

The Gaussian fits to the spectral fluctuations (shown as σ in Fig. 6.19) are also indicated. These examples clearly show that the fluctuations are not symmetrically distributed but reveal at least two spectrally different (by less than 50 meV separated) components. They are not spectrally resolved, but there relative contribution varies as a function of PL intensity. This is schematically illustrated in Fig. 6.20(bottom). Again, we observe a red shift of PL energy with decreasing intensity. The overall range of spectral fluctuations is up to 50 meV, while the width of spectral fluctuations at each PL intensity is about 30 meV (see Fig. 6.21). We arrive at the following picture. States 1 and 2 are separated by about 10 meV with a spectral width of typically 30 meV, which is considerably larger than the separation of electronic states. Each of these states shifts (in the course of a blinking time trace) by up to at least –50 meV with decreasing intensity.

Similar to the procedure described in Section 6.2.1 we determined also the PL decay time in the blue and red detection channels for each intensity separately. The results are shown in Fig. 6.21 for a CdSe/ZnS QD in comparison to the spectral shifts, spectral widths, and intensity-level density. Results are qualitatively similar for CdSe/CdSe/CdS/PMMA QDs. PL decay data have been fitted by the sum of two exponential decay functions. A triexponential fit (see Section 6.2.1) was not possible due to the reduced sensitivity caused by the splitting in two spectrally separated channels. Note that absolute decay times are partly larger compared to the ones given in Section 6.2.1, which is related to the use of a different batch of QDs with obviously different QYs.

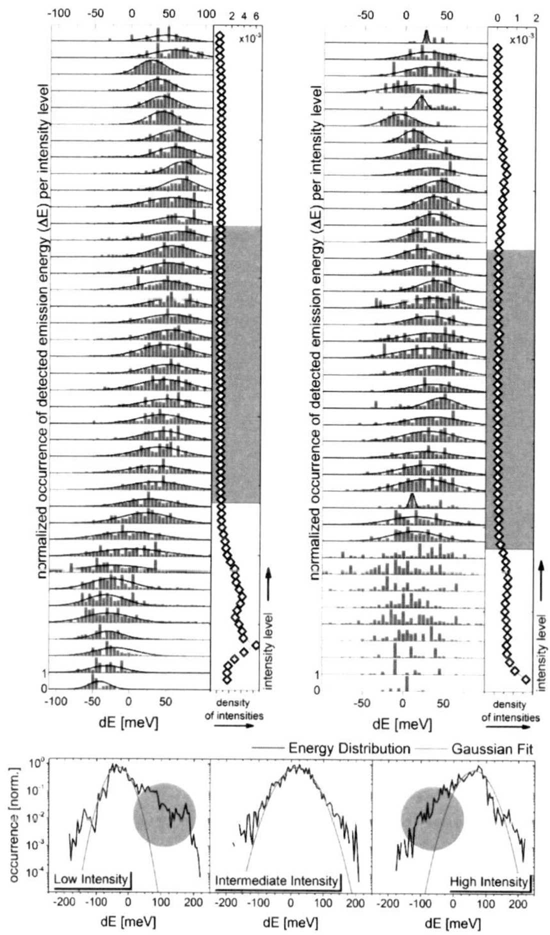

Figure 6.20 Top: Spectral shifts ΔE with respect to the dichroit as a function of PL intensities (given according to intensity numbers with increasing PL intensity from bottom to top) for CdSe/ZnS (left) and CdSe/CdS/PMMA (right). The density of intensities is also given. Gray ranges correspond to intermediate intensities. Curves correspond to the respective Gaussian fit through the distribution of spectral shifts at a given intensity. Bottom: Detailed presentation of spectral fluctuations ΔE at low (left), medium (middle), and high (right) PL intensities for CdSe/ZnS (see upper-top spectral series on the left) on a normalized logarithmic PL intensity scale. The gray curve corresponds to a Gaussian fit to the spectral fluctuations. Major deviations from the Gaussian fit are encircled. The full-width at half-maximum (FWHM) is broadest in the range of intermediate intensities.

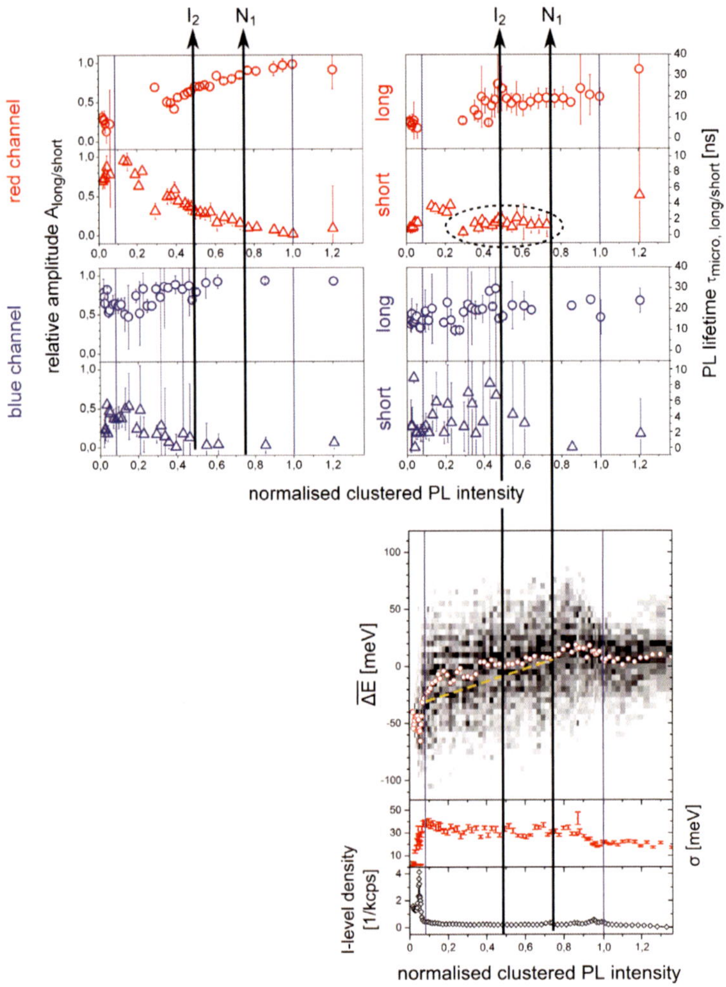

Figure 6.21 Comparison of PL decay times (top, right) and related amplitudes (top, left) with spectral shifts ΔE (open circles correspond to average spectral shifts), spectral fluctuations σ (obtained by Gaussian fitting of ΔE), and intensity-level densities (bottom) as a function of clustered and normalized PL intensities for a single TOPO-capped CdSe/ZnS QD. The arrow I_2 marks the inset of the intensities assigned to state 2. The black arrow N_1 marks up to which decreasing intensity the PL decay remains monoexponential, though intensities are decreasing (see text). The broken oval marks intensity-independent short decay times in the red channel. The broken yellow line indicates a shift in ΔE of ~40 meV. The blue lines mark the L–I and N–S intersections.

Time resolved data show complex features, which we will discuss step by step. Considering the large fitting errors we have only included those decay components where the related relative amplitudes A_i are larger than 0.1. According to the density of intensities we have indicated two lines: The line N_1 indicates the intensity at the transition from the intensity range N to range I (see Section 6.2.1). The line I_2 has been set to $0.6 \cdot N_1$ according to the highest intensity of the PL decay component of state 2. This value has been independently deduced from nonspectrally resolved decay data (see Fig. 6.11). Due to spectral diffusion, each individual PL intensity can contribute to both detection channels according to the magnitude of its spectral width σ and the steepness of the dichroit.

First we discuss the results in the blue detection channel. For intensities larger than N_1 the decay is monoexponential with a decay time of about $\tau_1 \approx 22$ ns at the highest intensities where it remains constant across a considerable intensity range. Below N_1 we find a second decay close to $\tau_2 \approx 8$ ns becoming shorter down to 3 ns at the lowest intensity. The contribution A_2 (relative amplitude of the short component) is relatively increasing with decreasing intensity but becomes at the most as large as A_1 (relative amplitude of the long component), while in the L range (left from the first/left blue line vide infra), no components shorter than 3 ns are observed. All these features are qualitatively similar to those already reported in Section 6.2.1, though both the experimental setup and QD batch are different. Therefore we assign A_2 to state 2 and A_1 to state 1 (see the previous section).

Now we turn to the red detection channel. Here we also observe a component $\tau_1 \approx 22$ ns, being constant at the largest intensities and then gradually decreasing with intensity down to ~8 ns. However, we do (at high intensities) not observe the second relatively long component τ_2 which has been assigned to state 2. We take this as a further indication that state 2 is somewhat higher in energy compared to state 1. Surprisingly, in the intermediate intensity range we find an intensity-independent short component $\tau_3 \approx 2$ ns (dashed ellipse) but no strong indication of τ_2. For such a short time we expect a very low intensity according to Eq. 6.1, which is obviously not the case. There are also indications that τ_3 contributes (visible only at high intensities) in the blue channel but cannot be identified any more as soon as the τ_2 comes into play at about I_2. How can a constant decay time at varying intensities be explained?

Again, as we have already discussed with respect to τ_1 in the previous section, the relationship can be explained by fluctuating (blinking) population routes followed by a fast but constant deactivation rate. The most effective population of this state is obviously in the low-intensity range since the relative contribution of τ_3 is increasing with decreasing intensity.

It has to be pointed out, however, that there might be a distinct and resonance-like second short decay component in the L range both in the red and in the blue detection channel. Obviously the contribution of this resonance-like state is more pronounced in the red detection channel since at the specific intensity it completely dominates the decay in the red channel, while it contributes only up to 50% in the blue channel (see Fig. 6.21, top left). We would like to mention, however, that an alternative explanation of the resonance-like behavior might be related to an S-type blinking, which would offer a strong population peak, as shown in Fig. 6.7.

Relating the kinetic data with the spectral shifts in the lower part of Fig. 6.21 reveals that decreasing decay times are accompanied by continuously red-shifted PL, while the spectral width σ increases on average only slightly with decreasing intensity. However, two peculiarities are observed. Firstly, we find a jump in σ, which corresponds to the maximum in the blue shift that is the highest PL energy. We have assigned this energy to the (silanol) OH group–related state, as discussed intensively in the previous section. Carrying the speculation, which we have posed earlier, further on this intensity might be related to the nonquenched exciton, giving rise to an increased density of intensities (see also Section 6.2.1) [Krasselt et al., 2011; Schmidt et al., 2014] and is obviously accompanied by a further blue shift of about 10 meV. Secondly, σ experiences a sudden increase close to the one of ΔE at the L–I intersection (indicating that an enlarged number of states is involved during spectral diffusion). This can be also concluded from the presentation of σ in Fig 6.20 (bottom left), which shows a (weak) side band shifted to the blue by about 50 meV from the principal band. At the same intensity we find a relative maximum of the short decay component τ_3 both in the blue and in the red channel (see Fig. 6.21, top left). We take this as a further confirmation of the presence of a third decay component similar to the one shown in Fig. 6.11.

σ further decreases abruptly at very low intensities (close to where intensities of state 3 are expected) followed by a sharp narrowing in the range of the high density of PL intensities in the L range. The dashed yellow line in Fig. 6.21 marks a range of spectral shifts $\Delta E < 50$ meV, which has to be compared to the spectral width σ of maximal 40 meV in the same range. Since these values are of similar magnitude it is not surprising that all decay times show up in both detection channels with comparable magnitude. Only in the S range σ was found to be as narrow as 20 meV, which corresponds to the typical PL linewidth of a single QD at 300 K [Curie et al., 2014]. At very low PL intensities σ becomes much smaller, which we therefore assign to noise fluctuations.

Figure 6.22 shows the low-intensity range of the blue channel. Here, we find (similar to what is seen in the red channel not shown here) across a narrow intensity range a very short-lived PL component of <2 ns. This behavior is close to what we have reported recently [Schmidt et al., 2014] and confirms again the presence of a very short decay time and a related narrow intensity (energy) distribution, as discussed in Section 6.2.1.

That a very short-lived state can be found in the blue spectral range is demonstrated in Fig. 6.21, which shows on an expanded scale a low-intensity band blue-shifted by ~20–30 meV. Such a band has already been expected from the sudden increase of the spectral width σ at low intensities shown in Fig. 6.21. The PL band at $\Delta E \approx 0$ has a similar energy as the one at the very low intensity in the L range. We therefore tentatively assign the latter to the strongly quenched exciton.

As the signal-to-noise ratio limits the accuracy to determine the complete set of all kinetic data for a single QD, we plot the decay times detected in the blue or red detection channel biexponentially fitted for 22 QDs of both types (CdSe/ZnS and CdSe/CdS) in Fig. 6.23. The comparison of decay times in the two channels and for the two different types of QDs immediately reveals that the manifold of decay times is very similar. However, decay times < 2 ns are predominantly found in the red channel, while the longest decay times (>12 ns) are predominantly in the blue channel. It is interesting to note that the grouping of times is very similar to the grouping found recently [Petrov et al., 2006; Zenkevich et al., 2015].

Additional support with respect to the assignment of PL decay times to PL energies stems from ensemble experiments [Petrov et al., 2006; Zenkevich et al., 2015]. There we found that those energy components related to τ_3 and τ_2 contribute effectively more at higher

PL energies than the one assigned to τ_1. At first glance this seems to be a contradiction, since we find, from Fig. 6.21, shifts to lower energies with decreasing decay times (Pl intensities). However, this is not necessarily a contradiction since the (red) low-energy channel contains contributions from both high- and low-energy states due to spectral diffusion (see Fig. 6.20, bottom left). As the dichroit is not necessarily at the center of the (averaged) PL band, absolute PL intensities in the respective channels cannot be determined easily.

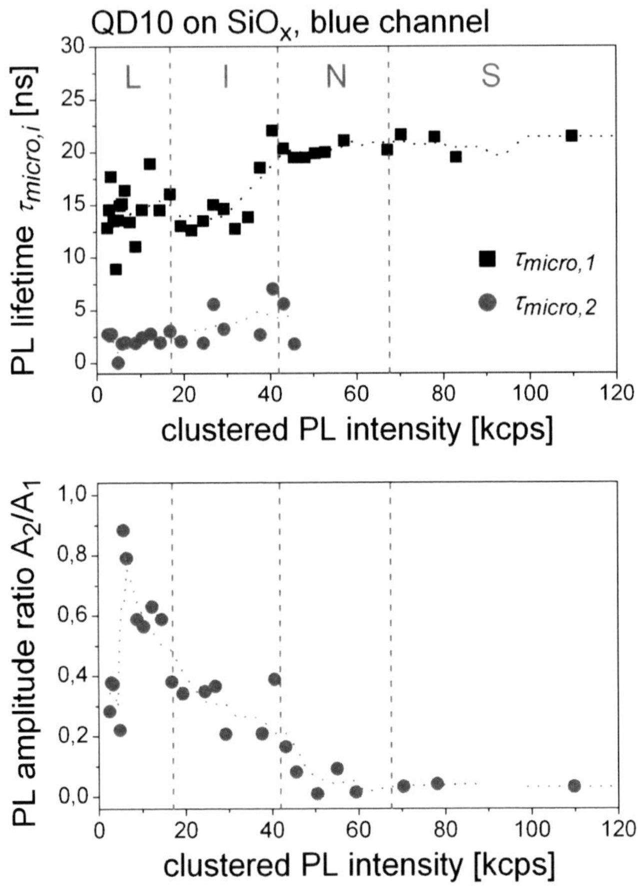

Figure 6.22 Biexponentially fitted PL decay times $\tau_{\mathrm{micro},i}$ and amplitude ratio A_2/A_1 of a CdSe/ZnS QD on SiO_x as a function of clustered intensities in the detection channel denoted by solid circles.

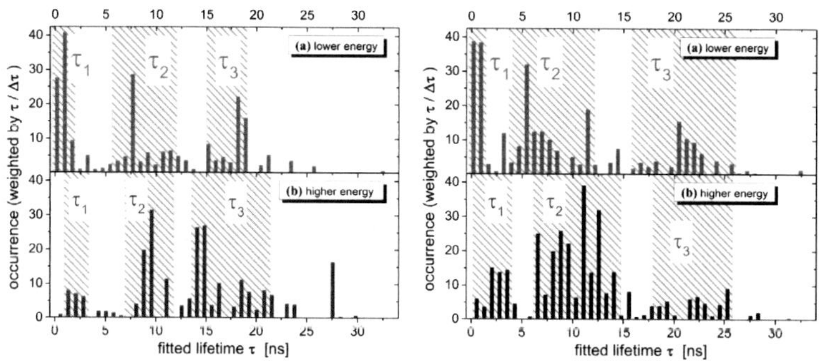

Figure 6.23 Distribution of biexponentially fitted PL decay times for CdSe/ZnS (left) and CdSe/CdS/PMMA (right) in the (a) red and (b) blue detection channel. Times have been grouped into three classes (shaded areas).

It is somewhat surprising that the differences (spectral fluctuations and PL decay components) in general findings for the spectrally resolved single-QD experiments on CdSe/CdS/PMMA are comparable to those found for CdS/ZnS [Göhler; 2014]. Since the energy barriers induced by the CdS shell to the CdSe core are quite low in the conduction band, one would expect much larger differences [Cui et al., 2014]. Only with respect to blinking we found that CdS/CdS/PMMA exhibits somewhat longer "on" times and shorter "off" times compared to CdSe/ZnS [Göhler, 2014]. Possibly the PMMA shell is responsible for this finding.

Besides the distribution of PL energies and the uniquely related intensity distribution one can directly determine the spectral fluctuations *dE*, as has been reported in several publications [Müller et al., 2005; Neuhauser et al., 2000; Martin et al., 2012; Schmidt et al. 2014]. The respective CPA procedure has been described in Chapter 4. In Fig. 6.24 we show the distribution of *dE* jumps for a typical single (a) TOPO-capped CdSe/ZnS and (b) TOPO-capped CdSe/CdS/PMMA QD. By setting appropriate intensity thresholds we discriminate between spectral diffusion for "on," "intermediate," and quasi-"off" intensities. It is evident that the related widths of distributions of *dE* are by about 10 meV narrower for "on" intensities compared to the other ones. This finding depends slightly but not significantly on the type of QD (CdSe/ZnS or CdSe/CdS/PMMA).

However, the distribution of *dE* cannot be described by a Gaussian function. To provide a rough estimate of the width of the distributions we have fitted the distribution by a Lorentzian function and determined the ratio of the widths between fluctuations of *dE* in the "on" compared to the "off" intensities. The ratio is shown in Fig. 6.24 (right). We find a wide distribution of such ratios which is centered at about 0.6. How can such a distribution be approximated more quantitatively?

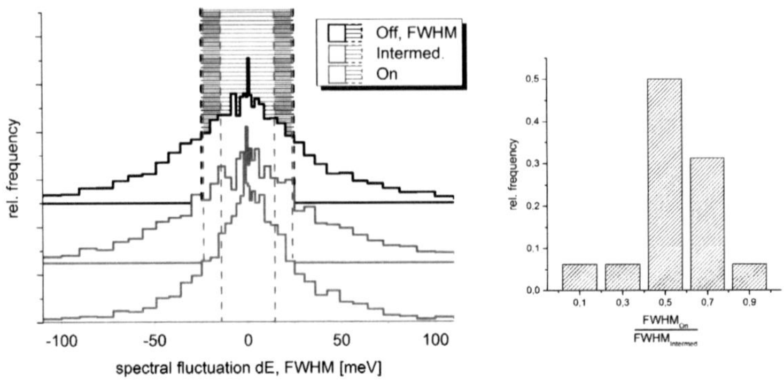

Figure 6.24 Left: Probability distribution of spectral fluctuation *dE* (in a given intensity range) for a single CdSe/ZnS QD and corresponding line width FWHM of the distribution as a function of the PL intensity range "off," "intermediate," and "on" as obtained by appropriate threshold settings. Right: Probability distribution of the ratio of $FWHM_{on}/FWHM_{off}$ (Lorentzian approximation) for 20 QDs.

In an analogy to PL intensity jumps (blinking) shown in Fig. 6.9, we expect that three different classes of energy jumps are needed. One of them is centered at $\Delta E = 0$; the other two are split nearly symmetrically into two bands shifted each below and above the central band. The fit results are shown in Fig. 6.25, where we depict the shifts *dE* for the respective threshold-based intensities for 20 QDs again with the related (Gaussian) widths' full-width at half-maximum (FWHM) obtained during a blinking time trace.

We find for all intensities a narrow distribution close to $dE \approx 0$, which is probably related to noise. The second distribution is close to $dE = \pm(20\text{–}40)$ meV, while a third very weak one is $dE \geq 80$ meV. Note that for "on" intensities there is an indication of a jump of *dE* (diamonds) of $\sim\pm 20$ meV close to $dE = 0$. According to Fig. 6.19 "on" intensities contain predominantly contributions from state 1 and OH-related states. Therefore $dE \approx 20$ meV corresponds to the energy separation of these two states.

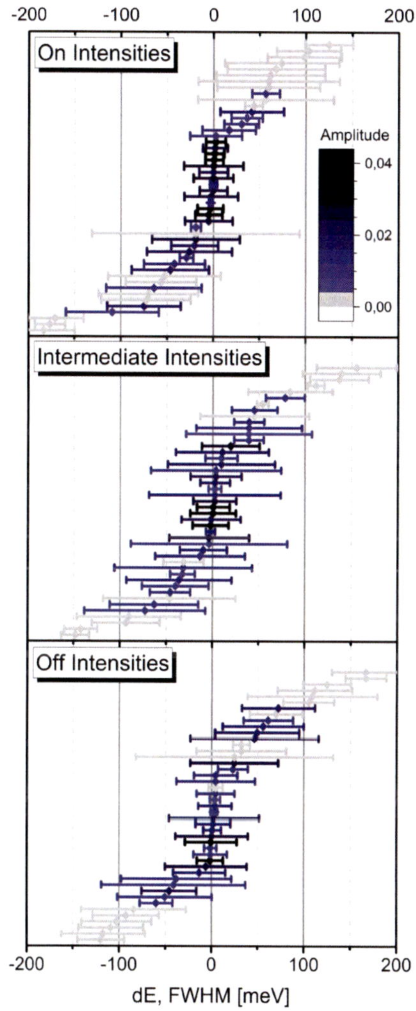

Figure 6.25 Distributions of spectral jumps *dE* sorted by PL intensities "on" (top), "intermediate" (middle), and quasi-"off" (bottom) for 20 CdSe/ZnS QDs on SiO_x during a blinking time trace of 20 min. Up to 5 Gaussian bands are fitted for each QD. The center of each band is marked by a solid diamond and the related FWHM by a colored horizontal bar. The color code indicates the probability (amplitude) of occurrence. Data for each fit have been vertically shifted by arbitrary units. The color code represents the respective probabilities of *dE*. Note that there is an indication for a jump of *dE* (diamonds) of ~±20 meV close to *dE* = 0 for "on" and of ~ ± 40 meV for "off" intensities.

"Off" intensities show more clearly side bands at ±40 meV, which resembles energetically jumps between the lowest-intensity state at L and the high intensities at the N–S intersection. The energy 40 meV is close to ΔE = 50 meV, as assigned by the yellow line in Fig. 6.21 between the two blue lines marking the intermediate-intensity range. Intermediate intensities, on the other hand, smear out the behavior of "on" and "off" contributions since now state 2 comes into play, which is energetically between state 1 and the OH-related state.

We have suggested (see the discussion related to Fig. 6.19) that this energy difference might correspond to the energy difference of states 1 and 2. This assignment is supported by recent findings from deconvolution of PL spectra [Zenkevich et al., 2015] and the data shown in Fig. 6.13. Consequently, spectral fluctuations contain a convolution of small jumps in energy between two basically different electronic states and additional spectral fluctuations typically larger than ±40 meV. All three intensity ranges show a strong (narrow) distribution close to dE = 0, which is probably strongly influenced by noise. As indicated by the color code the magnitude of dE is increasing with decreasing intensity. In other words, the largest spectral jumps happen during "off" periods.

All findings are in good agreement with the distribution of intensity jumps ΔI in Fig. 6.9. The difference is that the absolute values of ΔI depend on the absolute intensity of the PL of a single QD, while dE varies only slightly from QD to QD. According to the correlation diagram of Fig. 6.15 S-type blinking is accompanied by large intensity jumps. However, these jumps do not show up as clear side bands in Fig. 6.9, which might, however, be due to the smaller occurrence of S-type blinking in comparison with the other blinking types. An increase of large intensity jumps ΔI is observed in the presence of OH groups (see Fig. 6.9) which do not play such an eminent role in the presently described QDs. For this reason we do not observe such prominent side bands for dE. Since the contribution of these OH-related states varies from QD to QD, we also obtain a considerable variation of the relative contributions of the spectral jumps (see Fig. 6.24, right).

The conclusion is that small spectral fluctuations (also containing noise-related contributions) occur at all intensities, but large spectral jumps are predominantly observed during "intermediate" or "off" intensities. The broad distribution of spectral jumps dE observed for intermediate intensities is partly artificially imposed since we collect all intermediate states in the analysis which exhibit, according to Fig. 6.20, a distribution of PL energies in the order of also (40–50) meV across the total intensity range. If the sensitivity would be high enough we would probably observe the same features as for intensity jumps, namely that spectral jumps among intermediate intensities occur with relatively small probabilities.

Summarizing, we find clear indications that the states 3 (L), 1, and 2 are shifted to the red with respect to the OH-related state by about −40, −20, and −10 meV, respectively. A more exact energy resolution is hampered by two effects, namely noise fluctuations in the order of 10 meV and spectral diffusion of energies in the same order of magnitude.

Finally, we would like to comment on spectral fluctuations in the S range of PL intensities. These fluctuations are included in Fig. 6.24 in the analysis of "on" intensities, but according to the relatively low densities of these states (see Fig. 6.20) they do not contribute significantly neither to the spectral nor to the intensity jump distributions. However, as can be directly seen from Figs. 6.19 and 6.21 we find that in this intensity range $\sigma \approx 30$ meV is the narrowest distribution across the total intensity range (except in the lowest L range). We would like to note that intensity jumps in the S range are quite large, which seems at first glance to be in contradiction to the observed small range of spectral fluctuations. However, we like to remind that the S range is governed by different population rates, while the same electronic state is emissive (constant PL decay time). Thus $\sigma \approx 30$ meV corresponds to the convolution of homogeneous width and the remaining spectral fluctuations in this state with the highest PL intensity. Taking into account recent estimations of the homogeneous width at 300 K [Cui et al., 2014], spectral fluctuations are negligible for the highest intensities. We therefore conclude that spectral fluctuations take predominantly place (i) in small jumps among intermediate intensities with different electronic energies and (ii) in large jumps following an "off" period, as has been reported previously [Neuhauser et al., 2000].

To summarize, large spectral jumps occur predominantly passing through a (nearly) "off" intensity, while spectral jumps occurring during periods of "on" intensities are on average by about a factor of 2 smaller (see also Fig. 6.24, right). Moreover, the clear 1:1 relation between energies and PL intensities allows relating energy fluctuations with (intensity-dependent) "on" times shown in Fig. 6.7. From this comparison emerges that small spectral jumps occur on timescales by at least a factor of 2 faster than for large spectral jumps. The distributions of spectral jumps do not depend significantly on the type of QD (and embedding matrix) but the diffusion dynamics do. This finding is identical with that for intensity jumps.

6.2.3 Modeling PL Intensity and Lifetime Blinking

After having discussed blinking dynamics in relation to PL intensities, decay times, and energies we are ready to set up a comprehensive blinking model. We would like to emphasize that such a model is limited to the presently investigated CdSe/ZnS QDs. However, various features will be also valid for other types of colloidal QDs. As we have outlined in Chapter 4 several basically different blinking models have been developed and it might be difficult to rely only on one of them. Basically two alternative approaches are at hand: (i) the charging model via Auger-type processes [Efros and Rosen, 1997] and (ii) the multiple recombination center (RC) model [Frantsuzov et al., 2009]. While the charging model has been questioned recently [Zhao et al., 2010; Rosen et al., 2010], other reports demonstrate the validity of this model [Galland et al., 2012] or show evidence of a combination of different blinking types [Galland et al., 2011]. Previously, Gómez et al. (2009) presented also a non-power-law-type blinking, assuming three relevant states, including a positively charged trion. Very recently Osad'ko et al. (2015) reported a complex model combining these two types of blinking. However, the underlying experimental results do not take into account sufficient experimental parameters such as PL decay times, PL energies, and the influence of the environment, including ligands to substantiate the ambitious model. In the following paragraphs we prefer to present a recombination-type model which we consider to be most appropriate in the presence of strong disorder as manifested by a broad distribution of intermittent intensities and PL decay times. The basic assignments and identifications of blinking classes are given in Section 6.2.1.4. We would like to emphasize that applying a charging model implies that charges are photogenerated and the PL of the subsequently excited exciton is quenched. This approach will result in well-defined clear "on-off" intensities. Our model follows the idea that exciton excitation is followed by intermittent charge localization, which creates states which are still emissive to a certain degree.

Most surprising in the present investigation is the appearance of blinking at high PL intensities in combination with a constant PL decay time, the so-called *S*-type blinking. A comprehensive illustration of the *S*-type blinking behavior is shown in Fig. 6.26. As discussed recently, it can be explained by an intermittent population from energetically higher states which involves the presence of hot electrons [Schmidt et al., 2014]. The existence of such a "bypass" is supported by the comparison of absorption and PL excitation spectra in Fig. 6.27. Here we find (as we did for all types of CdSe QDs) that the excitation into high-energy exciton states results in a reduction of the PL QY compared to excitation into the lowest one. Two principally different alternatives exist upon high-energy excitation: Excited electrons (or holes) (i) relax radiationless to the lowest exciton state, arriving as hot electrons (holes) from where they can escape (right scheme in Fig. 6.26), or (ii) immediately relax radiationless to the ground state (left scheme in Fig. 6.26). Statistic fluctuations between these basically different routes will result in variations of the population routes to the emitting and blinking states. At the same time PL decay dynamics and energies remain unaffected. We call these fluctuations *S*-type blinking since it appears experimentally as intense intensity spikes (S) in the course of a blinking time trace.

An outstanding observation is that the distribution of clustered PL intensities in the *S* range behaves completely different from the ones in the *I* and *L* ranges. While the density of intensities in the *S* range is high at low intensities (upper end of the *N* range), it scales down to isolated intensities (spikes) at the high-intensity range of the respective distribution. That this is not caused by the clustering algorithm can be seen from the appearance of intensities spikes also in conventional binned time traces in Fig. 6.1a.

Moreover, there is a clear correlation of intensities and related "on" times in the *S* range: Rare high intensities correspond with short "on" times (see Fig. 6.26) compared to low intensities close to the *N* range. This observation might be explained in a detailed model for

S-type blinking, also shown in Fig. 6.26, which needs a kind of switch closing frequently the radiationless bypass most efficiently for short times (left scheme) and less effectively for longer "on" times (right scheme).

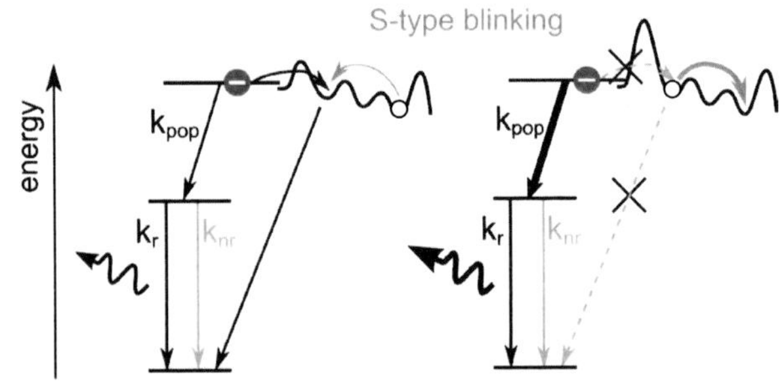

Figure 6.26 Schematic presentation of varying (populating) excitation pathways responsible for *S*-type blinking. Two limiting situations for hot-electron trapping are shown. Left: Low tunneling barrier for hot-electron trapping followed by radiationless deactivation. Right: High tunneling barrier. The open circles dots mark the fluctuating population within the multilevel energy landscape of centers blocking the escape of hot electrons. Equilibrated population of low-energy blocking states results in long "on" times for *S*-type blinking and vice versa.

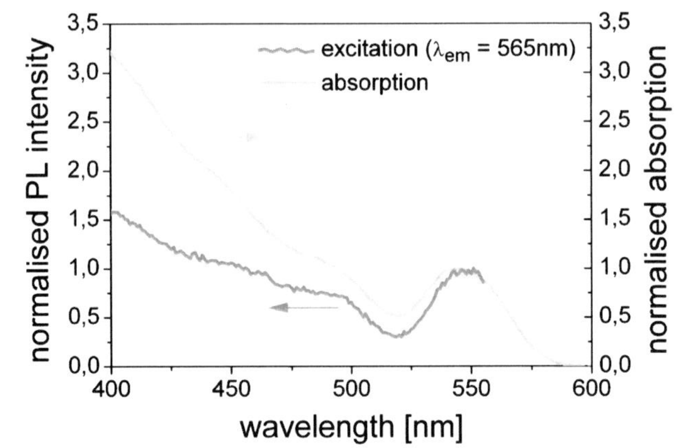

Figure 6.27 Absorption and PL excitation spectra of TOPO-capped CdSe/ZnS QDs in toluene. PL emission is centered at 565 nm. PL has been normalized to the maximum of the lowest absorption band. The PL quantum yield is reduced by up to a factor of 2 upon high-energy excitation.

"Closing" corresponds to an increased population of the emitting state(s) and thus increases PL intensity (but a constant PL decay time). This scheme does not only apply to the strong emitting state but also applies to other states which are populated with intermittent efficiency. Indications for a similar process are discussed in Section 6.2.2.

Such a close relationship of switching time and effectiveness needs a hierarchy of processes, as suggested in Fig. 6.26, where it has been schematically realized via a distribution of two-level systems which are characterized by a hierarchy of energies able to accept a photogenerated electron. Such a hierarchy guarantees that the highest (but most effectively Coulomb blocking) level is less probably populated than the lower ones. Since all "on" times depend critically on the polarity of the surroundings, (partial) charges are probably involved. We speculated that blocking the radiationless bypass (for the hot electron) is realized via charging the manifold of two-level systems. Alternatively, variations of the local density of polarons will have similar effects as localized electrons. Only a finite small number (<6) of levels is needed to reproduce the observed density of states in the *S* range (see Fig 6.7). This finding corresponds nicely with the model of multiple RCs [Frantzusov et al.,2009; Schmidt et al., 2014].

We suggest that the switching mechanism in the bypass is a kind of Coulomb blockade in the sense that the excited (hot) electron can only escape from the excitation cycle to the RC(s) in the case of realization of appropriate conditions for (self-) trapping of the hot electron. That charges (or alternatively polarons) are involved is concluded from the observed strong dependence of switching times on solvent polarity and the influence of alternating and static external electric fields [Krasselt, 2015]. A similar influence within an electrochemical cell has been reported by Galland et al. (2011).

The timescale of the *S* blinking spans a larger range than the one of the *I* or *L* blinking. We suggest that this is related to tunneling among the distribution of two-level systems coupled to the RC. Via a blinking analysis applying only the lowest possible excitation energy it could in principle be checked if in fact hot electrons are mandatory to explain *S*-type blinking. Such an experiment has, to the best of our knowledge, not been performed.

As will be discussed in more detail in Section 6.3 several investigations have examined the influence of excitation energy on blinking statistics and report controversy observations. Knappenberger et al. report that the cutoff time of power-law-distributed "on" times shortens with increasing excitation energy [Knappenberger et al., 2007], while in the study of Stopel et al. no influence of excitation energy on blinking statistics was observed [Stopel et al., 2013]. These apparent differences might be related to the presence or absence of *S*-type blinking. In case *S*-type blinking is absent one might not expect an influence of excitation energy, since trapping of hot electrons is absent. In case *S*-type blinking is present excitation energy and thus the "temperature" of the hot electrons might control the bypass to a relaxation center. However, the question related to the microscopic nature of the RCs remains still open.

I-type blinking shows a rather flat distribution of "on" times. It is related to relatively small intensity and energy jumps basically among states 1 and 2.

"On" times of *L* blinking are considerably longer than the ones in the *I* and *S* ranges. However, "on" times in the *L* range are biexponentially distributed since there are also very short "on" times present. Experimentally it is difficult to assign these by up to 1 order of magnitude different "on" times to a well-defined PL intensity since the density of intensities is very high in the *L* range. The biexponentiality probably reflects the superposition of two completely different classes of emitting states, the one related to A_3 (state 3) which is predominantly present in this *L* intensity range (see also the discussion in Section 6.2.2) and the other ones related to A_1 (state 1) and A_2 (state 2) contributions. The latter we have recently assigned to two types of trapped hole states (A_1 and A_2) [Schmidt et al., 2014]. Basically, there is a high probability for small intensity jumps between zero intensity and probably state 3 (related to A_3). Since A_3 is the dominant component in the *L* range, we assign the long "on" time to this state.

Recently, we have assigned this state to the strongly (due to hole trapping) quenched excitonic state [Schmidt et al., 2014]. However, we cannot completely exclude that this intensity belongs to an Auger-type quenched state, as has been suggested previously [Nirmal et al., 1996; Kraus et al., 2005; Peterson and Nesbit, 2009; Rosen et al., 2010; Amecke and Cichos, 2011]. However, there is strong evidence that our assignment is correct, since we observe a partly blue shift of PL related with this low intensity. Moreover, intensity jumps from state 3 to zero indicate that there is also a completely quenched state present which might be related to the Auger-type quenching process. If this is the case direct quenching and thus blinking processes originate from an already extensively quenched exciton state to a completely quenched state. Taking this suggestion a step further, zero intensity might appear as a consequence of detrapping the hole into the QD core, enabling Auger-type quenching processes. A schematic view of all three blinking types is shown in Fig. 6.28.

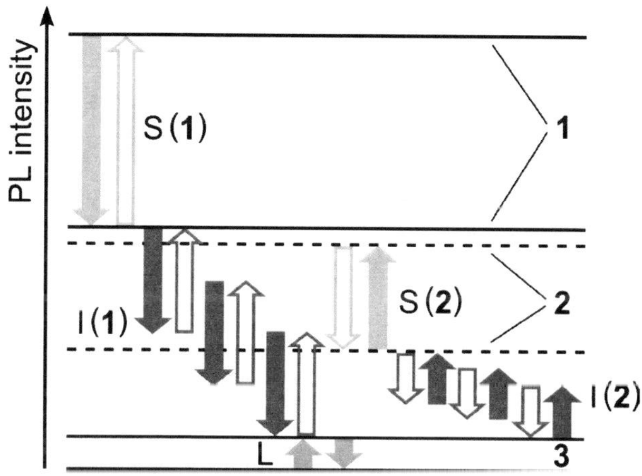

Figure 6.28 Schematic presentation of the distribution of blinking events in the intensity ranges *L*, *I*, and *S*. We assume two principally different electronic states (1) (solid horizontal lines) and (2) (broken horizontal lines) which are subject to the same blinking dynamics. *S*-type blinking—*S*(1) and *S*(2)—contains large intensity jumps across the respective intensity range, while *I* type blinking takes place predominantly via intermediate intensity jumps. The upward and downward intensity jumps in this approximation are not completely symmetrically with respect to their probability, as indicated by solid and open arrows. This is possible due to the presence of a third pathway via the *L* blinking connecting state (3) and the zero intensity level. *L*-type intensity jumps are closely spaced. An assignment of the symmetry with respect to up and down jumps is not feasible.

Of central importance is the finding that every PL intensity is related to at least two differently decaying states, which causes an additional type of blinking, namely lifetime blinking [Galland et al.,

2012]. In the *I* range, this lifetime blinking occurs monoexponential predominantly among states 1 and 2.

We have shown that the PL decay times τ_{micro}, absolute PL energies, spectral jump widths, and relative intensity jumps do not depend significantly on the matrix (PVA or PS), whereas τ_{macro} (which in fact corresponds to "on" times) depends strongly on the embedding environment. Most importantly, "on" times can now be uniquely assigned to PL decay times and energies. The independence of spectral diffusion on solvent properties has previously been reported [Gómez et al., 2006] and can be now understood in the light of our elaborated investigations. The strong dependence of the "on" times on the properties of the surrounding (such as solvent, substrate, and ligands) has been demonstrated in a whole bunch of publications [Issac et al., 2005; Cichos et al., 2007; Krasselt et al., 2011; Issac et al., 2012; Schmidt et al., 2012; Schmidt et al., 2014].

According to our findings, this implies that the dynamics of blinking reflect the properties of the surrounding. Since "on" times are not (mono-) exponentially distributed we make the conjecture that there are small fluctuations in a bath of two-level systems surrounding the QD governed by tunneling processes since dynamics are slow and nearly temperature independent [Shimizu et al., 2001]. We suggest two simplified variants for the nature of RCs. Firstly, electrons or holes of the excited QDs are trapped in the slowly fluctuating (tunneling) manifold of two-level systems. These two-level systems model the inhomogeneous (local) distribution of local charge densities at or close to the interface of the QD with the environment. These charge densities can in principle be explained by the combined dielectric properties of matrix and QD surface. They are controlled by the heterogeneous solvent and ligand shell (and can further be modified through external electrical fields) and influence the nanoscopic distribution of band-edge states of the QD. Kilina et al. (2009) have shown in recent calculations that band-edge-state energies depend critically on the type of solvent. We name this approach the *self-trapping charging model*. Secondly, since ligands have also a considerable influence on blinking dynamics (vide infra in Section 6.4), fluctuations in ligand positions (binding) on (to) the QD surface (tunneling) dynamics will also considerably modify band-edge energies, as has been calculated recently [Kilina et al., 2009]. Position-dependent mobility of ligands has also recently been proposed [Voznyy, 2011; Voznyy et al., 2014]. Naturally, formation dynamics of ligand bonds will depend on the dielectric environment [Schapotschnikow et al., 2009]. We name this approach the *ligand tunneling model*. Probably a realistic assumption is a combination of these two models.

As a remark aside we like to point out that other factors such as the interaction with OH groups impose both (well-defined) new exciton states and blinking dynamics, which, however, we do not want to discuss at this place since they do not impose qualitatively new aspects. Nevertheless, they are important since the related interactions might suppress blinking [Fomenko and Nesbitt; 2008; Odoi et al., 2009].

We suggest that the results of the CPA approach can be cast on the threshold-based one in the following way. The longest "on" times at high PL intensities (in the *S*-type blinking regime—typically 30 ms in PS and 10 ms on PVA) can be identified with the truncation time of the "on" time power law found in the threshold-based analysis, while the long "on" times at very low PL intensities (in the *L*-type regime—typically (50–100) ms in PS and 6 ms on PVA) can be identified with the truncation time in the threshold-based "off" time analysis. The latter identification seems to be reasonable since a threshold set at the lowest intensity level will automatically identify low-intensity "on" times as "off" times. The question as to why cutoffs for "on" times are always faster than for "off" times remains.

At least two open questions remain:

- Why is the timescale of fluctuations in the submillisecond to several hundred millisecond range? Partly, this might be due to the limited time resolution both of the threshold- and CPA-based approach. Different techniques indicate much faster fluctuations [Verberk et al., 2002; Fomenko and Nesbitt, 2008], which, however, will not discard our present discussion and interpretation.
- It has been shown that "on" time as well as "off" time fluctuations are light induced [Baker et al., 2009]. In Section 6.3.1 we demonstrate the influence of excitation rates on blinking, especially on "on" times. However, such experiments are not conclusive to identify the impact of optical excitation on blinking dynamics uniquely. No doubt, "off" to "on"

relaxation takes place also in the dark [Cichos et al., 2004; Chung et al., 2004; Schuster et al., 2007]. Additionally, "off" time distributions can be influenced by variation of excitation power [Cichos et al., 2004; Schuster et al., 2006] (see also Section 6.3).

Finally, one of the most intriguing questions about the intrinsic nature of the emitting states is still not yet completely answered. We identified at least three different states (named 1, 2, and 3). These states have been (somewhat modified) identified also in ensembles of CdSe/ZnS QDs in liquid solution, on substrates, and embedded in polymers [Petrov et al., 2006; Zenkevich et al., 2015]. State 1 is characterized by the longest PL decay times of (14–22) ns and exhibits the highest PL intensity. It has to be assigned either to the (band-edge) exciton or to a strongly emitting trapped hole or surface state. State 2 is blue-shifted by about 10 meV with respect to state 1 but with a shorter PL decay times of (5–10) ns accompanied by a radiative rate about two times higher as the one of state 1. We have recently assigned this state also to a trapped hole state [Schmidt et al., 2014]. Its radiative and nonradiative rates are faster by about a factor of 2 as compared to state 1. An alternative explanation for these two states might be related to an interchange of fine structure levels (see Chapter 2) and as has been recently suggested [Salman et al., 2009]. Considering a very similar explanation, well-defined relations among two states with nearly equal intensities (blinking suppression) have been described by Fomenko and Nesbitt (2008).

Both states are intermittent among themselves and among each other. They fluctuate between low and high intensities and correspondingly between fast and slow decay times. Energy shifts occur in parallel to lower energies with decreasing PL intensity by up to 50 meV, as is observed in form of spectral diffusion. Spectral diffusion as a total is constituted from these jumps among all these states and from additional fluctuations in a way that a well-defined intensity (state) is subject to fluctuations of its energy. We suggest that these combined shifts in intensity, lifetime, and energy are due to fluctuating interactions of (trapped) electrons and (trapped) holes [Schmidt et al., 2014]. The interrelation of states 1 and 2 with respect to their spectral features is schematically shown in Fig. 6.29.

Blue shifts in the range of 5 to 10 meV have been theoretically assigned to positively charged trions which are generated by trapping electrons [Troparevsky and Franceschetti, 2005; Califano et al. 2007] but with a decreased radiative rate, which is opposite to what we find for state 2. Strong red shifts of 10 to 40 meV have been predicted theoretically [Troparevsky and Franceschetti, 2005; Califano et al. 2007] and observed experimentally [Shimizu et al., 2002; Fernée et al., 2009] for negatively charged trions (trapped holes). Radiative rates have been calculated to be similar to those for the uncharged exciton [Califano et al. 2007]. PL decay times are (in the case of CdSe/CdS QDs) shortened moderately [Spinicelli et al., 2009; Gomes et al., 2009] or drastically [Jha and Guyot-Sionnest, 2009; Cohn et al., 2013]. Taking PL decay times and energy (red shifts) into account we suggest that the intensities related to state 3 are a superposition of a negatively charged trion (very low PL energy shifted by 50 meV to the red) and to a lesser extent by a quenched exciton (at high PL energy; see Fig. 6.20, bottom left). Intermediate PL intensities (uniquely accompanied by PL energies and decay time) are assigned to fluctuations of relative distances of positive and negative charges (quantum-confined Stark effect), as we have discussed recently with respect to spectral fluctuations in Si nanocrystals [Martin et al., 2012].

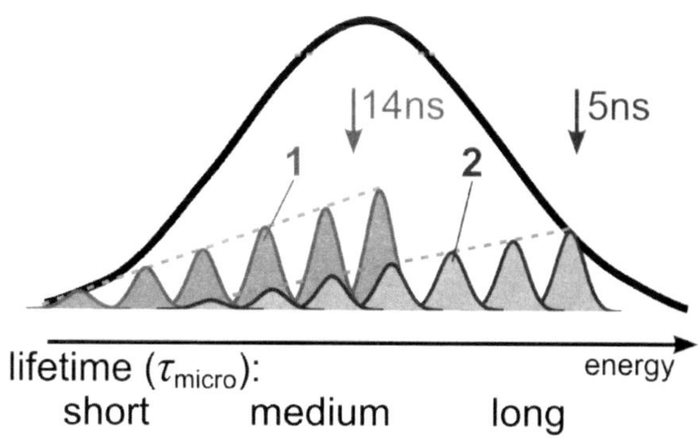

Figure 6.29 Schematic presentations of the energy distributions (with intrinsic FWHM) of state 1 and 2, assuming for a single QD. The respectively longest decay times of typically 14 ns and 5 ns are indicated. PL decay times (and intensities) decay (differently) toward lower PL energies. The overlay of the two spectral distributions results in the observation of a short decay at high and a longer one at lower energies, both decreasing with further decreasing energy.

State 3 is with a decay time of (1–2) ns unambiguously only detected at very low intensities. However, it has contributions both at low and high energies (see also Petrov et al. [2006] and Zenkevich et al. [2015]) which are of nearly equal magnitude according to those experiments. This makes us believe that there are in fact two different short-lived states. One of them is the result of the continuous Stokes shift of states 1 and 2 to low energies due to a (quasi-) charging process. The other one is in fact a high-energy state, which could be either a strongly quenched exciton or an Auger-type state with a localized electron and a hole in the remaining CdSe core.

Though we are (not yet) able to unravel all blinking features our model provides a most comprehensive approach to intermittency, including intensity, energy, and lifetime blinking under complex and heterogeneous environmental conditions. Somewhat opposite to many statements in the literature which have their origin in solid state physics of semiconductors, we followed more closely a molecular approach since many features of QD blinking are similar to those observed for single molecules [Cichos et al., 2007]. Probably both approaches have their own merits. However, we believe that the strongly disordered interfaces (including shells, ligands, and embedding matrix) might be more adequately described by borrowing ideas from disordered organic materials.

6.3 Excitation Intensity and Energy-Dependent Blinking

6.3.1 Influence of Excitation Intensity

It is well documented that blinking dynamics might depend on excitation power [Knappenberger et al., 2007; Peterson and Nesbitt, 2009; Goushi et al., 2009; Gómez et al., 2009; Cordones et al., 2011a]. In case a dependency is found the most pronounced influence is observed for the cutoff time of the "on" time distribution [Peterson and Nesbitt, 2009], while "off" times are nearly unaffected. The remaining influence might be related to the fact that the most commonly used threshold-based analysis cannot in all cases correctly separate "on" and "off" times. For example, for very high thresholds, "on" times might be assigned to "off" times, resulting in an apparent influence of excitation power. Another observation is that cutoffs in "on" time distributions are shorter by up to 2 orders of magnitude when determined by confocal microscopy compared to wide-field microscopy. This can be readily explained (see later) by the fact that if cutoff times are shortened with increasing laser power this influence will be more pronounced with confocal excitation since the power/(excited area) is typically several orders of magnitude higher in comparison to wide-field microscopy. This implies that the power dependence has to be carefully analyzed with respect to the concrete experimental conditions. These circumstances might explain why there is some controversy in the literature.

In the following paragraphs we will describe via wide-field measurements the power-law dependence of CdSe/ZnS QDs capped by TOPO or a mixture of TOPO/hexadecylamine (HDA) ligands in two different environments, namely on a SiO_x substrate and embedded in films of the organic hole transport material *N,N'*-diphenyl-*N,N'*-bis(3-methylphenyl)-(1,1'-biphenyl)-4,4'-diamine (TPD) [Trenkmann, 2015]. Figure 6.30 shows a comparison of blinking time traces for the two materials, each at two different excitation powers.

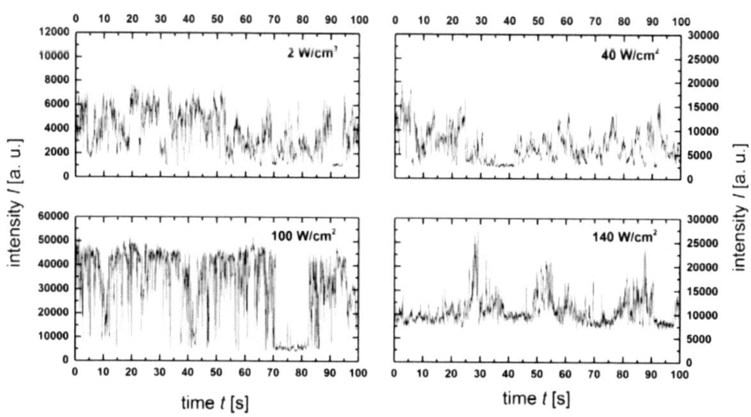

Figure 6.30 Excitation power-dependent blinking time traces of (TOPO/HDA)-capped CdSe/ZnS on SiO_x (left) and in a TPD film (right) at 300 K excited at 2.6 eV excitation energy.

It is immediately evident that blinking varies between different kinds of materials. Especially, blinking in TPD at high excitation power results in a kind of bunching of the "on" intensities. These differences become more evident when analyzing the power dependence of the blinking cycles (frequencies) between "on" and "off" intensities, which is given by the (identified) number N_{on} (N_{off}) of "on" and "off" times, respectively, and the related observation times $t_{\text{on},i}$ and $t_{\text{off},i}$ by

$$\text{cycles} = \frac{N_{\text{on}} + N_{\text{off}} - 1}{\sum_i t_{\text{on},i} + \sum_i t_{\text{off},i}} \tag{6.3}$$

and the "on" time fraction Φ_{on}

$$\phi_{\text{on}} = \frac{\sum_i t_{\text{off},i}}{\sum_i t_{\text{on},i} + \sum_i t_{\text{off},i}} \tag{6.4}$$

The power dependence of these blinking parameters is shown in Fig. 6.31.

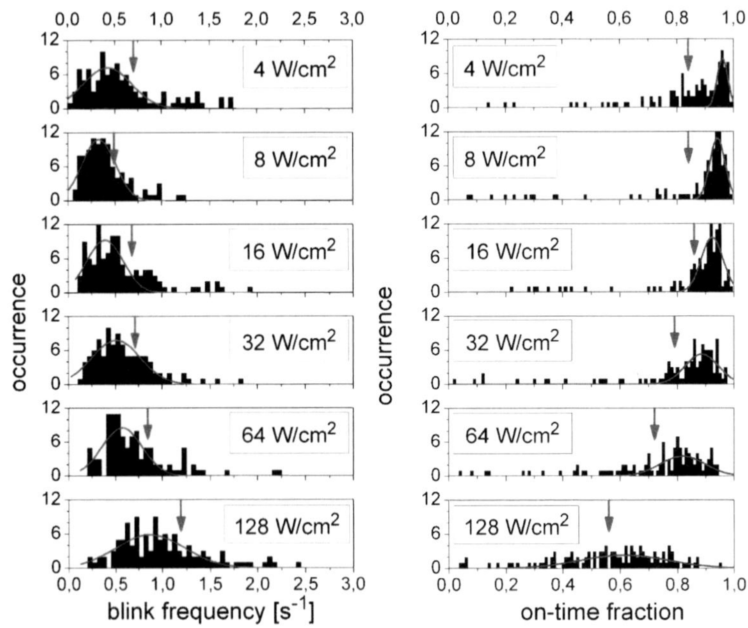

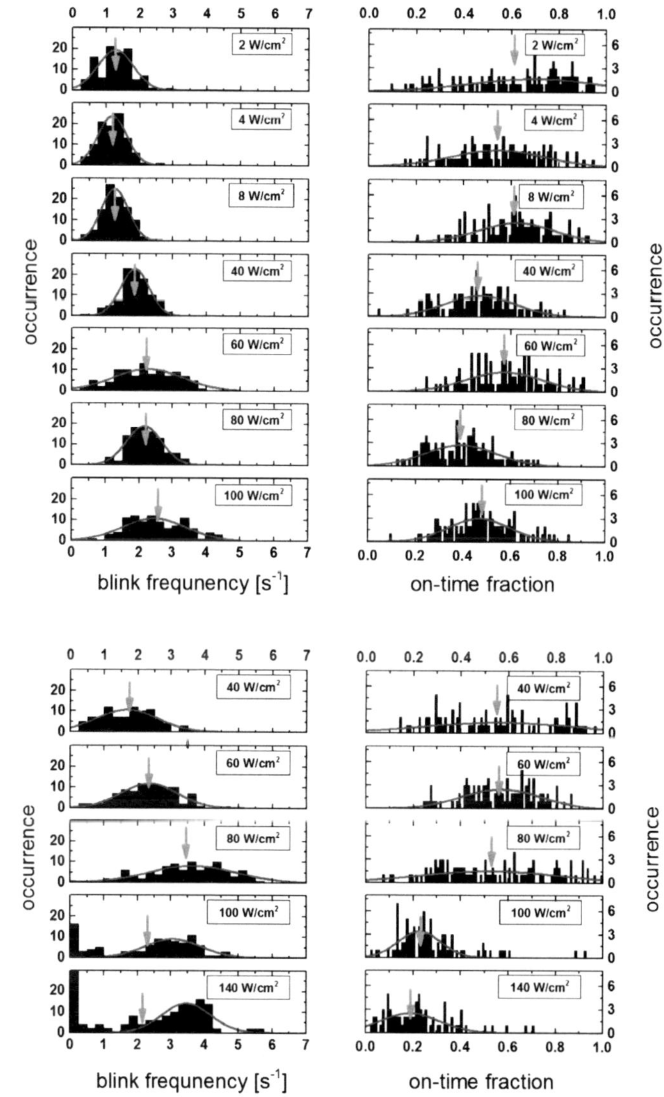

Figure 6.31 Dependence of blinking parameters, blinking frequency and "on" time fraction, on excitation power for three different QD/ligand systems. The curves are Gaussian fits to the respective distributions, while the arrow marks the arithmetic average. Top: TOPO-capped CdSe/ZnS on SiO_x. Middle: (TOPO/HDA)-capped CdSe/ZnS on SiO_x. Bottom: (TOPO/HDA)-capped CdSe/ZnS in TPD. Left column: Blinking frequency (see Eq. 6.3). Right column: "On" time fractions Φ_{on} (see Eq. 6.4).

We observe initially an increase of blinking activity and a broadening of the related distribution with increasing excitation power. Remarkably, at about 100 W/cm^2 we observe in the case of the TPD matrix at least two distribution peaks. At the same excitation power we find a sudden decrease of the "on" time fraction, while the latter one shifts continuously to slightly lower values in the case of SiO_x.

A conventional threshold-based analysis has been applied to all sets of data to obtain "on" and "off" times as a function of excitation power. The results are shown in Figs. 6.32 and 6.33. Since PL of QDs is weaker in TPD, excitation power starts only at 40 W/cm^2. "Off" times can be described by a truncated power law (Fig. 6.32, bottom). The power-law exponent is slightly increasing (decreasing) on SiO_x (in TPD), while the cutoff times are slightly decreasing [Trenkmann, 2015]. However, both effects are small and might be related to incorrect threshold settings.

"On" times are more sensitive to excitation power variation, as shown in Fig. 6.32 (top). "On" times have to be described at least on SiO_x by Eq. 6.2 that is a superposition of a truncated power law and an exponential function (for details see Section 6.2.1). The presence of the exponential function is less pronounced in TPD. It was not possible to get a satisfying and systematic dependence of the parameters of Eq. 6.2 at all excitation power levels [Trenkmann, 2015] since there are too many parameters to be fitted on a rather smooth power dependence. Moreover, the QDs are capped by a mixture of ligands. As ligands have a noticeable influence on blinking [Gómez et al., 2006] (see also Section 6.4), this increases the complexity of the blinking behavior. Additionally, we observe in the case of TPD the inset of a second type of kinetics of the power dependence in Fig. 6.31, which adds another degree of complexity. However, qualitatively, "on" times become overall shorter with increasing laser power for SiO_x, which is not immediately evident as in the case of TPD. To get a more quantitatively reliable fit we kept the power-law exponent constant at 1.4 during power variation, since a constant exponent has been reported in most of the published data [Bruhn et al., 2012; Knappenberger et al., 2007; Gómez et al., 2009]. We observe for both matrices that the cutoff time (rate) is becoming shorter (higher) with increasing laser power. As can be seen from

Fig. 6.32 (top) the dependence is more pronounced in the case of SiO_x.

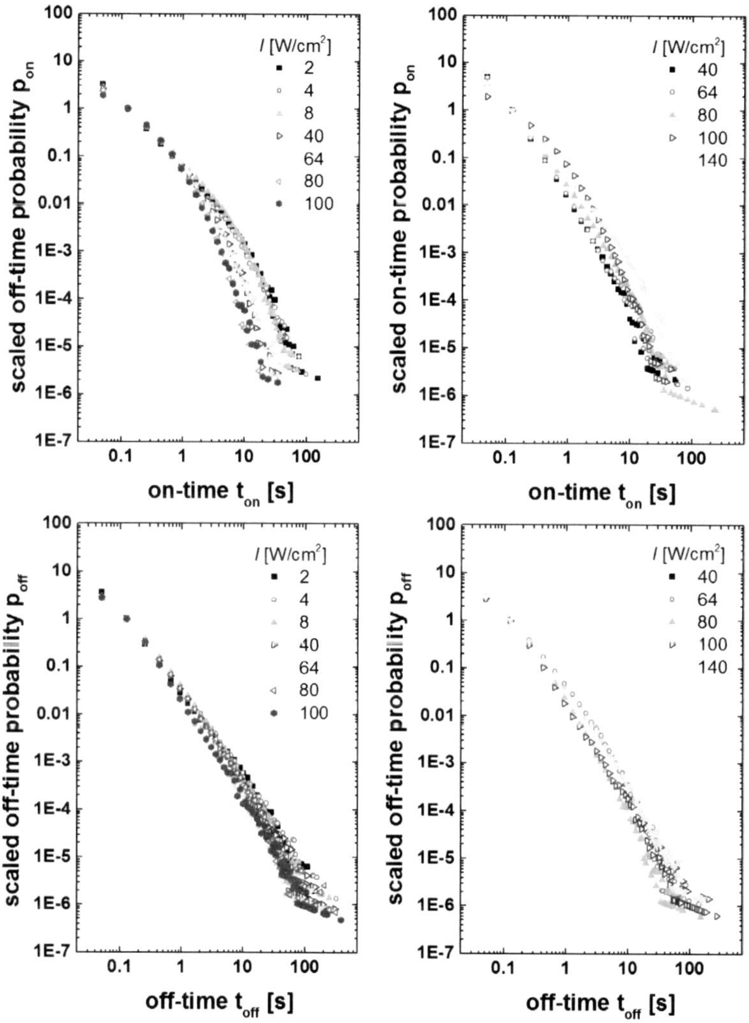

Figure 6.32 Analysis of blinking of (TOPO/HDA)-capped CdSe/ZnS on SiO_x (left) and in TPD (right) for "on" times (top) and "off" times (bottom).

Blinking data and analysis for TOPO-capped CdSe/ZnS QDs on SiO_x are shown in Fig. 6.33. As will be discussed in Section 6.4 TOPO-capped QDs show "on" time blinking according to Eq. 6.2, which complicates analysis since Γ_{on} and γ_{on} cannot easily discriminated against each other and the relative amplitude ratio B/A of the exponential versus the power-law part is changed by a factor of 2 upon excitation power variation. This results in a large uncertainty of α_{on}. Nevertheless, Γ_{on} (and γ_{on}) increases with increasing excitation power. Again, "off" time dynamics are much less dependent on excitation power.

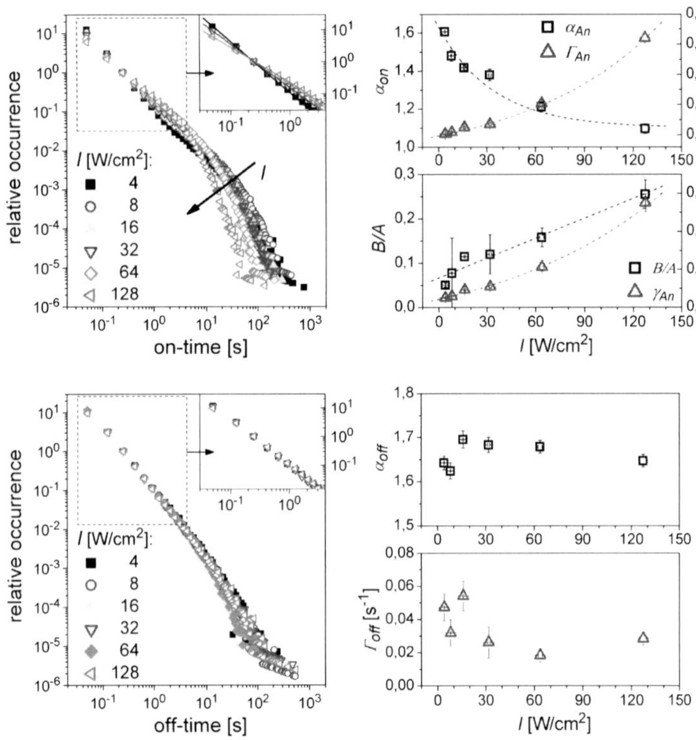

Figure 6.33 Power dependence of blinking of TOPO-capped CdSe/ZnS on SiO_x for "on" times (top) and "off" times (bottom). Blinking data are given on the left and the analysis according to Eq. 6.2 is given on the right.

To obtain a more quantitative dependence on cutoff rates we plot in Fig. 6.34 the truncation (cutoff) rate as a function of excitation power both on a linear and on a logarithmic scale. Though the range of data is limited there are three regions of the dependence of Γ on the absorption rate. These QDs on SiO_x are tentatively assigned (indicated by three dashed lines) to an absorption rate–independent ($m = 0$), linear (m = 1), and quadratic (m = 2) behavior. Two of these three dependencies have been reported in the literature [Knappenberger et al., 2007; Wang, et al., 2006; Early and Nesbitt, 2013]. A power-independent behavior has not yet been reported. The quadratic dependence is an indication for the inset of two-photon processes, for example, biexciton excitation [Peterson and Nesbitt, 2009; Early and Nesbitt, 2013]. Usually this latter process will dominate in experiments with confocal detection, since the typical power in such experiments is higher than 100 W/cm^2. The absorption rate dependence does not show a clear inset of two-photon processes. The interpretation of these findings will be explained at the end of this section.

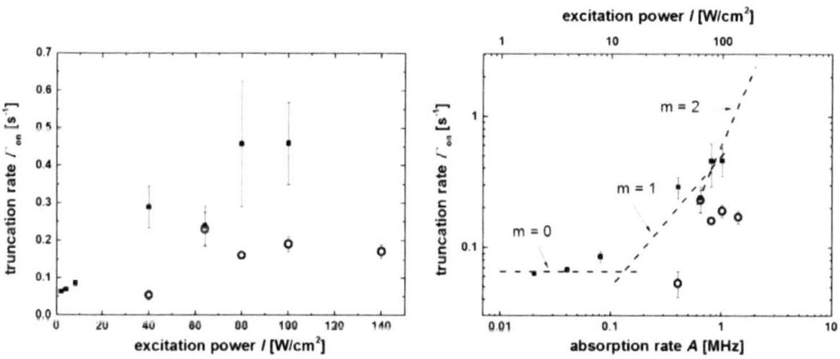

Figure 6.34 Power dependence of the cutoff rate Γ (left) of "on" times for CdSe/ZnS (TOPO/HDA) on SiO_x (solid squares) and in TPD (open circles). The same data on a double-logarithmic scale (right) where the excitation power (on the top scale) has been converted into the absorption rate A (corresponding to excitation power multiplied by the absorption cross section, divided by excitation energy and corrected for the illuminated area of the sample). The dashed lines mark the order of absorption rate dependence: independent (m = 0), linear (m = 1), and quadratic (m = 2).

The results for excitation power dependence of blinking can in agreement with the literature [Early and Nesbitt, 2013] be summarized as follows:

- Blinking activity increases with excitation power accompanied by a decrease of the "on" time fraction by up to a factor of 2.
- Blinking in TPD at elevated power is accompanied by the set-in of a second blinking process at intermediate laser power. Moreover, at the same power level α_{on} increases from about 1.4 to 2.0 at the highest power level.

- "Off" times are almost not influenced by the power variation. Small variations might be due to insufficient threshold discrimination between "on" and "off" times.
- "On" times show a noticeable shortening of the cutoff time with increasing excitation power on SiO_x. Such a shortening is much less pronounced in TPD.
- In the case of superposition of a truncated power law and a superimposed exponential distribution function (see Eq. 6.2) both parts depend differently on excitation power in the case of SiO_x, which can, however, not reliably be discriminated at all power levels. Fixing α_{on} to a constant value, for example, $\alpha_{\text{on}} = 1.4$, results in an increase of B/A.
- α_{on} becomes smaller with increasing excitation power. This is probably an artifact related to the increase of blinking activity during which "on" times become shorter, not being any more properly resolved due to long binning times.
- The main difference between the two matrices SiO_x and TPD is the appearance of two blinking processes clearly separated in time for TPD (accompanied by a switch-over from a truncated power law to a truncated power law superimposed by an exponential function) and the reduced influence of the excitation power on the cutoff time of "on" times.

Comparison with the literature [Knappenberger et al., 2007; Peterson and Nesbitt, 2009; Cordones et al., 2011a; Early and Nesbitt, 2013; Bharadwaij and Novotny, 2011] shows a high similarity, though one has to keep in mind that most of the results are obtained with pulsed and/or confocal techniques which correspond to a higher-power regime. Goushi et al. (2009) find a dependence on the matrix and ascribe it to a different polarizability. Additionally, comparing results for TOPO- and TOPO/HDA-capped CdSe/ZnS shows that the concrete variation of blinking with laser power depends not only on the matrix but also on the type of ligands, as will be discussed in Section 6.4.

Our data also show that the increase of excitation power influences the fitting parameters in the sense that the increasing blinking activity is accompanied by shorter "on" (and "off") times. This will result in a reduced time resolution for short times as a function of the binning time. As a consequence the probability to identify short times is reduced, giving rise to an apparent lower power law. This effect is clearly seen in Fig. 6.33 (left). Surprisingly, "off" times, though on the same timescale, are seemingly not hampered by a decreased time resolution. This is a further indication that "off" times are almost not influenced by excitation power, which implies that the blinking-inducing microscopic process is not changed by the continued absorption of the QD. If the recovery from a dark to a bright state is predominantly light induced, as has been reported in the literature [Baker et al., 2009], the related recovery process does not depend significantly on excitation power.

Anyway, experiments show that a comparison of blinking dynamics under varying conditions needs the knowledge of excitation power. Such a comparison is also shown in Figs. 6.32 and 6.33 for the presently investigated QDs. Moreover, excitation power has to be scaled to the excitation rate of the QD, depending on the (size-dependent) absorption cross section. Generally speaking, power-law coefficients and cutoff times obtained by confocal techniques become smaller and shorter, respectively, compared to wide-field techniques.

6.3.2 Influence of Excitation Energy

Another possible parameter influencing blinking dynamics is the excitation energy. Already low-power excitation shows that the QY is reduced when exciting into energetically high QD absorption bands compared to excitation into the lowest band, as shown in Fig. 6.35 (and Fig. 6.27). Therefore it is intriguing whether this also influences blinking dynamics. Knappenberger et al. (2007) and Gómez et al. (2009) have reported an influence of excitation energy on the cutoff times, but Stopel et al. (2013) did not observe such effects.

We have investigated blinking TOPO-capped CdSe/ZnS QDs on a SiO_x surface at three different excitation wavelengths with a wide-field microscope under continuous wave excitation and a power density of 16 W/cm^2, as shown in Fig. 6.36. One of the excitations takes place in the lowest absorption band, whereas the others populate the two next higher absorption bands, as indicated in the left part of Fig. 6.36. Excitation power has been experimentally adapted to the same excitation rate, taking the appropriate absorption cross section into account. The analysis of fit parameters according to Eq. 6.3 shows that only minor changes take place (upon decreasing excitation energy we find slightly decreasing power-law exponents, decreasing cutoff times, and an increasing exponential contribution). However, as the direct comparison of the blinking data shows, changes are not really significant.

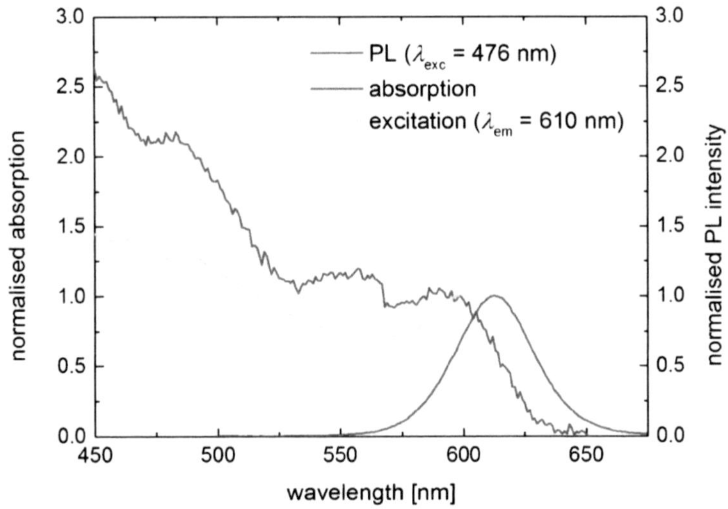

Figure 6.35 Comparison of PL excitation (at λ_{em} = 610 nm) and absorption spectra of TOPO/HDA-capped CdSe/ZnS QDs in toluene. Additionally, the PL spectrum excited at 476 nm is shown by the red line. PL excitation has been normalized to the intensity of the lowest absorption band. This comparison is similar to the one shown in Fig. 6.27 for a slightly smaller and differently ligated QD.

Our experimental conditions are very similar to those of Knappenberger et al. (2007) but the results are different in the respect that we observe (for slightly smaller QDs) no dependence on excitation energy but only a dependence on excitation power. One important difference is that we observe a coupling to specific silanol (OH) groups, which is absent in the reported results.

No influence on excitation wavelength has been reported by Stopel et al. (2013) and Ma et al. (2012). However, Stopel et al. (2013) have excited only into the energetically lowest first excitonic absorption band, where Knappenberger et al. (2007) found the blinking to be independent of the excitation energy. But Knappenberger et al. (2007) have not adjusted the excitation power according to the strongly varying absorption cross section. These limitations induce some uncertainties with respect to the reported results. Moreover, part of their analysis shows deviations from a truncated power law which could have been analyzed by Eq. 6.3. This is not surprising since they have performed the experiments in PVA, which often shows very strong deviations from a truncated power law [Schmidt et al., 2014].

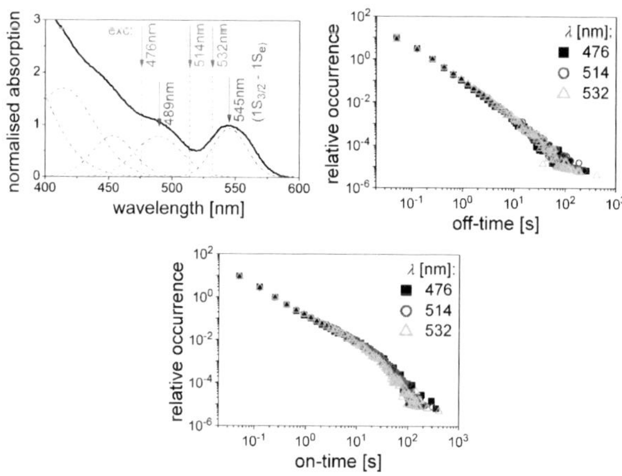

Figure 6.36 Comparison of excitation energy-dependent blinking dynamics of TOPO-capped CdSe/ZnS QDs. Top: Deconvolution of the absorption spectrum of an ensemble of QDs in toluene by five Gaussian-shaped absorption bands. Experimental excitation wavelengths are indicated by green arrows. Middle: Distribution of "off" times as a function of excitation wavelength. Bottom: Distribution of "off" times as a function of excitation wavelength.

6.3.3 Excitation-Dependent Blinking Model

In Fig. 6.37 we present a scheme which takes into account the most important excitation routes which might influence blinking differently with respect to the excitation rate k_0 and the excitation energies. The basic underlying assumptions are that photoexcitation results in charging of the QD via hole and/or electron trapping [Schmidt et al., 2014]. In this way, the QD remains emissive but charged. The overall behavior can be described within the framework of multiple RC_i's [Frantzusov et al., 2009]. Additionally,

Auger processes might quench the PL completely or at least very strongly [Amecke and Cichos, 2011].

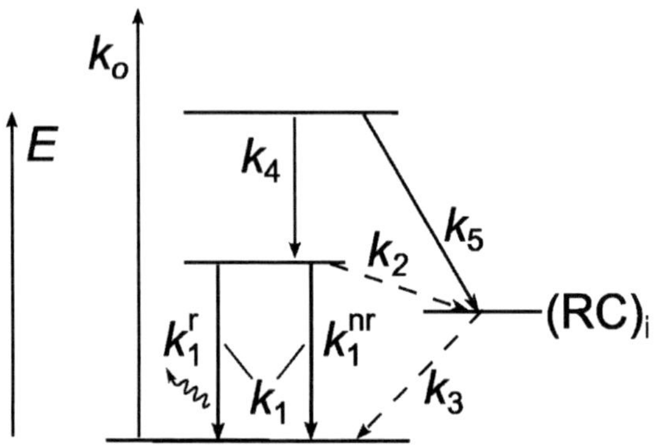

Figure 6.37 Excitation scheme assuming besides the ground state two electronically excited states with related deactivation rates. For definition of rates see text. Multiple recombination centers (RC)$_i$'s are assumed to control blinking dynamics by creating dark or dim emissive states. Naturally, all rates are in fact broadly distributed. In this model, k_2 and k_3 have to be identified with inverse "on" and "off" times, respectively.

The PL QY depends obviously on the kind of excited QD state, as shown in Figs. 6.27 and 6.35 and by Hoheisel et al. (1994) via a comparison of absorption and PL excitation spectra. The specific excited states which might have different barriers for activation of (specific) RC$_i$'s which effect blinking phenomena (especially the cutoff rate of "on" times when varying excitation energies) [Hoheisel et al., 1994; Cordones et al., 2011a; Knappenberger et al., 2008].

According to the excitation scheme in Fig. 6.37 the population of energetically high electronic states is deactivated by two competing decay rates k_4 and k_5. In case $k_4 >> k_5$ we do not expect a dependence of blinking on excitation energy. Accordingly, RC$_i$'s become effective via the combined deactivation cycle ($k_4 + k_2$).

The increase of excitation power (rate) basically shortens the cutoff of "on" times [Bruhn et al., 2014; Cordones et al., 2011a]. "On" times correspond to k_2^{-1}. Assuming for the moment the simplest case that we find only one "on" time we borrow ideas from the model of fluorescence fading [Fischer et al., 1983]. Roughly speaking, in case the excitation rate k_0 is comparable to or faster than the (slow) decay rate k_3 fluorescence dynamics is overcome by k_0 and a "fading" rate k_F is observed according to [Fischer et al., 1983]

$$k_F = k_3 + k_2 k_0/(k_1 + k_2). \tag{6.5}$$

The general procedure is very similar to the intensity autocorrelation function reported for single-molecule detection involving a (nonemissive) triplet state [Orrit and Bernard, 1990]. Transferring this (certainly oversimplified) model to QD blinking we have to assume a distribution of "on" times [Schmidt et al., 2014] whereby the longest "on" time can be identified by the cutoff time of an (apparent) power-law distribution. We identify this cutoff rate with k_2. With increasing excitation rate k_0 will replace successively the respective longest "on" time, resulting in shorter and shorter cutoff times, as experimentally observed. We identify the smallest cutoff rate Γ_{on} (k_2) with the excitation power–independent rate $k_2 = \Gamma_{on} \approx 7 \times 10^{-2}$ s^{-1} (see Fig. 6.34 for QDs on SiO_x), the smallest cutoff rate Γ_{off} with $k_3 \approx 3 \times 10^{-2}$ s^{-1} (see Fig. 6.33 and Trenkmann [2015]), and the excitation rate with $k_0 \approx 10^6$ s^{-1}. With these values and $k_1 \approx 5 \times 10^7$ s^{-1} we expect the excitation power–induced cutoff rate to be $k_F \approx 3.14 \times 10^{-2}$ s^{-1}, which is in the range of "on" time excitation power–dependent cutoff rates of larger than 7×10^{-2} s^{-1}. (For the selected excitation rate we observed a cutoff rate of 30×10^{-2} s^{-1}). Though these results are qualitatively in agreement with experimental findings, uncertainties in the effective absorption rate will have strong impact on quantitative results. This simple model explains qualitatively the independent and linearly dependent parts of the power dependence of cutoff rates shown in Fig. 6.34.

We like to remark that charging of QDs might also occur via accumulation of abundant charges in the matrix, which probably will leave blinking properties unaffected from excitation power and energy.

In case $k_5 > k_4$ we might observe changes in cutoff times for two reasons: (i) k_5 might be faster than the population limiting rate k_2 and (ii) excitation into elevated electronic states might open the access to additional (fast) RCs.

In conclusion we have shown that "on" time blinking does not depend on excitation power since wide-field excitation at low incident power levels does not interfere with intrinsic blinking dynamics.

6.4 Influence of Substrate Surfaces and Ligands on Blinking

There are ample reports in the literature that organic ligands influence optical energies [Smith and Nie, 2010; Brown et al., 2014], PL decay times [Odoi et al., 2007; Yuan et al., 2007; Fomenko and Nesbitt, 2008; Cui et al., 2014], photostability [Munro and Ginger, 2008; Hammer et al., 2007], QY [Ji et al., 2008; Boldt et al., 2011; Shen et al., 2015], blinking dynamics [Gómez et al., 2006], and spectral diffusion [Gómez et al., 2006]. This is not surprising since they saturate dangling bonds related to hole or electron states [Frenzel et al., 2007]. Detailed calculations show that even replacement of single ligands at specific surface sites influences noticeably QD energies and decay dynamics [Kilina et al., 2009 Voznyy, 2011; Gómez-Campos and Califano, 2012]. We have recently argued that these effects can be made visible by attaching dye molecules to the QD surface [Zenkevich et al., 2015]. In a solid state approach (see also Chapter 2) this is due to the extension (tunneling) of the excitonic wavefunction beyond the QD core [Blaudeck et al., 2008]. It is far beyond the present contribution to review even only the basic findings. Some more aspects will be presented in Zenkevich and Borczyskowski (2016) and Trenkmann (2015). Here we will concentrate on some not yet reported findings with respect to blinking in combination with the influence of the embedding matrix or supporting substrate. We suggest that the influence of ligands can in some respect be treated the same way as the influence of the dielectric properties of a solvent upon blinking [Cichos et al., 2007].

6.4.1 Influence of Substrate Modifications on Blinking

We recently reported that blinking related "on" times depend in some cases critically on the OH group concentration on SiO_x substrates [Krasselt et al., 2011]. The corresponding influence shows up as a pronounced monoexponential contribution superimposed to the otherwise broad (power-law-like) distribution of "on" times (Eq. 6.2). Similar observations are reported in the literature but have not been discussed with respect to the related origins [Bharadwaij and Novotny, 2011]. The manifestation of such effects depends critically on the type of ligand and the surface treatment, as qualitatively shown for some selected blinking time traces in Fig. 6.38 for TOPO-capped CdSe/ZnS QDs.

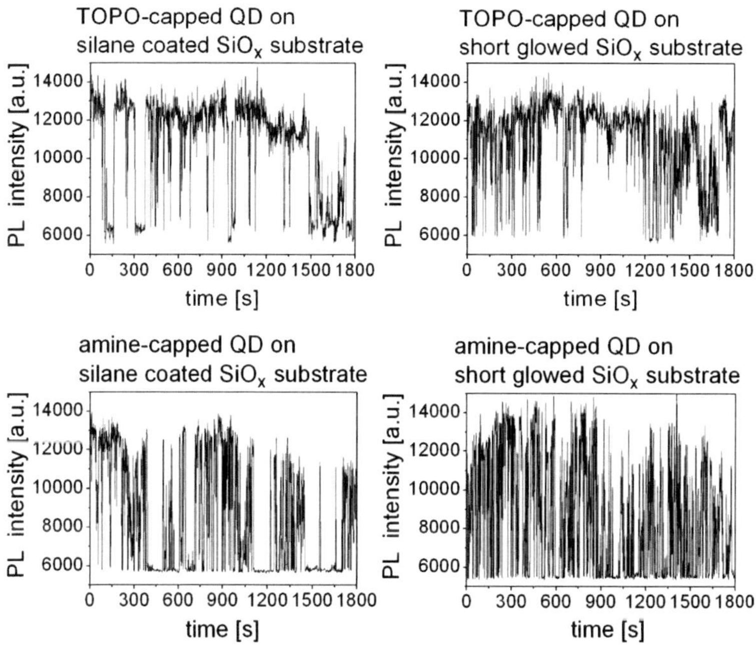

Figure 6.38 Blinking time traces for CdSe/ZnS QDs capped by TOPO (top) or long-chain amines (bottom) on differently treated SiO_x substrates. Left column: Octadecylsiloxan (ODS)-coated substrates. Right column: At 700 K for 10 min heated substrates to remove OH groups [Krasselt et al., 2011].

It is immediately evident that blinking is considerably influenced, as qualitatively shown in Fig. 6.39, by evaluation of the "on" times, while the "off" times (not shown here) remain nearly unaffected by surface modifications of this kind. Figure 6.39 (left) shows that "on" times have to be described by Eq. 6.2. However, the monoexponential contribution is strongly suppressed by lowering the OH group (silanol) concentration on the surface. This finding has been explained by formation of hole trap states [Krasselt et al., 2011]. Under identical conditions amine-capped CdSe/ZnS QDs do not show the monoexponential contribution, as can be seen from Fig. 6.39 (right).

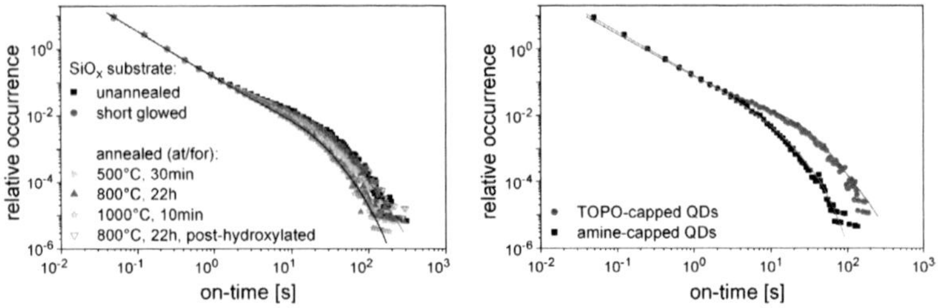

Figure 6.39 "On" time distribution for TOPO-capped CdSe/ZnS for various surface treatments (left). Solid lines are fits according to Eq. 6.2. Comparison of "on" times for TOPO-capped and amine-capped QDs under otherwise identical conditions (right).

To understand the influence of surface treatments on QD blinking, we will take a closer look at the chemical nature of a SiO_x surface. Figure 6.40 shows a scheme of the surface coverage with OH groups or a silanized one. OH groups add a polar contribution to the surface. They can be removed via annealing at high temperatures [Krasselt et al., 2011]. Such OH groups (silanols) are not homogeneously distributed across the surface but form clusters, as will be also investigated and discussed by Taeuber and Borczyskowski (2016). Silanized surfaces, however, are expected to remove OH groups and are much less polar.

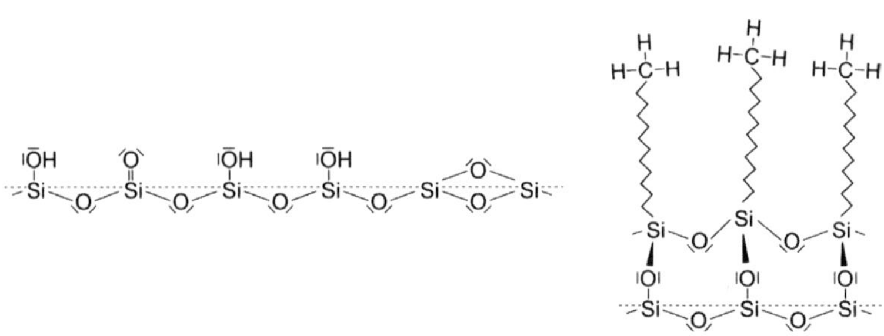

Figure 6.40 Scheme of a polar SiO_x surface with silanol groups (left) and by ODS silanized one (right).

We tentatively explain that the contribution of silanol groups to the "on" time distribution is nearly monoexponential since it is a well-defined interaction of polar OH groups with the QD surface (dangling bonds). We have previously [Cichos et al., 2007] and in Chapter 4 argued that power-law-type distributions have their origin in disorder, which is obviously not important or even absent for interactions with OH groups. We also found a strong monoexponential contribution for CdSe/ZnS QDs in a PVA film [Schmidt et al., 2012; Schmidt et al., 2014]. Some examples are also shown in Section 6.2 of this chapter. Since PVA also contains OH groups these might be the origin of the additional strong monoexponential component.

In case a locally polar surface is the origin of a deviation from a (truncated) power law, silanization should remove the monoexponential contribution. However, Fig. 6.41 shows that this is not the case, though contact angle experiments show a considerable reduction of polarity. We explain this by the fact that despite silanization a few OH-containing sites remain at which QDs preferentially bind. Such type of processes is described by Taeuber and Borczyskowski (2016) and Trenkmann (2016).

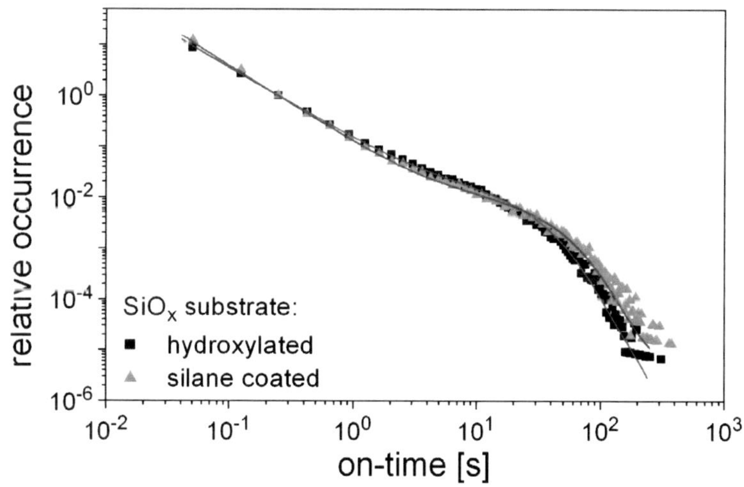

Figure 6.41 Comparison of the "on" time distribution for 100 TOPO-capped CdSe/ZnS QDs on a hydroxylated SiO_x and by an ODS-silanized surface. Solid lines are fits to Eq. 6.2.

Figure 6.42 compares "on" and "off" times for CdSe/ZnS QDs capped by amine ligands. As can be seen there is hardly any monoexponential contribution to the "on" time distribution. Also in this case no influence of silanization is observed. We speculate that this is due to the fact that the surface coverage is due to steric reasons [Blaudeck et al., 2012; Kowerko et al., 2010] much smaller for TOPO ligands compared to amine ligands by roughly a factor of 3, while the binding energies are considerably larger for TOPO compared to amine ligands [Schapotschnikow et al., 2009]. Moreover, for the amine-capped QDs surface treatment does not change blinking dynamics, as is shown exemplary for the "on" times.

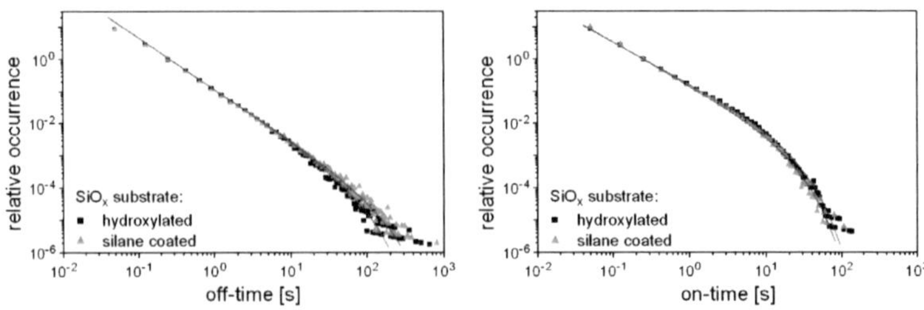

Figure 6.42 Left: "Off" time distributions for amine-capped CdSe/ZnS QDs on a hydroxylated and silanized SiO_x surface. Solid lines are fits to a truncated power law. Right: Influence of surface modification on "on" times. Solid lines are fits to Eq. 6.2.

6.4.2 Impact of Ligands on Blinking

In this section we present some new results of the influence of different ligands on blinking dynamics. A typical comparison has been reported previously for similar ligands but less detailed [Yuab et al., 2007]. We will address the question of whether this influence can be modeled the same way as the influence of the dielectric properties of an embedding matrix [Cichos et al., 2004; Issac et al., 2005; Issac et al., 2012]. Naturally, when varying ligands we also have to take the influence of the embedding matrix into account.

Figure 6.43 shows the "on" and "off" time distribution of CdSe/ZnS QDs capped by a mixture of TOPO and HDA in various matrices. We have also chosen besides others the matrix TPD since it is a (hole) conducting organic semiconductor. Details of sample preparation are given in Trenkmann (2015). The data for QDS in TPD are both for "on" and "off" times qualitatively different in a way not showing a clear truncated power law.

Table 6.1 collects the results obtained by fitting "on" and "off" time distributions by a truncated power law, neglecting small contributions from superimposed exponential terms (see Eq. 6.2).

Data are tabulated according to the increasing dielectric constant ε_M of the embedding or supporting matrix. All fitting parameters depend on the type of ligand and matrix. Care has to be taken if comparing different series of experiments, since the experience is that various unidentified effects influence absolute values of the parameters, including those which are related to the threshold setting, excitation power, energy (see the previous section), and sample preparation.

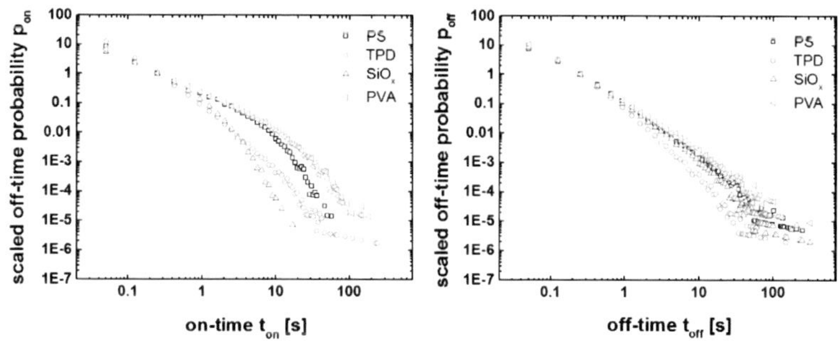

Figure 6.43 Distribution of "on" (left) and "off" (right) times for TOPO/HDA-capped CdSe/ZnS QDs in various matrices: PS, TPD, SiO_x, and PVA. The dielectric constants are increasing from PS to PVA. Data have been fitted by a truncated power law (Eq. 6.2) and are collected in Table 6.1.

Table 6.1 Fitting parameters for blinking of CdSe/ZnS QDs in various matrices and with different ligands

Matrix	ε_M	Ligands	α_{on}	$\Gamma^{-1}{}_{on}$(s)	α_{off}	$\Gamma^{-1}{}_{off}$(s)
p-Terphenyl	2.12	TOPO[a]	1.42	11.1	1.77	130
PS	2.53	TOPO	1.44	9.9	1.70	130
		TOPO/HDA	1.25	12.5	1.71	48
TPD	2.95	TOPO/HDA	1.69	3.75	1.66	36
PMMA	3.40	TOPO	1.43	8.5	1.65	130
SiO_x	3.80	TOPO	1.42	7.6	1.61	120
		TOPO/HDA	1.10	1.25	1.71	18
PVP	4.80	TOPO	1.44	2.4	1.58	100
NaCl	5.30	TOPO	1.43	2.1	1.54	0
PVA	14.00	TOPO	1.14	1.9	1.46	90
		TOPO/HDA	1.14	18.1	1.55	15

[a]Data for TOPO are taken from Issac et al. (2012).

To compare data more consistently we plot the parameters according to the Onsager model [Issac et al., 2005; Issac et al., 2012]. The results for the power-law exponent $\alpha_{\rm off}$ are shown in Fig. 6.44 separately for TOPO- and TOPO/HDA-capped QDs. Data are fitted linearly and show a different increment for the two types of ligands. To explore whether this different increment can be cast on the same functional dependence, we attempt to describe the influence of ligands within the framework of the dielectric properties of the ligands. The related approach is schematically shown in Fig. 6.45. Basically this model is a kind of (continuous medium) shell model with shells of different dielectric constants. In some respect it follows a previously suggest approach by Leatherdale et al. (2001).

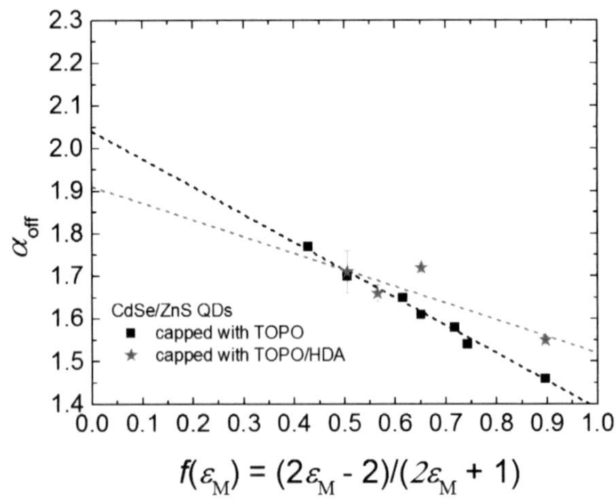

Figure 6.44 Dependence of $\alpha_{\rm off}$ for the two types of ligands (TOPO and TOPO/HDA) on the dielectric constant ε_M of the matrix following the Onsager model. Fits have been obtained by linear regression of the data.

Within this model we assume that ligands form a homogeneous (dielectric) shell around the QD, which is characterized by the respective dielectric constant and which is further embedded in the dielectric medium of the matrix. In the simplest approach we further assume that blinking occurs via trapping an electron in either of these two shells. As we have argued recently [Issac et al., 2005; Issac et al., 2012] the electron becomes stabilized in the respective dielectric medium following the Onsager approach. The influence of trapping the electron now depends on the spatial position of the trapped electron and the resulting Onsager function $f(\varepsilon_e)$ for QDs capped with TOPO according to

$$f(\varepsilon) = x_{CdSe/ZnS} f(\varepsilon_{CdSe/ZnS}) + x_{matrix} f(\varepsilon_M) + x_{TOPO} f(\varepsilon_{TOPO}) \quad (6.6a)$$

and with TOPO/HDA according to

$$f(\varepsilon) = x_{CdSe/ZnS} f(\varepsilon_{CdSe/ZnS}) + x_{matrix} f(\varepsilon_M) + x_{TOPO} f(\varepsilon_{TOPO}) + x_{HDA} f(\varepsilon_{HDA}) \quad (6.6b)$$

with the relative fraction of each material x_i, $\varepsilon_{TOPO} = 2.5$, $\varepsilon_{HDA} = 2.25$ [Wawrzynczyk et al., 2013], $\varepsilon_{CdSe} = 9.3$, and $\varepsilon_{ZnS} = 8.25$ [Zubritskii, 1997].

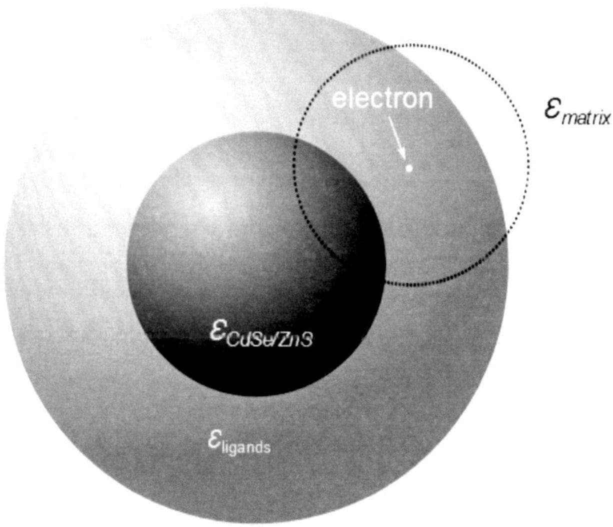

Figure 6.45 Schematic presentation of dielectric shells and the position of a trapped electron in the ligand shell.

To apply the model we have estimated that the coverage of the QD surface by ligands is close to 28% (TOPO capped) and 23% (TOPO/HDA capped) [Trenkmann, 2015]. We estimated the interaction radius of the electric field of the trapped electron to the molecular diameter of the ligand and the matrix [Leatherdale and Bawendi, 2001], which is about 1 nm. Now we can calculate dependence of

α_{off} on $f(\varepsilon_e)$ following Eq. 6.6. The results are shown in Fig. 6.46 for three values of d, which is the distance of the trapped electron from the CdSe/ZnS surface. According to the graphs we obtain a satisfying agreement for all α_{off} for the two types of ligands with $d \approx 0.5$ nm. Taking the oversimplifications of the model into account we suggest that differences of blinking parameters caused by different ligands can basically be approximated by the dielectric properties of the ligands. This model reveals that matrices and ligands influence blinking dynamics in very similar way, namely via their respective dielectric functions. We take this as a further indication that blinking is accompanied by charge separation or polarization.

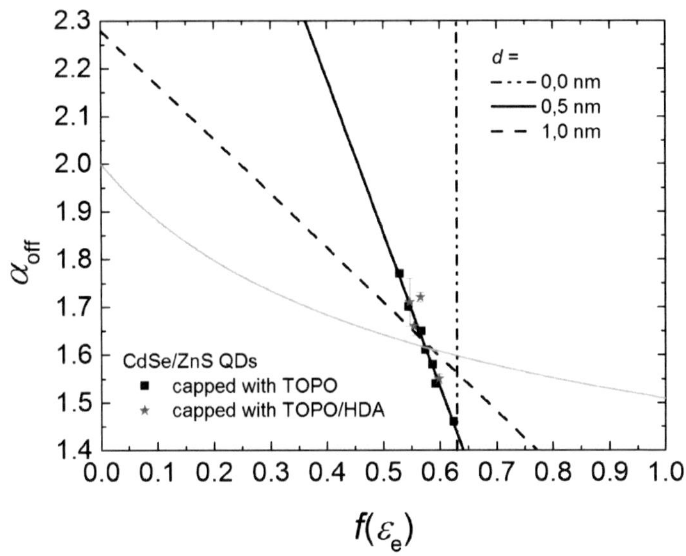

Figure 6.46 Dependence of α_{off} on the dielectric function $f(\varepsilon_e)$ for TOPO- and TOPO/HDA-capped CdSe/ZnS QDs calculated according to Eq. 6.6. The parameter d is the (average) distance of the trapped electron from the QD surface. The bended curve is a calculation according to Eq. 6.7 proposed for the self-trapping model [Issac et al., 2005].

The bended green curve in Fig. 6.46 follows the self-trapping model suggested previously by Issac et al. (2005) according to

$$\alpha_{off} = 1 + \sqrt{\frac{\Delta E_v}{\Delta E_v + E_0 f(\varepsilon_M)}}. \tag{6.7}$$

The fitting parameters apply to a tunnel barrier of $\Delta E_V = 0.26$ eV and a stabilization energy of $E_0 = 0.74$ eV. The comparison shows that the presently applied Onsager model describes more accurately both the influence of ligands and matrices compared to the self-trapping model.

The cutoff time Γ^{-1}_{off} of the "off" time power-law distribution also depends according to Table 6.1 on the combined dielectric Onsager function $f(\varepsilon)$, as shown in Fig. 6.47 separately for the two types of ligand configurations. In comparison, Γ^{-1}_{off} is by about a factor of 3 longer for TOPO-capped QDs compared to TOPO/HDA-capped ones. Moreover, both dependencies show across over (sudden shortening of Γ^{-1}_{off}) close to $f(\varepsilon) \approx 0.55$–$0.6$, while they seem to be constant below and above this value. Presently we do not have a consistent explanation for this finding. However, since $\varepsilon_{TOPO} = 2.5$ is somewhat larger than $\varepsilon_{HDA} = 2.25$, this might explain the longer trapping times in the case of TOPO ligands. The constant long trapping time might be related to the fact that there is a physically determined longest trapping time.

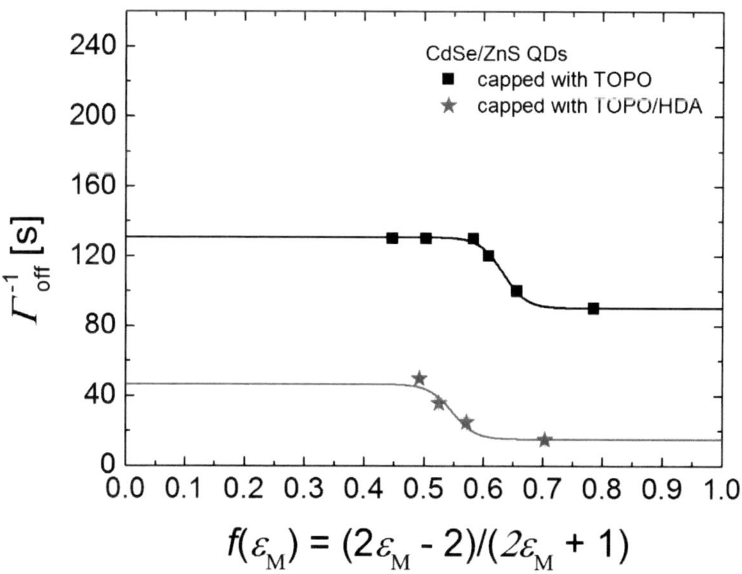

Figure 6.47 Dependence of Γ^{-1}_{off} on the dielectric function $f(\varepsilon_M)$ of the matrix for TOPO-capped (top) and TOPO/HDA-capped CdSe/ZnS QDs. Data for TOPO-capped QDs are from Issac et al. (2012).

Finally we would like to address the influence of ligands and matrices on the "on" times in a similar way as has been published recently [Issac et al., 2012]. Corresponding blinking data are collected in Table 6.2 for TOPO/HDA-capped CdSe/ZnS QDs in various matrices. Data have been evaluated according to either a truncated power law or Eq. 6.2. While the power-law exponent α_{on} does not depend critically on which approximation has been used, the cutoff times Γ^{-1}_{on} become shorter when applying Eq. 6.3. The fitted results are shown in Fig. 6.48 together with recently reported data of Issac et al. (2012) for TOPO-capped CdSe/ZnS QDs as a function of the Onsager function $f(\varepsilon_M)$ of the matrix.

Table 6.2 Dependence of "on" time blinking parameters of TOPO/HDA-capped CdSe/ZnS QDs as a function of matrices and fitting equation according to Eq. 6.2 or a truncated power law

Matrix	ε_M	Eq.	α_{on}	Γ^{-1}_{on} (s)	γ^{-1}_{on} (s)	B/A
PS	2.53	6.2	1.25	12.5		
		6.3	1.13	4.12	5.45	0.22
TPD	2.95	6.2	1.69	3.75		
		6.3	1.69	3.75	5.89	0.019
SiO_x	3.80	6.2	1.10	1.25		
		6.3	1.10	1.25	3.81	0.004
PVA	14.00	6.2	1.14	18.1		
		6.3	1.03	5.60	66.7	0.16

While data for TOPO-capped QDs are either with respect to $f(\varepsilon_M)$ independent for α_{on} (besides PVA which provokes a considerably large error) or nearly linearly dependent for Γ_{on}, a similar behavior emerges for TOPO/HDA-capped data. With respect to Γ^{-1}_{on} we observe (besides TOPO/HDA-capped QDs in PVA) a nearly linear behavior with a similar increment in parallel with TOPO-capped QDs but by about a factor of 1.5 faster, which is qualitatively similar to corresponding "off" time data. The deviation for PVA might be explained by the observation that in this matrix a truncated power-law behavior of "on" times is quite unrealistic (see Section 6.2).

The power-law exponent α_{on} is constant within experimental error besides data for TPD. Moreover, the absolute value of α_{on} is considerably smaller (~1.1) compared to TOPO-capped QDs

(~1.4), while it is much larger (~1.7) in TPD. This finding is most surprising and might be related to the fact that TPD is an organic (hole) conductor. While absolute values of power-law exponents have not to be taken too seriously when comparing data obtained under different experimental conditions (e.g., caused by different threshold settings, excitation power, or energy), a deviation under identical experimental conditions has to be taken quite seriously. An increased α_{on} indicates according to the self-trapping model [Issac et al., 2005; Issac et al., 2012] a reduced tunneling barrier for charge (hole) ejection. Similar results have been obtained by Bixby et al. (2012), who observe for TOPO-capped CdSe/ZnS QDs in TPD α_{on} = 1.6 and in PMMA α_{on} = 1.43. We speculate that in the case of TPD a hole is ejected, while it is an electron in all other cases.

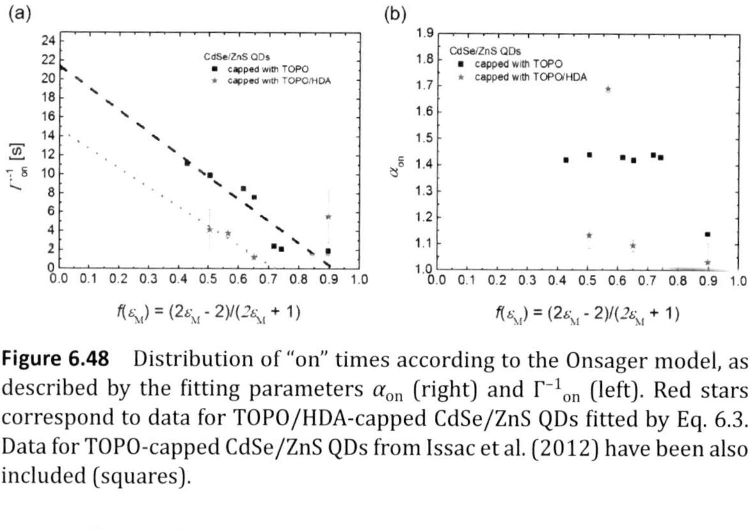

Figure 6.48 Distribution of "on" times according to the Onsager model, as described by the fitting parameters α_{on} (right) and $\Gamma^{-1}{}_{on}$ (left). Red stars correspond to data for TOPO/HDA-capped CdSe/ZnS QDs fitted by Eq. 6.3. Data for TOPO-capped CdSe/ZnS QDs from Issac et al. (2012) have been also included (squares).

Finally we like to comment on the behavior of the exponential contribution as characterized by $\gamma_{on}{}^{-1}$ which we have assigned to the influence of OH groups. While the ratio B/A will depend on the concentration of OH groups (depending on sample preparation, $\gamma_{on}{}^{-1}$ should be an intrinsic property of the QD-matrix system. Krasselt et al. (2011) have not observed in the case of only HDA ligands such a contribution, while it was strong for TOPO ligands with $\gamma_{on}{}^{-1}$ in the range of 10.0–15.4 s, which is about a factor of 2 slower compared to TOPO/HDA ligands (see Table 6.2). TOPO/HDA-capped QDs in PVA show $\gamma_{on}{}^{-1}$ = 66.7 s, but TOPO-capped QDs in PVA revealed ~0.02s [Schmidt et al., 2012] (and 0.25 s in PS). However, it has to be mentioned that the latter experiment has been performed under confocal and pulsed laser excitation. Concluding, under the same experimental conditions γ_{on}^{-1} is of the same order of magnitude for TOPO/HDA and TOPO ligands but seems to be much faster under high excitation fluence. Whether there is a systematic behavior of γ_{on}^{-1} on matrices and ligands remains presently open.

6.5 Conclusion

We have intensively investigated blinking single CdSe/ZnS QDs as one of the prototypes of emitters in disordered environments. Though blinking quantum systems have been investigated intensively for almost two decades, a microscopic understanding has only been partly achieved. Basic models have been presented shortly after the first experimental reports on these phenomena. They basically rely on photoinduced charge separation followed by Auger-type luminescence quenching of the sequentially excited excitons. Certainly this is the most direct approach, which, however, does not take into account the complex interface of colloidal QDs. Surface and interface effects governed by ligands, embedding matrices, defects of the morphology, and impurities will give rise to a manifold of contributions to charge generation, charge diffusion or localization, fluctuations of electron–phonon coupling, and Stark effects.

In this chapter we presented a comprehensive approach combing a PL intensity–related (CPA) investigation of spectral and dynamic properties of CdSe/ZnS QDs and the influence of several external influences such as excitation power and energy, ligands, and matrix variation. We aimed at a detailed understanding of the interrelation of these variables on the blinking dynamics. Without aiming to present a review on what has been published in the literature, we made appropriate references, whenever necessary.

We proved that power-law behavior is not a general feature of PL intermittency. Many indications for this statement can be identified upon careful re-inspection of reported investigations on blinking. The without any doubt often seemingly power law is a result of the strong structural and energetic disorder inherent to colloidal

QDs. We find a strong correlation of PL energies, decay times, and intensities. While energies and decay times depend only moderately on external parameters such as ligands and embedding matrices, blinking dynamics depend strongly on these variables. Variation of excitation conditions influence or control the effectiveness of different relaxation pathways.

We were able to identify three independent classes of blinking and at least three involved and basically different states which constitute two of the blinking classes. The third one is related to blinking processes in the excitation cycle which is related to a kind of Coulomb blockade for hot electrons. The application of CPA allows us to unravel the complex stepwise interconnections of the involved states identified by their intensity, PL decay, and energy.

Moreover, we also observed lifetime blinking, since for a given PL intensity at least two or even three PL decay times are observed.

We take the identified interrelations as a strong indication that blinking statistics themselves are controlled by the environment and its inherently related disorder, while PL energies and decay dynamics are predominantly inherent QD properties. As has been shown experimentally coupling of the excitonic states to the environment can be suppressed by increasing order to the interface via shell engineering (reducing strain by gradual shell construction or giant shell formation confining the excitonic wavefunction nearly completely to the ordered core/shell). Fluctuations in the environment should in the future be modeled by (tunneling) ligands, defects, or surface atoms in combination with the coupled landscape of two-level systems describing the features of the amorphous environment, especially its dielectric properties.

This is supported by the recent investigation of blinking CdSe/ZnS QDs in a conducting polymer [Al-Attar et al., 2015] which shows that the conducting polymer favors the presence of charged excitons, giving rise to changes of the blinking dynamics with increased correlation between neighboring QDs.

References

Al Attar, N., Kennedy, Eamonn, Kelly, G., and Rice, J. H. (2015). Photoluminescence blinking from single CdSe/ZnS quantum dots in a conducting polymer matrix, *J. Phys. Chem. C*, **119**, 6278–6287.

Amecke, N., and Cichos, F. (2011). Intermediate intensity levels during the emission intermittency of single CdSe/ZnS quantum dots, *J. Lumin.*, **131**, 375–378.

Baker, T. A., Rouge, J. L., and Nesbitt, D. J. (2009). Single molecule studies of quantum dot fluorescence intermittency: evidence for both dark and light-assisted blinking dynamics, *Mol. Phys.*, **107**, 1867–1878.

Bharadwaij, P., and Novotny, L. (2011). Robustness of quantum dot power-law blinking, *Nano Lett.*, **11**, 2137–2141.

Betzig, G. H., Patterson, G. H., Sougrat, R., Lindwasser, O. W., Olenych, S., Bonifacino, J. S., Davidson, M. W., Lippincott-Schwartz, J., and Hess, H. F. (2006). Imaging intracellular fluorescent proteins at nanometer resolution, *Science*, **313**, 1642–1645.

Biebricher, A., Sauer, M., and Tinnefeld, P. (2006). Radiative and nonradiative rate fluctuations of single colloidal semiconductor nanocrystals, *J. Phys. Chem. B*, **110**, 5174–5178.

Bixby, T. J., Cordones, A. A., and Leone, S. R. (2012). CdSe/ZnS quantum dots in N,N'-diphenyl-N,N'-bis(3-methylphenyl)-(1,1'-biphenyl)-4,4'-diamine (TPD), *Chem. Phys. Lett.*, **521**, 7–11.

Blaudeck, T., Zenkevich, E. I., Cichos, F., and Borczyskowski, C. von (2008). Probing wave functions at semiconductor quantum-dot surfaces by non-FRET photoluminescence quenching, *J. Phys. Chem. C*, **112**, 20251–20257.

Blaudeck, T., Zenkevich, E. I., Abdel-Mottaleb, M., Szwaykowska, M. K., Kowerko, D., Cichos, F., and Borczyskowski, C. von (2012). Formation principles and ligand dynamics of nanoassemblies of CdSe quantum dots and functionalised dye molecules, *ChemPhysChem*, **13**, 959–972.

Boldt, K., Jander, S., Hoppe, K., and Weller, H. (2011). Characterization of the organic ligand shell of semiconductor quantum dots by flurescence quenching experiments, *ACS Nano*, **5**, 8115–8123.

Brown, P. R., Kim, D., Lunt, R. L., Zhao, N., and Bawendi, M. G. (2014). Energy level modification in lead sulfide quantum dot thin films through ligand exchange, *ACS Nano*, **8**, 5863–5872.

Bruhn, B., Qejvanaj, F., Sychugov, I., and Linnros, J. (2014). Blinking statistics and excitation-dependent luminescence yield in Si and CdSe nanocrystals, *J. Phys. Chem. C*, **118**, 2202–2208.

Califano, M., Franceschetti, A., and Zunger, A. (2007). Lifetime and polarization of radiative decay of excitons, biexcitons, and trions in CdSe nanocrstal quantum dots, *Phys. Rev. B*, **75**, 115401.

Chung, I., and Bawendi, M. G. (2004). Relationship between single quantum-dot intermittency and fluorescence intensity decays from collections of dots, *Phys. Rev. B*, **70**, 165304.

Cichos, F., Martin, J., and Borczyskowski, C. von (2004). Emission intermittency in silicon nanocrystals, *Phys. Rev. B*, **70**, 115314.

Cichos, F., Borczyskowski, C. von, and Orrit, M. (2007). Power-law intermittency of single emitters, *Curr. Opin. Colloid Interface Sci.*, **12**, 272–284.

Cody, G. D. (1984). *Hydrogenated Amorphous Silicon, Semiconductors and Semimetals* (Pankove, J. I., ed), Academic, New York, Vol. 21, pp. 11–79.

Cohn, A. W., Rinehart, J. D., Schimpf, A. M., Weaver, A. L., and Gamelin, D. R. (2013). Size dependence of negative trion Auger recombination in photodoped CdSe nanocrystals, *Nano Lett.*, **14**, 353–358.

Cordones, A. A., Bixby, T. J., and Leone, S. R. (2011a). Evidence for multiple trapping mechanisms in single CdSe/ZnS quantum dots form fluorescence intermittency measurements over a wide range of excitation power, *J. Phys. Chem. C*, **115,** 6341–6349.

Cordones, A. A., Bixby, T. J., and Leone, S. R. (2011b). Direct measurement of off-state trapping rate fluctuations in single quantum dot fluorescence, *Nano Lett.*, **11**, 3366–3369.

Cui, J., Beyler, A. P., Bischof, T. S., Wilson, M. W. B., and Bawendi, M. G. (2014). Deconstructing the photon stream from single nanocrystals: from binning to correlation, *Chem. Soc. Rev.*, **43**, 1287–1310.

Dickson, R. M., Cubitt, A. B., Tsien, R. Y., and Moerner, W. E. (1997). On/off blinking and switching behaviour of single molecules of green fluorescent protein, *Nature*, **388**, 355–358.

Early, K. T., and Nesbitt, D. J. (2013). Size-dependent photoionization in single CdSe/ZnS nanocrystals, *Nano Lett.*, **13**, 4844–4849.

Efros, A. L., and Rosen, M. (1997). Random telegraph signal in the photoluminescence intensity of a single quantum dot, *Phys. Rev. Lett.*, **78**, 1110–1113.

El-Kork, N., Huisken, F., and Borczyskowski, C. von (2011). Dielectric effects on the optical properties of single silicon nanocrystals, *J. Appl. Phys.*, **110**, 074312.

Empedocles, S. A., and Bawendi, M. G. (1999). Influence of spectral diffusion on the line shapes of single CdSe nanocrystallite quantum dots, *J. Phys. Chem. B*, **103**, 1826–1830.

Fernée, M. J., Littleton, B. N., and Rubinsztein-Dunlop, H. (2009). Detection of bright trion states using the fine structure emission of single CdSe/ZnS colloidal quantum dots, *ACS Nano*, **3**, 3762–3768.

Fernée, M. J., Plakhotnik T., Louyer, Y., Littleton, B. N., Potzner, C., Tamarat, P., Mulvaney, P., and Lounis, B. (2012). Spontaneous spectral diffusion in CdSe quantum dots, *J. Phys. Chem. Lett.*, **3**, 1716–1720.

Fischer, U., Borczyskowski, C. von, Seiff, F., and Stehlik, D. (1983). Long-range energy transport in doped dibenzofurane crystals detected by intermolecular fluorescence fading, *Chem. Phys. Lett.*, **97**, 476–482.

Fisher, B. R., Eisler, H. J., Stott, N., and Bawendi, M. G. (2004). Emission intensity dependence and single-exponential behaviour in single colloidal quantum dot fluorescence lifetimes, *J. Phys. Chem. B*, **108**, 143–148.

Fomenko, V., and Nesbitt, D. J. (2008). Solution control of radiative and nonradiative lifetimes: a novel contribution to quantum dot blinking suppression, *Nano Lett.*, **8**, 287–293.

Frantsuzov, P. A., Volkán-Kacsó, S., and Jankó, B. (2009). Model of fluorescence intermittency of single colloidal semiconductor quantum dots using multiple recombination centres, *Phys. Rev. Lett.*, **103**, 2070402.

Frenzel, J., Joswig, J.-O., and G. Seifert, G. (2007). Optical excitations in cadmium sulfide nanoparticles, *J. Phys. Chem. C*, **111**, 10761–10770.

Galland, C., Ghosh, Y., Steinbrück, A., Sykora, M., Hollingsworth, J. A., Klimov, V. I., and Htoon, H. (2011). Two types of luminescence blinking revealed by spectrochemistry of single quantum dots, *Nature*, **479**, 203–207.

Galland, C., Ghosh, Y., Steinbrück, A., Hollingsworth, J. A. Htoon, H., and Klimov, V. I. (2012). Lifetime blinking in nonblinking nanocrystals quantum dots, *Nat. Commun.*, **3**, 908.

Göhler, C. (2014). *Schnelle Spektralanalyse des Blinkens von Halbleiternanokristallen*, Master Thesis, Technische Universität, Chemnitz, Germany.

Gómez-Campos, F. M., and Califano, M. (2012). Hole surface trapping in CdSe nanocrystals: dynamics, rate fluctuations, and implications for blinking, *Nano Lett.*, **12**, 4508–4517.

Gómez, D. E., Embden, J. van, Jasieniak, J., Smith, T. A., and Mulvaney, P. (2006). Blinking and surface chemistry of single CdSe nanocrystals, *Small*, **2**, 204–208.

Gómez, D. E., Embden, J. van, and Mulvaney, P. (2006). Spectral diffusion of single semiconductor nanocrystals: the influence of the dielectric environment, *Appl. Phys. Lett.*, **88**, 154106.

Gómez, D. E., Embden, J. van, Mulvaney, P., Fernée, M. A., and Rubinsztein-Dunlop, H. (2009). Exciton-trion transitions in single CdSe-CdS core-shell nanocrystals, *ACS Nano*, **3**, 2281–2287.

Goushi, K., Yamada, T., and Otomo, A. (2009). Excitation intensity dependence of power-law blinking statistics in nanocrystal quantum dots, *J. Phys. Chem. C*, **113**, 20161–20168.

Hammer, N. I., Emrick, T., and Barnes, M. D. (2007). Quantum dots coordinated with conjugated organic ligands: new nanomaterials with novel photophysics, *Nanoscale Res. Lett.*, **2**, 282–290.

Hell, S. W., and Wichmann, J. (1994). Breaking the diffractive resolution limit by stimulated emission: stimulated-emission-depletion fluorescence microscopy, *Opt. Lett.*, **19**, 780–782.

Hoheisel, W., Colvin, V. L., Johnson, C. S., and Alivisatos, A. P. (1994). Threshold for quasicontinuum absorption and reduced luminescence efficiency in CdSe nanocrystals, *J. Chem. Phys.*, **101**, 8455.

Hohng, S., and Ha, T. (2004). Near-complete suppression of quantum dot blinking in ambient conditions, *J. Am. Chem. Soc.*, **126**, 1324–1325.

Issac, A., Borczyskowski, C. von, and Cichos, F. (2005). Correlation between photoluminescence intermittency of CdSe quantum dots and self-trapped states in dielectric media, *Phys. Rev. B*, **71**, 161302.

Issac, A., Krasselt, C., Cichos, F., and Borczyskowski, C. von (2012). Influence of the dielectric environmenton the photoluminescence intermittency of CdSe quantum dots, *ChemPhysPhysChem.*, **13**, 3223–3230.

Jha, P. P., and Guyot Sionnest, P. (2009). Trion decay in colloidal quantum dots, *ACS Nano*, **3**, 1011–1015.

Jha, P. P., and Guyot-Sionnest, P. (2010). Electrochemical switching of photoluminescence of single quantum dots, *J. Phys. Chem. C*, **114**, 21138–21141.

Ji, X., Copenhaver, D., Sichmeller, C., and Peng, X. (2008). Ligand bonding and dynamics on colloidal nanocrystal at room temperature: the case of alkylamines on CdSe nanocrystals, *J. Am. Chem. Soc.*, **130**, 5726–5735.

Kilina, S., Ivanov, S., and Tretiak, S. (2009). Effect of surface ligands on optical and electronic spectra of semiconductor nanoclusters, *J. Am. Chem. Soc.*, **131**, 7717–7726.

Klimov, V. I. (2003). Charge carrier dynamics and optical gain in nanocrystal quantum dots: from fundamental photophysics to quantum dot lasing, in *Semiconductor and Metal Nanocrystals: Synthesis and Electronic and Optical Properties* (Klimov, V. I., ed.), Marcel Dekker, New York, pp. 159–214.

Knappenberger, K. L., Wong, D. B., Romanyuk, Y. E., and Leone, S. R. (2007). Excitation wavelength dependence of fluorescence intermittency in CdSe/ZnS core/shell quantum dots, *Nano Lett.*, **7**, 3869–3874.

Kraus, R. M., Lagoudakis, P. G., Mueller, J., Lupton, J. M., Feldmann, J., Talapi, D. V., and Weller H. (2005). Interplay between Auger and ionization processes in nanocrystal quantum dots, *J. Phys. Chem. B*, **109**, 18214–18217.

Kowerko, D., Schuster, J., Amecke, N., Abdel-Mottaleb, M., Dobrawa, R., Würthner, F., and Borczyskowski, C. von (2010). FRET and ligand related non-FRET processes in single quantum dot-perylene bisimide assemblies, *Phys. Chem. Chem. Phys.*, **12**, 4112.

Krasselt, C., Schuster, J., and Borczyskowski, C. von (2011). Photoinduced hole trapping in single semiconductor quantum dots at specific sites at silicon oxide interfaces, *Phys. Chem. Chem. Phys.*, **13**, 17084.

Krasselt, C. (2015). *Dynamik der Photo-Lumineszenz-Unterbrechung von Halbleiter-Nanokristallen in elektrischen Feldern*, PhD Thesis, Technische Universität, Chemnitz, Germany.

Leatherdale, C. A., and Bawendi, M. G. (2001). Observation of solvatochromism in CdSe in colloidal quantum dots, *Phys. Rev. B*, **63**, 165315.

Ma, X. Tan, H., Kipp, T., and Mews, A. (2012). Optical spectroscopy of single semiconductor nanocrystals close to gold nanoparticles, *Proc. SPIE*, **8232**, 82321I–82321I.

Martin, J., Cichos, F., and Borczyskowski, C. von (2012). Spectral diffusion of quasi localized excitons in single silicon nanocrystals, *J. Lumin.*, **132**, 2161–2165.

Müller, J., Lupton, J. M., Rogach, A. L., Feldmann, J., Talapin, D. V., and Weller, H. (2005). Monitoring surface charge migration in the spectral dynamics of single CdSe/CdS nanodot/nanorod heterostructures, *Phys. Rev. B*, **72**, 205339.

Munro, A. M., and Ginger, D. S. (2008). Photoluminescence quenching of single nanocrystals by ligand adsorption, *Nano Lett.*, **8**, 2585–2590.

Negele, C., Haase, J., Budweg, A., Leitenstorfer, A., and Mecking, S. (2013). Stable single-photon emission by quantum dot/polymer hybrid particles, *Macromol. Rapid Commun.*, **34**, 1145.

Neuhauser, R. G., Shimizu, K. T., Woo, W. K., Empedocles, S. A., and Bawendi, M. G. (2000). Correlation between fluorescence intermittency and spectral diffusion in single semiconductor quantum dots, *Phys. Rev. Lett.*, **85**, 3301–3304.

Nirmal, M., Dabbousi, B., Bawendi, M., Macklin, J. J., Trautman, J. K., Harris, T. D., and Brus, L. E. (1996). Fluorescence intermittency in single cadmium selenide nanocrystals, *Nature*, **383**, 802–804.

Odoi, M. Y., Hammer, N. I., Early, K. T., McCarthy, K. D., Tangirala, R., Emrick, T., and Barnes, M. D. (2009). Fluorescence lifetimes and correlated photon statistics from single CdSe/ologo(phenylene vinylene) composite nanostructures, *Nano Lett.*, **7**, 2769–2773.

Orrit, M., and Bernard, J. (1990). Single pentacene molecules detected by fluorescence excitation in a *p*-terphenyl crystal, *Phys. Rev. Lett.*, **65**, 2716.

Osad'ko, I. S., Eremchev, I. Y., and Naumov, A. V. (2015). Two mechanisms of fluorescence intermittency in single core/shell quantum dot, *J. Phys. Chem. C*, **119**, 22646–22652.

Peterson, J., and Nesbitt, D. J. (2009). Modified power law behaviour in quantum dot blinking: a novel role for biexcitons and Auger ionization, *Nano Lett.*, **9**, 338–345.

Petrov, E. P., Cichos, F., and Borczyskowski, C. von (2006). Intrinsic photophysics of semiconductor nanocrystals in dielectric media: formation of surface states, *J. Lumin.*, **119–120**, 412– 417.

Plakhotnik, T., Fernée, M. J., Littleton, B., Rubinsztein-Dunlop, H., Potzner, C., and Mulvaney, P. (2010). Anomalous power laws of spectral diffusion in quantum dots: a connection to luminescence intermittency, *Phys. Rev. Lett.*, **105**, 167402.

Riley, E. A., Hess, C. M., and Reid, P. J. (2012). Photoluminescence intermittency from single quantum dots to organic molecules: emerging themes, *Int. J. Mol. Sci.*, **13**, 12487–12518.

Rosen, S., Schwartz, O., and Oron, D. (2010). Transient fluorescence of the off state in blinking CdSe/CdS/ZnS semiconductor nanocrystals is not governed by Auger recombination, *Phys. Rev. Lett.*, **104**, 157404.

Salman, A. Al, Tortschanoff, A., van der Zwan, G., van Mourik, F., and Chergui, M. (2009). A model for the multi-exponential excited-state decay of CdSe nanocrystals, *Chem. Phys.*, **357**, 96–101.

Schlegel, G., Bohnenberger, J., Potapova, I., and Mews, A. (2002). Fluorescence decay time of single semiconductor nanocrystals, *Phys. Rev. Lett.*, **88**, 1374011.

Schmidt, R. (2010). *Zeitaufgelöste Konfokalmikroskopie an einzelnen CdSe/ZnS-Halbleiternanokristallen,* Diploma Thesis, TU Chemnitz, Germany.

Schmidt, T., Chizhik, A. I., Chizhik, A. M., Potrick, K. Meixner, A. J., and Huisken, F. (2012). Radiative exciton recombination and defect luminescence observed in single silicon nanocrystals, *Phys. Rev. B*, **86**, 125302.

Schmidt, R., Krasselt, C., and Borczyskowski, C. von (2012). Change point analysis of matrix dependent photoluminescence intermittency of single CdSe/ZnS quantum dots with intermediate intensity levels, *Chem. Phys.*, **406**, 9–14.

Schmidt, R., Krasselt, C., Göhler, G., and Borczyskowski, C. von (2014). The fluorescence intermittency for quantum dot is not power-law distributed: a luminescence intensity resolved approach, *ACS Nano*, **8**, 3506–3521.

Schuster, J., Cichos, F., and Borczyskowski, C. von (2005). Influence of self-trapped states on the fluorescence intermittency of single molecules, *Appl. Phys. Lett.*, **87**, 051915.

Schuster, J. F. Cichos, F., and Borczyskowski, C. von (2006). Variation of power-law dynamics caused by dark state recovery of fluorescence intermittency of a single quantum system, *Proc. SPIE*, **6258**, 625804.

Schuster, J., Brabandt, J., and Borczyskowski, C. von (2007). Discrimination of photoblinking and photobleaching on the single molecule level, *J. Lumin.*, **127**, 224.

Schapotschnikow, P., Hommersom, B., and Vlugt, T. J. H. (2009). Adsorption and binding of ligands to CdSe nanocrystals, *J. Phys. Chem. C*, **113**, 12690–12698.

Shen, Y., Tan, R., Gee, M. Y., and Greytak, A. B. (2015). Quantum yield regeneration: influence of neutral ligand binding on photophysical properties in colloidal core/shell quantum dots, *ACS Nano*, **9**, 3345–3359.

Shimizu, K. T., Neuhauser, R. G., Leatherdale, C. A., Empedocles, S. A., W. K. Woo, W. K., and Bawendi, M. G. (2001). Blinking statistics in single semiconductor nanocrystal quantum dots, *Phys. Rev. B*, **63**, 205316.

Shimizu, K. T., Woo, W. K., Fisher, B. R., Eisler, H. J., and Bawendi, M. G. (2002). Surface-enhanced emission from single semiconductor nanocrystals, *Phys. Rev. Lett.*, **89**, 117401.

Smith, A. M., and Nie, S. (2010). Semiconductor nanocrystals: structure, properties, and band gap engineering, *Acc. Chem. Res.*, **43**, 190–200.

Spinicelli, P., Mahler, B., Buil, S., Quélin, X., Dubertret, B., and Jean-Pierre Hermier, J.-P. (2009). Non-blinking semiconductor colloidal quantum dots for biology, optoelectronics and quantum optics, *ChemPhysChem.*, **10**, 879–882.

Taeuber, D., and Borczyskowski, C. von (2016). Identification of heterogeneous surface properties via fluorescent probe molecules, in *Self-Assembled Organic-Inorganic Nanostructures: Optics and Dynamics* (Zenkevich, I., and Borczyskowski, C. von, eds.), Pan Stanford, Singapore.

Trenkmann, I. (2015). *Mikroskopie und optische Spektroskopie an heterogenen Nano- und Mikrostrukturen: Halbleiter-Nanokristalle, molekulare Farbstofffilme und funktionalisierte Hybridstrukturen*, PhD Thesis, TU Chemnitz, Germany.

Trenkmann, I. (2016). Selective surface binding of dye molecules and hybrid structures for humidity sensors, in *Self-Assembled Organic-Inorganic Nanostructures: Optics and Dynamics* (Zenkevich, I., and Borczyskowski, C. von, eds.) Pan Stanford, Singapore.

Troparevsky, M. C., and Franceschetti, A. (2005). Radiative recombination of charged excitons and multiexcitons in CdSe quantum dots, *Appl. Phys. Lett.*, **87**, 263115.

Vela, J., Htoon, H., Chen, Y., Park, Y.-S., Ghosh, Y., Goodwin, P. M., Werner, J. H., Wells, N. P., Casson, J. L., and Hollingsworth, J. A. (2010). Effect of shell thickness and composition on blinking suppression and the blinking mechanism in 'giant' CdSe/CdS nanocrystal quantum dots, *J. Biophot.*, **3**, 706–717.

Verberk, R., van Oijen, A., and Orrit, M. (2002). Simple model for the power-law blinking of single semiconductor nanocrystals, *Phys. Rev. B*, **66**, 233202.

Voznyy, O. (2011). Mobile surface traps in CdSe nanocrystals with carboxylic acid ligands, *J. Phys. Chem. C*, **115**, 15927–15932.

Voznyy, O., and Sargent, E. H. (2014). Atomistic model of fluorescence intermittency of colloidal quantum dots, *Phys. Rev. Lett.*, **112**, 157401.

Wang, S., Querner, C., Emmons, T., Drndic, M., and Crouch, C. H. (2006). Fluorescence blinking statistics from CdSe core and core/shell nanorods, *J. Phys. Chem. B*, **110,** 23221–23227.

Watkins, L. P., and Yang, H. (2005). Detection of intensity change points in time-resolved single molecule measurements, *J. Phys. Chem. B*, **109**, 617–628.

Wawrzynczyk, D., Bednarkiewicz, A., Nyk, M., Strek, W., and Samoc, M. (2013). Ligand-dependent luminescence of ultra-small Eu(3+)-doped $NaYF_4$ nanoparticles, *J. Nanopart. Res.*, **15,** 1707.

Yuan, C. T., Chou, W. C., Chen, Y. N., Chou, J. W., Chuu, D. S., Lin, C. A. J., Li, J. K., Chang, W. H., and Shen, J. L. (2007). Study of fluorescence enhancement of colloidal cdse/zns quantum dots bound to hexadecylamine by single-molecule measurements, *J. Phys. Chem. C*, **111**, 15166–15172.

Zenkevich, E. I., Stupak, A., Kowerko, D., and Borczyskowski, C. von (2012). Influence of single dye molecules on temperature and time dependent optical properties of CdSe/ZnS quantum dots: ensemble and single nanoassembly detection, *Chem. Phys.*, **406**, 21–29.

Zenkevich, E. I., Stupak, A., Göhler, C., Krasselt, C., and Borczyskowski, C. von (2015). Tuning electronic states of a CdSe/ZnS quantum dot by only one functional dye molecule, *ACS Nano*, **9**, 2886–2907.

Zenkevich, E. I. and Borczyskowski, C. von (2016). Structural and energetic dynamics in QD-dye nanoassemblies, in *Self-Assembled Organic-Inorganic Nanostructures: Optics and Dynamics* (Zenkevich, I., and Borczyskowski, C. von, eds.), Pan Stanford, Singapore.

Zhang, K., Chang, H., Fu, A., Alivisatos, A. P., and Yang, H. (2006). Continuous distribution of emission states from single CdSe/ZnS quantum dots, *Nano Lett.*, **6**, 843–847.

Zhao, J., Nair, G., Fisher, B. R., and Bawendi, M. G. (2011). Challenge to the charging model of semiconductor-nanocrystal fluorescence intermittency from off-state quantum yields and multiexciton blinking, *Phys. Rev. Lett.*, **104**, 157403.

Zubritskii, V. V. (1997). Phonon focusing in CdSe, ZnS, ZnO crystals, *Tech. Phys.*, **42**, 639–643.

Index